CONQUERING NATURE

The Environmental Legacy of Socialism in Cuba

PITT LATIN AMERICAN SERIES

Billie R. DeWalt, *General Editor*

Reid Andrews, *Associate Editor*

Catherine Conaghan, *Associate Editor*

Jorge I. Domínguez, *Associate Editor*

CONQUERING NATURE

The
Environmental Legacy of
Socialism in Cuba

Sergio Díaz-Briquets
AND
Jorge Pérez-López

UNIVERSITY OF PITTSBURGH PRESS

Published by the University of Pittsburgh Press, Pittsburgh, Pa. 15261
Copyright © 2000, University of Pittsburgh Press
All rights reserved
Manufactured in the United States of America
Printed on acid-free paper

10 9 8 7 6 5 4 3 2 1

Library of Congress Cataloging-in-Publication Data

Díaz-Briquets, Sergio.
    Conquering nature : the environmental legacy of socialism in Cuba/
Sergio Díaz-Briquets and Jorge Pérez-López.
        p.   cm. — (Pitt Latin American series)
Includes bibliographical references (p.   ) and index.
    ISBN 0-8229-4118-X (cloth : acid-free paper) — ISBN 0-8229-5721-3
(paper : acid-free paper)
    1. Cuba—Environmental conditions. 2. Environmental degradation—
Cuba. 3. Environmental policy—Cuba. 4. Socialism—Cuba.
    I. Pérez-López, Jorge F. II. Title. III. Series.
    GE160.C9D53   1999
    363.7'02'097291—dc21   99-050680

For Beatriz and Kathleen
and in memory of our fathers,
Alberto Díaz Salazar (1903–1995) and
René G. Pérez-López (1914–1995)

Contents

List of Tables and Maps

Preface

OVER TEN years ago, when we began an innocent conversation about the lack of information and analysis on the environment in Cuba and the imperative to introduce environmental concerns in analyses of the current situation and its transition to a democratic, free-market economy, little did we know that we would be embarking on a major project to document the environmental legacy of socialism on the island. Our interest was piqued by the information that began to flow in the late 1980s and early 1990s about environmental disasters in the (then) Soviet Union and socialist countries of Eastern Europe and the overall recognition of the critical importance of sustainable development—a development process that enhances or protects, rather than damages or depletes, resources, including human resources.

What, we wondered, is the environmental situation on the island? Is the Cuban environmental situation toward the end of the 1990s an unmitigated disaster, similar to that faced by the Soviet Union and Eastern Europe after the fall of the Berlin Wall? Has there recently been a "greening" of Cuban socialism, as sharp cuts in imports of chemicals, oil, and equipment have forced changes in farming techniques? Are these changes genuine and long lasting? What are the environmental implications of the Cuban socialist government's rush to attract foreign investment to develop the tourism and mining sectors? Will the accumulated environmental disruption be a drag on post-socialist Cuba?

While we believe that our effort to examine the socialist environmental legacy has been rewarding, we also wish to acknowledge the many difficulties and frustrations we have experienced, in part due to the fragmentary and often contradictory nature of much of the data we have been able to assemble, as well as our growing humbleness as we have come to appreciate the enormous complexity of environmental–socioeconomic interactions. The data sets we have examined contain many discrepancies; where appropriate, we have highlighted data reliability problems. We have also chosen to briefly summarize some of the most basic environmental–socioeconomic relationships in the introductory sections of each chapter to assist those not familiar with these matters to interpret the environmental issues we have discussed and make this work accessible to a wider readership: readers familiar with nuclear issues may not be as familiar with those related to forestry, and vice versa.

Our primary purpose in writing this book, therefore, is to provide an overview of environmental trends in Cuba under socialism, highlighting some of the most obvious environmental issues. When we speak of the environmental legacy of socialism in Cuba, we are referring to the environmental consequences of development policies pursued under the socialist regime that has ruled the country for four decades. The design and implementation of these policies were colored by ideological notions that inevitably deepened their environmental consequences, as documented throughout this book. Whether or not comparable environmental outcomes would have occurred under a different socioeconomic and political system will never be known, although it is very likely that Cuba's environmental record, while in some respects resembling that of other Caribbean island nations currently confronting mounting environmental problems, would not have been as alarming absent the systemic flaws inherent in the socialist development model and their environmental underpinnings.

It is our hope that specialists on these topics may find our efforts a useful starting point from which to begin more systematic analyses of the environmental issues we have identified and the development of policies to address the more serious ones, including comparative analyses between Cuba and other developing countries.

The fact that neither one of us is a physical or environmental scientist (our disciplines are demography and economics, respectively) has meant that we have relied very heavily on the work done by others. We are indebted to many friends and colleagues who assisted us in obtaining information, understanding some of the concepts involved, or sharpening the analysis. The list is very long and, at the risk of missing important contributors, includes Manuel Acevedo, José Alvarez, Amparo Avella, Teo Babún, Rafael Barba, Jonathan Benjamin-Alvarado, Sergio Besú, René Costales, Larry Daley, Nancy Diamond, María Dolores Espino, Antonio Gayoso, José Ramón Gonzalez, José Carlos Lezcano, Manuel Madrid-Aris, Daniel B. Magraw, Federico Poey, Armando H. Portela, José Oro, Roberto Orro Fernández, Arturo Pino, Héctor Sáez, Joseph Scarpaci, María C. Werlau, and several anonymous reviewers of the manuscript.

We are also grateful to the Association for the Study of the Cuban Economy (ASCE), and in particular to former ASCE president Armando Lago, for providing a forum for the discussion of environmental issues in today's as well as in a post-socialist Cuba. Through the annual meeting of the ASCE, we were able to present parts of our research and elicit very helpful comments and leads on additional information. Early drafts of several of the chapters of this book were published in ASCE's annual proceedings *Cuba in Transition*.

We also appreciate the assistance of William Nelson and Ian Canda with the maps and figures, and of the dedicated staff of the University of Pittsburgh Press—Eileen Kiley, Ann Walston, Mark Jacobs, Hope Kurtz, and Emily Walker—for shepherding the manuscript through the editing and publication process.

Last, but not least, we are grateful to our families for their continuing support in juggling family, work, and research.

Sergio Díaz-Briquets
Jorge Pérez-López
July 1999

CONQUERING NATURE

The Environmental Legacy of Socialism in Cuba

☙ 1 ❧

Socialism in Cuba and
the Environment

EVEN BEFORE the collapse of socialism, there was considerable concern in the West about the environmental consequences of decades of centrally managed economic policies in the former Soviet Union and Eastern Europe. Concern deepened as environmental issues acquired an increasingly prominent role in the global economic and political agenda and as a growing number of observers began to voice alarm about the general disregard toward the environment of socialist development policies. Environmental deterioration was not supposed to occur under socialism. According to conventional Marxist-Leninist dogma, environmental deterioration was precipitated by the logic of capitalism and its relentless pursuit of profits. Exploiting workers and the natural world were inevitable sequelae of capitalist accumulation. Socialism could only beget environmentally benign economic development policies. Guided by "scientific" principles, socialism's goal was a classless and bountiful society, populated by men and women living in harmony with each other and the environment.

In reality the socialist environmental record proved to be, as in so many other political and economic realms, far different from the utopian view. The magnitude of the environmental disaster in these countries that became apparent on the fall of socialism surpassed expectations of even the most pessimistic observers. As Murray Feshbach and Alfred Friendly (1992, 1) observed with regard to the Soviet Union: "When historians finally conduct an autopsy of the

1

Soviet Union and Soviet Communism, they may reach the verdict of death by ecocide. . . . No other great industrial civilization so systematically and so long poisoned its land, air, water, and people. None so loudly proclaiming its efforts to improve public health and protect nature so degraded both. An no advanced society faced such a bleak political and economic reckoning with so few resources to invest toward recovery."

The situation in Eastern Europe, and the diagnosis of its cause, was very similar: "The legacy of our polluted continent [Europe] can partly be blamed on the policies adopted by the socialist Communist states over the last four decades. The Eastern bloc countries never admitted to pollution problems during the first two decades of the post–Second World War era. In spite of Stalinist and post-Stalinist heavy industrialization policies, pollution of any kind was, according to their propaganda, only to be found in the West where capitalist profit motive was the cause of their environmental degradation problems. Hindsight has now proved the fallacy of such claims, but does not solve the way forward in these countries" (Carter and Turnock 1993, 189).

It would have been surprising if Cuba, a country that eagerly embraced the socialist development model from the early 1960s to the late 1980s, could have escaped unscathed from the systemic environmental failings of socialism. In this book we seek to provide a preliminary overview of the environmental legacy of socialism in Cuba based on the examination of secondary sources, informed by the study of development and environmental trends in other socialist countries as well as in the developing world. Socialist Cuba is a closed society, and independent research within the country is not an option for researchers who reside abroad (such as the authors of this volume) or on the island.[1] Most of our inferences and conclusions have been corroborated in published materials and discussions with scientists who emigrated from Cuba in recent years.[2]

Any issue related to Cuba is controversial, and the environment is no exception. Witness the following two descriptions of the environmental situation in Cuba made at about the same time, the first by a U.S. professor of environmental law who visited Cuba in the early 1990s and the second by a Cuban scientist who defected from the island in 1992 and currently resides in the United States:

> There is an island in the Caribbean where at certain moments you feel that you are wandering through the pages of "Ecotopia," Ernest Callenbach's novel about an environmental and egalitarian utopia. There are few cars and no smog. There are no graffiti, no commercial billboards and signs. The streets are cleaned, trash is picked up. Everything is recycled, nothing wasted: the truck that delivers a box of fruit and

vegetables to the hotel restaurant takes away a box of banana peels and vegetable trimmings from the day before. Some communities get electricity from windmills and cow-dung slurries that generate combustion methane. Small dairy herds have been established and new fields planted to make the island self-sufficient in agriculture and break its dependence on cash crops for export. . . . The island is Cuba. (Benson 1992)

The Cuban scientist Dr. José Oro, former Director General of the Department of Natural Resources of Cuba's Ministry of Basic Industry, in contrast, described the situation as grave and deteriorating further. Commenting on references to "Cuba's favorable environmental conditions" by President Fidel Castro at the UN Conference on the Environment and Development, the so-called Earth Summit, in Rio de Janeiro in 1992, Oro said: "It was funny to listen to Castro's speech in Rio. The environmental degradation [in Cuba] is rapid. We have the industrial production of Honduras and the pollution of East Germany" (Dunleavey and Penenberg 1993, 14).

We have made a deliberate effort to avoid extremes in this book. For this reason, we precede our review of socialist environmental trends with a brief examination of the pre-socialist situation. Through these historical reviews, we seek to establish what the environmental conditions in the country were before 1959 so that we can assess the likely environmental consequences of socialist development policies as opposed to conditions that socialist Cuba may have partially or fully inherited from the country's republican past. In Chapter 2 we provide basic information on the natural, demographic, and economic setting essential to provide a context for the discussion that follows.

Our analysis also recognizes that Cuba is a developing country. As such, in its attempt to overcome economic backwardness, it may have adopted economic practices whose long-term environmental consequences may not have been appreciated several decades ago. From this perspective, many of socialist Cuba's developmental policies gave rise to environmental consequences not very different from those observed in other developing countries, whether in the Caribbean, South America, or elsewhere, that followed a capitalist development path.

Conditions that are generally blamed for environmental deterioration in other developing countries, however, were not present in socialist Cuba between the early 1960s and the late 1980s—the period with which we are mostly concerned—and thus Cuba differs from these other countries in crucial respects. Among the most important mediating pathways to environmental decline in developing countries are population growth, poverty, and unequal access to resources. In addition much blame for environmental deterioration has been placed in recent decades on the introduction of the capital-intensive agricultural technologies associated with the "green revolution" (Vosti and Reardon 1997).

Population growth is regarded as a primary driver of environmental degradation, because it conditions the human–environment relationship by constantly increasing the pressure on the natural resource base (see, among many others, Livernash and Rodenburg 1998). Poverty magnifies the environmental effects of population pressure because "The poor often cannot afford the capital and other nonlabor inputs needed to protect the soil as agricultural intensification proceeds or the off-farm inputs needed to make land improvements not directly associated with intensification" (Vosti and Reardon 1997, 340).

Unequal access to resources is generally a concomitant of poverty and environmental deterioration since it leads to competition for limited resources among the poorest who, in their struggle for survival, and while prevented from gaining access to productive farm land, are often forced to abuse the natural resource base while disregarding the long-term consequences of their actions. More often than not, population growth, poverty, and unequal access to resources interact in a perverse cycle with agricultural modernization; as the introduction of capital-intensive technologies by the wealthiest landowners further deepens rural poverty (by reducing labor demand and blocking the access of peasants to farm land), environmental pressures intensify. William Durham (1979), in his study of the poverty-inequality-environment links in El Salvador, vividly illustrates what has come to be accepted as a classic paradigm.

The socioeconomic changes brought about by the 1959 Cuban Revolution essentially did away with extreme poverty—whether in rural or urban areas—for over a quarter of a century, first, through a radical redistribution of the country's wealth and, second, by the design of employment and income policies that for most of this period resulted in one of the most equitable income distribution regimes in the world. Furthermore, the rural structural transformations brought about by the revolutionary regime in the late 1950s and early 1960s obliterated the traditional "latifundio–minifundio" dichotomy found in much of Latin America that often acts as the nexus of the poverty–unequal access to resources–environment crucible. The Cuban peasantry became proletarized (except for a small peasant sector that was grudgingly allowed to survive, and that accounted for less than 10 percent of the rural work force) as employees of mechanized large-scale farms in which they worked for a salary and were provided with basic social services (e.g., health, education, social security), often in urbanized rural communities expressly built for that purpose. Moreover, by the late 1960s Cuba was following a demographic trajectory that in a few years would lead the country's population into below-replacement fertility, thus mitigating the population–environment connection, a tendency accentuated by the emigration of close to 10 percent of the country's population and continued urbanization. Already by the early 1970s, the absolute size of Cuba's rural population was on the decline.

Absent several of the foremost determinants of environmental decline in rural areas of developing countries, the only remaining explanation for environmental deterioration is agricultural intensification and in particular the capital-intensive practices associated with the socialist agricultural development model that incorporated most elements of the green revolution. And contrary to the situation in other countries in which the consequences of agricultural modernization are compounded by rural poverty and inequality, in Cuba the environmental effects of capital-intensive agriculture can be viewed in isolation. We contend, therefore, that environmental deterioration in Cuba over more than three decades of socialist rule responded to specific conditions not usually found in developing countries—in fact, some of the standard preconditions were absent—but were present in the former Soviet Union and the former Eastern European socialist countries. In Cuba, as in other socialist countries, central planning was oblivious to local environmental circumstances, lack of ownership rights led to the improper use and neglect of natural resources, the all-powerful state did not allow for citizenship involvement in decision making, and the manner in which rewards for individual managers was determined depended on complying with leadership directives, regardless of results. These systemic problems of socialism, together with centralization of political power, were aggravated in Cuba by Castro's personalistic governing style, his utter control over decision making, his whim to follow whatever technological fads he fancied, and his meddling in Cuba's development policies, which in many instances have led to the implementation of poorly assessed economic initiatives with adverse environmental consequences. In Chapter 3 we describe Cuba's legal structure of environmental protection and the harsh reality of poor implementation.

Not to be ignored either was Cuba's wasteful use of vast amounts of Soviet foreign assistance. A significant share of these resources was allocated to mega public works projects, particularly in agriculture, to achieve, in Soviet parlance, the "rapid and inexorable transformation of a backward economy." In other words, in Cuba's environmental realm, the systemic failings of socialism were compounded by its inefficient use of considerable investment resources and a personalistic style of government in which most major decisions ultimately were the responsibility of an infallible "maximum leader," capable of meddling in all facets of political, economic, and social life, and who is not known to shy away from difficult decisions, whether or not he is technically qualified to make them.

We have been equally cognizant of the fact that from a technological perspective, the socialist development model did not differ radically from the model pursued in capitalist countries, or that in capitalist countries, until very recently, environmental consequences were largely ignored in the selection and application of industrial and agricultural technologies. Global environmental deterioration has largely resulted from this disregard for environmental consequences, giving rise to ecological disasters of the magnitude of Love Canal in the United States, Miamata Bay in Japan, and other environmental calamities in Western Europe.

In agriculture, for example, socialist practices were inspired by North American agricultural methods at the turn of the century that were greatly admired by Lenin. For much of the twentieth century, capitalist and socialist agriculture have shared common features such as extensive mechanization, large-scale application of chemical inputs, and widespread irrigation, often with comparable detrimental environmental consequences. What we learned from the Soviet Union's experience is that the systemic characteristics of socialism aggravated the environmental consequences of modern capital-intensive agriculture. The Cuban experience simply serves to corroborate what many other researchers have found in their analyses of socialist agriculture. The same applies to industrial pollution, where the issue is not solely technological, since it is mediated by the economic framework and institutions in which the technologies are applied and by their detrimental effect on the environment.

Cuba's Environmental Ideological Discourse

In Cuba, as in the other former socialist countries, the dominant ideological discourse tended to minimize the adverse environmental consequences of socialist development policies, while claiming that capitalism is at the root of the global environmental deterioration. A representative example of this view was expressed by President Fidel Castro:

> Among the greatest harm that capitalism has inflicted on humanity ... is the deterioration of nature, the destruction of the environment, the mismanagement of forests and soils, the contamination of seas and the atmosphere. Capitalism has created the problems with the ozone layer and the greenhouse effect, which many scientists believe is irreversible. .... In barely 100 years, capitalism has exhausted most of the

fossils fuels on earth, coal and oil, and sometime in the future humanity will remember with horror these 100 years of capitalist development and how it has mistreated nature, how it has poisoned everything and has created situations in which deserts are expanding, agricultural land is shrinking, soils are being affected by salinization, and natural resources are scarce. (Castro 1992, 13–14)

With the fall of socialism, the nature of the discourse metamorphosed somewhat. Since then, the distinction is no longer between capitalist and socialist societies, but rather between consumer societies and the Third World. President Fidel Castro, in remarks at the 1992 Earth Summit, for example, did not point to capitalism directly for causing environmental deterioration, but instead attributed it more vaguely to "consumer societies" (which may or may not include the former Soviet Union and the industrialized Eastern European socialist countries), while drawing a clear distinction between the Third World and industrialized countries (Ministerio de Ciencia 1995, i).[3] According to this revisionist view, the Third World is blameless, since developing countries were colonies whose exploitation continues today under an unfair world economic order. Castro noted "that consumer societies are fundamentally responsible for the abject destruction of the natural environment. They arose out of the old colonial metropolises and imperial policies that, in turn, engender the backwardness and poverty that today afflict the vast majority of mankind." This same viewpoint is echoed by Cuban scholars writing on this topic (González 1992; Alvarez 1992).

Cuba's Environmental Problems

While denying responsibility, and arguing that "concern for protection and conservation of resources, considered the property of all the people, began in Cuba with the revolutionary victory of 1959" (Castro 1993, 44), the Cuban leadership admits that Cuba faces some environmental problems. Cuba's self-congratulatory report to the 1992 Earth Summit (COMARNA 1992), was very scant in terms of identifying environmental problems in the country, a fact duly noted by the UN staff in their compilation of national reports (United Nations 1992b, 118). However, Fidel Castro's long statement prepared for the conference, which emphasized the Cuban socialist government's commitment to preserving the environment and natural resources, made brief reference to some pressing environmental problems. He highlighted the following: (1) pollution of bays; (2) soil erosion and degradation, particularly in mining areas; (3) pollution of surface waters by waste products of the sugar industry; and (4) erosion of beaches and coastal areas and salinization of low-lying coastal lands (Castro, 1993, 46).

These problems were also implicit in Cuba's Programa Nacional sobre Medio Ambiente y Desarrollo (PNMAD, National Program on Environment and Development), a very lengthy document prepared by Cuba in 1993 containing 214 objectives and 816 actions to protect the environment and promote the rational use of natural resources (Vinculación 1993) and in a report presented by Cuba in 1994 to the UN Committee on Sustainable Development (Informe de Cuba 1994). Cuba's report to the Fifth Session of the UN Commission on Sustainable Development (Aplicación del Programa 1997), which met in New York in April 1997, and an Estrategia Ambiental Nacional (EAN, National Environmental Strategy), released in June 1997 (Ministerio de Ciencia 1997a) to coincide with a review session of the UN Conference on Environment and Development, also touch on environmental problems and steps taken by the Cuban government to address them. An environmental education strategy issued at the same time as the EAN spells out steps to enhance public consciousness of environmental problems, relying on formal education programs and informal approaches through the use of mass organizations (Ministerio de Ciencia 1997b).

The June 1997 EAN, although claiming some important achievements, provides a rather somber assessment of the country's environmental situation. Among the achievements cited are the eradication of extreme poverty; improvements in the population's environmental situation and in their quality of life within a social equity framework; increases in the country's forested land area; establishment of protected natural preserves and formulation of a proposal for integrating them into a national system; the systematic assessment of the national territory and environmental evaluation of priority investments; the application of scientific criteria for the assessment and development of technologies to address environmental problems; the gradual introduction of the environmental dimension in the national education system; and the overall strengthening of environmental concerns in the life of the nation (Ministerio de Ciencia 1997a, 1).

The causes for environmental problems are said to be insufficient awareness, knowledge, and education about environmental matters; poor management; limits in the introduction and broad application of science and environmental technology; inadequate attention to the environment in the design and implementation of development policies and plans; and the absence of a juridical system capable of integrating environmental controls in a coherent fashion. Furthermore, the scarcity of financial and material resources interfered with Cuba's ability to attain higher standards of environmental protection, a situation aggravated in the last few years by its economic situation following the loss of commercial relations with the former socialist camp and the sustained and strengthened economic "blockade" by the United States (Ministerio de Ciencia 1997a, 1). The principal environmental problems faced by Cuba, according to this document, were: soil degradation (such as erosion, poor drainage, salinity, acidity, and compaction); worsening of sanitary and environmental conditions in human settlements; contamination of terrestrial and marine waters; deforestation; and loss of biological diversity (9).

This list generally corresponds with the environmental problems we document in Chapters 4 to 9 of this book and testifies to the grave concerns emerging in Cuba regarding the country's environmental situation. Some of these environmental stresses were inherited from the past, but, as we show, some arose or were intensified by sectoral development strategies pursued by the socialist government. Urban pollution, for instance, could be partly traced to Cuba's extreme reliance on inefficient and highly contaminating factories and vehicles imported from the Soviet Union and Eastern Europe. In the agricultural sector a practice that resulted in environmental damage was the promotion of the Soviet-style, large-scale state farm production model (farm gigantism) based on widespread mechanization, heavy chemical inputs (e.g., fertilizers and herbicides), and extensive irrigation. The effect of large-scale mechanization on the compaction of soils has been reported as severe. The pollution of streams and coastal areas by organic waste discharges from the sugar industry has been a major concern for years. By the late 1980s, when sugar production was at its peak, the problem was considered so serious that to lower discharge rates, measures were instituted in more than ninety mills to fertilize sugarcane fields with organic waste.

Some of Cuba's bays became severely polluted because of human, industrial, and agricultural discharges but also by the runoffs associated with the deforestation from strip mining (e.g., in Moa). By the late 1970s, the UN Development Program was providing financial and technical assistance to the Cuban government to arrest the growing contamination of the Bay of La Habana. High levels of chemical and organic pollution were also present in the bays of Nipe, Chaparra, and Puerto Padre and more recently the Bay of Cienfuegos.

Numerous instances of soil salinization and erosion can be traced to waterlogging caused by poor irrigation and drainage practices, to excessive water extraction rates from coastal aquifers, and to schemes that led to the damming of low-volume streams and rivers that dried out during the dry season months. It is estimated that one million hectares, or about 14 percent of the country's agricultural surface, have excessive salt deposits. Of these, about 600,000 hectares are deemed to be affected by light to modest salinization levels and the rest by heavy salinization. The regions with the heaviest levels of salinization are in Guantánamo and the Cauto Valley.

The Cuban sugar industry, which controls the bulk of the industrial stock and the largest industrial enterprises in the island, is an important source of pollution. Sugar production generates very large amounts of air emissions and liquid industrial wastes. By-products of sugar production—especially torula yeast—emit toxic waste products that contaminate streams and eventually coastal areas as they are flushed out to the sea.

The Cuban nonsugar industrial sector is also a heavy polluter, discharging polluting agents into the atmosphere, the sea, or other ecological systems. Among the chief pollutants in the nonsugar industrial sector are: (1) the cement industry, a heavy generator of dust and smoke; (2) the chemical and metallurgical industries, producers of acid steams, smoke, and soot; (3) the steel and nonferrous alloy industries, also heavy producers of smoke and soot; (4) the sugarcane derivative industry, consisting of plants producing torula yeast, bagasse boards, paper, and so on, and generating a variety of air pollutants and solid wastes; and (5) the mining industry, especially nickel mining, which launches extremely large amounts of dust into the atmosphere and releases by-products into streams and the sea.

Who Is to Blame for Cuba's Environmental Problems?

While claiming that the source of some of these problems is Cuba's capitalist past and its exploitation by advanced industrial countries, especially the United States, Castro and some of his associates have recently been quite explicit in extending the blame for many of the country's most serious environmental concerns to the Soviet Union, Cuba's former socialist patron. A particularly clear instance of this line of reasoning was provided by Lionel Soto, former vice president of Cuba's Council of Ministers and ambassador to Moscow, when he declared in an interview—recorded in the Russian press—that the former Soviet Union had incurred a "debt" to Cuba of $20–25 billion (an amount roughly equal to Cuba's estimated financial debt to Russia) by exploiting its natural resources and contaminating its environment (Bai 1995). The implication is that the socialist development policies embraced by Cuba, presumably at the behest of the Soviet Union and its former Eastern European allies, are very much responsible for the environmental deterioration suffered by the country over the last several decades.

In a recent book a Cuban environmentalist associated with the University of La Habana observes that "It is a given that Cuba, with the assistance of the former socialist countries, increased the use of chemical fertilizers, pesticides, mechanization of cultivation and harvesting for example in the sugar industry, and irrigation, among others. This could be accomplished because the conditions had been created so that these advances could be incorporated into the educational system, and science, technology, and education worked closely. There is no exact estimate of the high economic-social cost of such 'advances'" (Cabrera Trimiño 1997, 182–83).

While the specific features of the socialist development model varied from place to place according to political, cultural, and national circumstances, the basic blueprint was inspired by the Soviet Union's historical experience. The essential characteristics of the agricultural organization model that emerged in the Soviet Union and was later adopted by other socialist countries, including Cuba, were the following:

▶ *Large-scale production units:* Farming units in the Soviet Union tended to be very large, presumably because large size facilitated the introduction of modern production techniques and maximized returns from mechanization; but, according to Nove (1965, 3), they also emerged because of administrative convenience. This pattern of large agricultural units has been referred to by Lazar Volin (1962, 254) as "farm gigantism." The Soviet proclivity toward large-scale operations, which extended to industry, has also been referred to as "gigantomania" (Gregory and Stuart 1974, 246).

▶ *Extensive cultivation:* With no rent charged for land use, it was sound decision making by farm managers to increase production by expanding the size of the farm units rather than by more intensely cultivating existing units (Raup 1990, 101). Examples of this tendency are the "virgin lands program" of the 1950s that brought large tracts of lands in Siberia and Kazakstan under cultivation. Much of the new lands brought under cultivation were marginal in terms of soil quality and, more important, subject to severe climatic conditions—dry, hot winds that blew into the virgin lands from the Central Asian desert, coupled with Arctic winds that brought snow as early as August and uneven rainfall (Willett 1962, 101).

▶ *Mechanization:* Lenin's "unbounded enthusiasm" for American tractors, coupled with the belief in the superiority of large-scale production in agriculture as well as in industry, made mechanization a critical part of the Soviet agricultural model (Volin 1962, 250). A large share of investment in the agricultural sector was devoted to the procurement of agricultural machinery and equipment.

▶ *Technological interventions:* Soviet authorities had a proclivity for relying on scientific and technological solutions to bottlenecks arising in the agricultural sector. The view that science and technology could "conquer" the problems of soil quality and unsuitable climate spread the myth of the unlimited agricultural resources of the Soviet Union and diverted attention from the management and incentives problems that were at the core of the agricultural production quagmire. Among the best documented of these technological interventions were the so-called Stalin Plan for the Transformation of Nature in the 1940s consisting of planting shelter belts and reforestation, crop rotations with perennial grasses and construction of ponds and reservoirs (Timoshenko 1953, 254), and a massive project designed to turn the semiarid lands of central Asia into a cotton-producing area through an irrigation scheme in the Aral Sea basin that drew water from the Syr Darya and Amu Darya Rivers, two of the main feeders of the Aral Sea (Akiner 1993, 256).

▶ *Use of agricultural inputs:* Faced with stagnating agricultural production, Nikita Khrushchev coined a new version of Lenin's slogan by declaring that "Communism is Soviet rule, plus electrification of the whole country plus 'chemicalization' of the economy" (Novak-Decker 1965, 193). Demand for chemical fertilizers and pesticides grew rapidly in the 1950s as a result of the expansion of land under cultivation pursuant to the virgin lands program. The drive to cultivate land more intensively and efficiently resulted in even higher usage of fertilizers and pesticides in collective and state farms.

While shifting the blame to others for development policies that increasingly appear to have taken a major environmental toll is a politically convenient rationale, it is not supported by the historical record. A question of considerable interest is whether Cuba could have avoided the environmental pitfalls that we now associate with socialism. Our conclusion is that this would have been unlikely.

During the 1960s, when the revolutionary leadership enthusiastically embraced the socialist development path, Cuban leaders and planners were in awe of the almost superhuman development plans envisioned by the Soviet Union and other socialist countries. In the water sector, for example, where great publicity was being given during the 1960s to the Soviet Union's plans to expand the amount of irrigated land, Cuban technical journals (and the media) published glowing accounts of what was to be achieved. According to one article, by the senior Soviet hydraulic advisor working in Cuba, in the 1965–1975 decade, the Soviet Union intended to increase the amount of irrigated land by 250 percent, or from 15 million hectares to 37–39 million hectares (Drovidech 1966, 48). Cuba was not to be left behind. In 1969 President Castro announced that in five years, thanks to Cuba's vast investments, 50 percent of Cuba's agricultural land would be irrigated (Nuevos cientos 1969, 2). This would have amounted to an extraordinary expansion in the amount of land irrigated (in the order of over 1,000 percent), since prior to the 1959 revolution, less than 4 percent of Cuba's agricultural land was irrigated. These claims were being made despite overwhelming evidence that the water reservoir and irrigation projects were running into major difficulties due to inadequate feasibility studies and poor construction practices.[4]

Conquering Nature

Even a cursory examination of the historical record suggests that the Cuban socialist leadership, since the early days of the revolution, eagerly promoted policies that in retrospect could only have had adverse environmental consequences.[5] In fairness many of the policies embraced were consistent with then-current thinking in developed capitalist and socialist countries (e.g., an agricultural development model that emphasized the widespread use of mechanization and chemical inputs to increase yields and large-scale hydraulic projects such as dams), but it would be disingenuous to claim that Cuba was pressured by the former Soviet Union to adopt such policies. Further, as in most of the world, including other socialist countries and the Soviet Union in particular, the thinking in Cuban leadership circles in the 1960s was dominated by the belief that nature could be conquered to serve humanity's needs. In a speech in 1966 Castro put the issue in near-epic terms: "We will struggle against the difficulties created by nature because, in the end, thus has been the story of mankind; to struggle to overcome the laws of nature, to struggle to dominate nature and to have it serve mankind. This is also part of the struggle of our people" (Castro 1992, 71).

The bigger the development projects, the better. This meant not only the establishment of immense, centrally managed state farms, but also the conceptualization and frequently the development of large-scale infrastructure projects that could not be justified in economic or environmental terms. The "gigantism" mentality, particularly in agriculture, coupled with a tendency to look at projects on a sectorial basis, are major factors behind environmental disruptions in Cuba (Coyula Cowley 1997, 59).

Some Cuban agricultural scientists and officials have recently claimed that they were victims of the imposition of a foreign development model. Two U.S. experts describe the line of argument put forth by these individuals:

> The Classical Model of conventional agriculture that developed during the first 30 years of the revolution was a model imposed from outside. They express resentment toward Soviet and other socialist bloc

advisors who were responsible for technology transfer to Cuba, and they are self-critical for having had a "colonized mentality." They believe that while the conventional model might be appropriate for Europe—where all of the expensive inputs are produced within each nation—for a developing country like Cuba it makes little sense because of the extreme dependency and external vulnerability that it promotes. (Rosset and Benjamin 1993, 22)

This claim belies the historical evidence and tracks with the tendency of Cuban officials to transfer blame to others, even more so if the erroneous policies emanated from the top leadership. The Cuban leadership, including Fidel Castro himself, eagerly embraced the socialist conception of how to bring about economic growth, whether in agriculture or industry. This was in keeping with Marxist-Leninist notions of the role nature ought to play in human development and the conviction that people could dominate it to serve their ends through technology. This view regarding the ability of the new socialist man to dominate nature paralleled the naive economic development perspective in vogue during the early days of the socialist revolution; it assumed that industrializing the country's economy in a few years would be a simple matter as long as the "capitalist" and "imperialist" yokes strangling Cuba were removed.

A lengthy quote from the leading geography textbook used to instruct generations of Cuban high school students, first published in the early 1960s and written by Antonio Núñez Jiménez (one of the country's most influential government officials, intellectuals, and members of the scientific community,[6] with close ties to President Fidel Castro), illustrates the leadership's conviction that under socialism Cuba was prepared to go to any lengths to conquer nature:

The Cuban [man] of socialism and communism not only restores the devastated forests, but also creates new ones; not only stops erosion, but creates new soils, terracing the sides of mountains to better manage the yield of the forests; other mountains will be demolished and their rocks taken to the depths of the seas to build dikes to transform these seas into productive soils; no river or subterranean stream will carry a single drop of potable water to the ocean; dams are being built to stop water courses, while longitudinal canals along the coasts capture waters from rivers to carry them where they are more needed; the endless energy of the sun will be used to desalinate sea waters; the winds will be trapped in powerful engines, and Cubans will dominate marine currents; the internal heat of the planet will be extracted through deeply dug perforations to power our industries of the future; to accomplish a profound transformation of nature, we will build atomic power plants as we construct powerful wind engines; we will learn to cultivate the bottoms of the sea, taking from submerged prairies cattle feed, edible algae, ultimately developing submarine agriculture; we will learn how to dominate hurricanes to capture their enormous energy; we will purify all industrial waters and recycle them; we will change and correct the course of rivers; we will control our variable climate, taking energy from the sun to temper it; we will create clouds and make rain according to agricultural needs. . . . Ultimately, the greatest challenge of man in communist society is to engage in a bloodless battle to transform nature. (Núñez Jiménez 1972, 289–90)

Núñez Jiménez described plans to build enormous dikes between mainland Cuba and nearby islets and cays to block the entry of seawater, remove the remaining water, and thus create new agricultural land. The magnitude of what was being envisioned can best be appreciated by studying the sketches reproduced in Map 1.1. The plans included not only constructing the dikes, but also draining the shallow sea area bound by them and filling the area with earth taken from Cuba's Zapata Swamp and other regions, or developing enormous potable water reservoirs, as was proposed at La Broa, next to the northwestern edge of Zapata Swamp in the southern part of the country, and in one of the arms of Nipe Bay in northeastern Cuba. One of these projects, draining the shallow waters between mainland Cuba and Isla de Pinos (currently called Isla de la Juventud), was claimed to have the potential to add 16,000 square kilometers (or 1.6 million hectares) of agricultural land to the country, constituting Cuba's own virgin lands program and increasing the land area by about 15 percent. In 1967 Castro himself, while recognizing the still hypothetical character of these ideas, revealed in one of his many speeches that National Academy of Sciences and School of Engineering staff were already at work on the project (Núñez Jiménez 1972, 299–300).

Plans were also being developed to change the natural course of the Toa River in eastern Cuba (Núñez Jiménez 1972, 302–303). One of the alternatives being discussed was to divert part of the Toa's flow to the arid Sabanalamar area in the Guantánamo region of southern Oriente Province. This would entail channeling much of the river's water away from its natural east-to-west flow, by connecting the Toa River through a tunnel or canal to the headwaters of the Sabanalamar River, five kilometers away. The engineering would have to take into account the rough topography of the region and the fact that the headwaters of the Sabanalamar were fifty meters below the bed of the Toa. Another option being contemplated by the Instituto Nacional de Recursos Hidráulicos

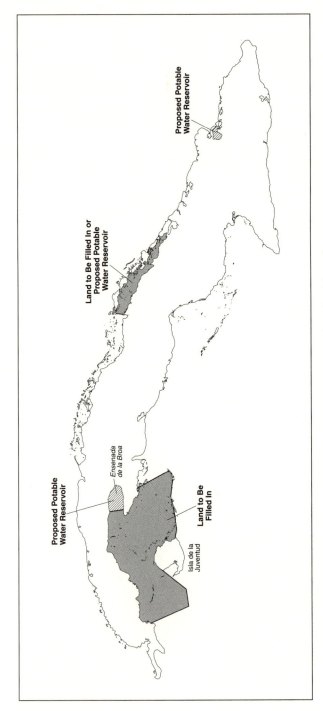

Map 1.1. Proposed projects to reclaim land from the sea

was to channel the Toa River to the Yateras River, which in turn was to have been redirected to the area of Guantánamo.

These projects, of course, never got off the ground, suffering the same fate as others that proposed, among other ideas, desiccating parts of the Zapata Swamp, Cuba's largest wetlands and a virtual natural treasure due to its biological diversity; this project, like many others, was obviously at variance with the avowed environmental preservation principles that many have suggested were at the heart of the development policies of the socialist government. A pilot project conducted at considerable cost led to the conclusion that this, like many other plans, was unfeasible or uneconomical, or both, and was quietly abandoned. Other grandiose projects, particularly in the agricultural sector, did proceed. Their detrimental environmental consequences, discussed in this volume, are just beginning to be appreciated. Suffice it to say at this point that the Cuban socialist discourse took pride in the endless repetition of slogans, most coined by Castro himself, which implied socialist Cuba's will (indeed, the phrase *voluntad hidráulica* or "hydraulic will" was coined by Castro to refer to the country's water policies[7]) to conquer nature: for example, "that not an inch of land should be left unused"; "that not a single drop of water be lost, that not a drop of water reach the sea . . . that not a single stream or river not be dammed." Speaking in 1970, President Castro (1992, 71) stated that the work ahead was "to complete the task of conquering rivers, complete the task of conquering floods, conquer nature. Unless we conquer nature, nature will conquer us."

This mentality has endured. As late as 1991, when the world was already well aware of the disastrous environmental consequences of socialist attempts to tamper with nature, the Cuban press continued to describe in glowing terms other ambitious hydraulic projects, such as the proposed Cauto–El Paso dam. This dam was said to be the "Baikal of Granma Province," an unfortunate reference to the once-pristine Russian lake that has been polluted by a pulp and paper plant and other industrial activities (En Cauto–El Paso 1991). Tourist development complexes, either recently completed or currently under construction, are further evidence of the same mentality, since there is a willingness to sacrifice the natural ecosystems if necessary to achieve pressing economic objectives.

The Cuban Experience Compared to That of Other Socialist Countries

To be sure, the environmental consequences of socialism in Cuba appear to differ in several important respects from those of the former Soviet Union and Eastern European socialist countries, although they share common systemic roots. One of the factors contributing to the divergent environmental paths between Cuba and the former Soviet Union and European socialist world was Cuba's initially weak industrial base and its assigned role within the global socialist division of labor: a producer of primary goods for the more industrialized economies. Thus, when one speaks about the environmental legacy of socialism in Cuba, attention immediately turns to agriculture and mining, sectors whose development strategy was guided by Cuba's natural resource endowment and place in the socialist international division of labor.

The most significant environmental legacy of socialism in Cuba will be in the agricultural sector. This should not come as a surprise since Cuba was, and still is, primarily an agricultural country. The zeal with which capital-intensive agriculture was implemented in socialist Cuba is widely believed to have led to the degradation of many of Cuba's soils. The socialist capital-intensive agricultural development model, with its mammoth-sized farms and great reliance on heavy equipment to work them, has led to soil compaction, and the excessive application of chemical inputs has contributed to a litany of maladies, ranging from contamination of soils and water bodies to problems with secondary pest infestations. There are alarming reports of widespread erosion, but assessing its extent, severity, and consequences must wait for carefully conducted studies of soil conditions in different Cuban regions. Chapter 4 includes a review of environmental concerns associated with agriculture. In Chapter 10 we discuss the alternative agricultural development model that began to be implemented in Cuba in the early 1990s partly in response to the inability to import foreign agricultural production inputs. This new agricultural model largely rests on the reintroduction of traditional peasant practices while giving added emphasis to modern organic farming methods.

The attempt to harness nature and increase agricultural yields—by constructing hundreds of large and small dams for irrigation—drove an ambitious water development program that has contributed to the salinization of the country's soils. Major culprits were inattention to proper drainage of irrigated fields, saltwater intrusions due to the excessive extraction of subterranean waters from aquifers, and tampering with the natural flow of rivers and streams. In some of Cuba's rivers, seawater reaches the walls of inland dams. Pollution is a major source of concern in many of Cuba's rivers and aquifers. There is also concern about the long-term consequences of some of the water development projects, since it has been posited that they could accentuate the destructive effects of natural disasters, like hurricanes.

Because the book is structured according to natural resource, some overlap occurs in the discussion of the relationship between water and environmental deterioration. A perspective of the environmental nexus of water and agriculture is provided in Chapter 4, and Chapter 5 focuses on broader water policies and their environmental impact, inclusive of some of their consequences in the countryside.

Cuba has been fortunate not to experience the monumental levels of industrial pollution recorded in many parts of the Soviet Union and Eastern Europe. Nevertheless, the bulk of the industrial infrastructure (e.g., plants that produce sugar derivatives, cement, and chemicals) that Cuba acquired under the tutelage of its socialist mentors exhibited an inattention to environmental matters not unlike industrial plants in socialist Europe. This disregard for the environmental consequences of development projects was also manifested and continues to be seen in some of the country's largest mining projects, especially in Eastern Cuba. In Chapter 7 we offer an overview of the environmental consequences of the Soviet-supplied industrial equipment and of open-pit mining practices initiated before the revolution, continued under socialism, and intensified today in partnership with foreign mining interests.

In some instances the differences in environmental outcomes between the former socialist countries and Cuba were the result of timing considerations and in others were accentuated by divergent paths in social policy. That Cuba managed to avoid (at least temporarily) the environmental threat posed by electricity-producing nuclear power plants has been largely fortuitous and caused by the country's inability to implement on schedule an ambitious nuclear energy program begun in the 1970s that called for the construction of three nuclear power plant complexes. When the Soviet Union collapsed, Cuba was nearing completion of its first nuclear facility at Juraguá, in south-central Cuba. The yet-to-be completed plant has been mothballed since 1992 because of Cuba's inability to obtain international financing for the remaining construction and equipment. In early 1997 President Castro announced that plans for completing the Juraguá plant had been indefinitely postponed. If the Juraguá plant is never completed, as now appears to be the case, it will become one of the largest failed development projects in the developing world, a white elephant in which Cuba wasted over one billion pesos. We discuss issues associated with Cuba's plans to develop nuclear power for electricity generation in Chapter 8.

Cuba also appears to differ from the former socialist bloc countries, especially the Soviet Union, regarding a posited cause-and-effect relationship between environmental deterioration and declining health standards. Numerous studies (see, for a review, Feshbach and Friendly 1992, 181–203) discovered a higher incidence of morbidity, rising infant and child mortality rates, and declining life expectancies in Soviet cities subjected to exceptionally high levels of air pollution. Many of these adverse morbidity and mortality trends have also been attributed to a contaminated supply of food and drinking water as well as to major problems supplying and managing the national public health sector. The evidence conclusively indicates that the Soviet Union grossly neglected the public health needs of its population.

For reasons noted earlier, Cuba managed to avoid the most egregious consequences of Soviet-led industrial development policies and allocated an inordinate amount of national resources to the health sector. National health care policy initiatives, subsidized by Soviet transfers and supported by equipment and medicine imports from the West, allowed Cuban health authorities to improve the health standards of the nation, at least up to the early 1990s, when the national economic crisis affected every sector of the national economy. A big question remains regarding the potentially adverse health effects that socialist development policies, particularly the use of chemical inputs in agriculture, may have had in contaminating the national water supply and through this on the nation's health. Recent health statistics and epidemiological studies are not available to assess potentially adverse trends (although some limited datasets suggest a deterioration of health trends). As will be shown, there is evidence that water pollution levels in the country are high enough to warrant the suspicion that they could be having a negative impact on the health of the Cuban people. These issues are addressed in Chapter 5, where we discuss water policies of the socialist government, as well as in Chapter 10, where we review the environmental consequences of the economic crisis of the 1990s.

While socialism's most lasting environmental legacy will be in the rural sector, some of the most visible and tangible manifestations of environmental decay are currently seen in urban Cuba, as reviewed in Chapter 9. La Habana, in particular, is a crumbling and environmentally aggrieved city, whose deterioration can be directly blamed on the leadership's decision not to allocate sufficient resources to maintain its housing stock and physical infrastructure and to control pollution, in order to promote a more balanced pattern of regional development. Its bay and surrounding beaches are heavily polluted, as are rivers and streams flowing through the metropolitan area. While visitors are appalled at the disrepair of the city's housing stock, they are much less aware of the chaotic conditions of La Habana's water distribution and sewerage systems. The city suffers from a severe shortage of potable water, mostly because of leaks in poorly maintained and obsolete distribution systems, and industrial and human effluents contaminate water resources. Industrial pollution is a major problem as well. The evidence suggests that the environmental situation in other large Cuban cities is as dismal as in La Habana.

Positive Environmental Trends Under Socialism

From an environmental perspective, not all socialist trends have been negative, however. Cuba's demographic growth has declined considerably, with the population currently growing at an annual rate well below 1 percent, and is rapidly approaching stabilization, if not population size decline. The economic crisis of the 1990s has further contributed to the long-term trend of fertility decline. Population size is expected to stabilize (or begin to decline) at fewer than 12 million people within the next few years, Cuba being the first Latin American country to achieve this distinction. If current fertility and emigration trends continue, Cuba's population will begin to contract by the early twenty-first century. Population density is likely to stabilize at around one hundred inhabitants per square kilometer, twice as high as in 1950 (when it had fifty-three inhabitants per square kilometer). Although this population density is well above the Latin American average, it is lower than for the smaller Caribbean island countries.

Socialist Cuba has also reversed a deforestation trend that had plagued the country throughout its modern history. Although it is apparent that the amount of land area covered by forests has expanded as a result of more than three decades of reforestation efforts, it is difficult to determine how successful these efforts have been from a broader environmental perspective, since they have not been sensitive to preserving biological diversity or conserving endemic species. There is also concern that some of the gains of nearly thirty-five years of reforestation policies may be lost as Cubans are forced to turn to the forests for lumber, firewood, and charcoal to address basic needs under the economic exigencies of the "special period," the label used by the socialist government to refer to the emergency economic policies implemented since 1990 following the collapse of the socialist world. We examine the forestry policies of the socialist government in Chapter 6.

The Special Period and Beyond

As we discuss in Chapter 10, the economic crisis of the 1990s has had major environmental consequences partly because Cuba has been forced to abandon many of the development policies it implemented under Soviet tutelage. With the financial support provided by Soviet subsidies and secure markets for its exports with the socialist bloc, Cuba for several decades pursued development policies characterized by the inefficient use of energy and other production inputs—particularly chemicals—in agriculture as well as in other sectors. With the disappearance of Soviet subsidies, Cuba has had no choice but to idle many of its inefficient industries because of its inability to purchase energy and other inputs and to reverse the course of three decades of mechanized and chemically dependent agriculture. Our assessment leads us to conclude that these changes, at least over the short term, will have beneficial environmental impacts since they have forced the reversal of many development policies that had, or could potentially have had, adverse environmental impacts. Their economic costs have been staggering, however, with levels of production in practically all sectors of the economy declining precipitously. Obvious examples of positive interactions between the special period and the environment are the partial abandonment of the capital-intensive agricultural development model and the mothballing of the Juraguá nuclear power plant.

On the other hand, the emergency economic policies instituted by the leadership to cope with the economic crisis pose other environmental challenges that have yet to be fully appreciated or studied. The most obvious policies, noted by some observers, have to do with the crash programs to develop the tourism sector that have been launched primarily in association with foreign investors. Some of the newer tourist projects have been sited in formerly pristine coastal locations that until now, and primarily because of the leadership's decision in the 1960s to limit the development of the international tourist industry, had remained largely in their natural state. There is evidence that in at least some of the tourism sites, environmental concerns have been sacrificed to economic expediency. There is also fear that the mining activities currently being encouraged, including off-shore oil exploration, may result in the further contamination of land or coastal areas.

Although revolutionary Cuba has developed an extensive legal infrastructure presumably designed to protect the environment (described in Chapter 3), our review of the evidence leads us to believe that there is a lack of vigor in implementing environmental laws and regulations, particularly during the special period. Some observers believe that the government lacks, in the face of a dire economic situation, the political will to rigorously enforce regulatory standards.

Perhaps more ominous is that under the current economic and political circumstances Cuba simply does not have access to the levels of financial resources that would be needed to implement aggressive restoration projects to reverse decades of environmentally damaging development policies. In the early 1980s Cuba defaulted on its loans to Western banks and official credit institutions, shutting off access to new funds; Cuba is not a member of the international financial institutions (e.g., International Monetary Fund, World Bank, International Finance Corporation, and Inter-American Development Bank) and therefore cannot draw on these sources of funds to finance environmental remediation. As long as U.S.-Cuban relations continue to be strained, this situation is not likely to change. For example, under Section 104 of the Helms-Burton Law, the United States is directed to vote against Cuba's admission to the international financial institutions, which are major potential sources of funding for environmental protection and restoration projects (Roy 1997, 82).

Particular areas of concern are related to the salinization of Cuba's soils and underground water resources, many of which resulted from the implementation of hydraulic development projects neglectful of complementary drainage infrastructure, and that may have for many decades changed the balance between surface and underground water sources. No less disturbing is the enormous waste of financial resources associated with the acquisition over several decades of an industrial and transportation infrastructure that proved to be as environmentally unfriendly as it was uneconomical to operate. These considerations suggest that over the short and medium term, priority should be assigned to preventing further damage to the environment, with environmental remediation to be tackled as financial resources become available.

Thanks to the educational policies pursued under socialism, which were in turn made possible by the generous economic subsidies and the scholarships awarded by the Soviet Union and other former European socialist states, socialist Cuba managed to train a large pool of professionals. This cadre of professionally trained personnel possesses the essential human capital to assess, under the right political and economic circumstances, Cuba's environmental situation and to design and implement appropriate remediation initiatives. This has yet to occur, however, since Cuba's socialist government continues to disregard technical advice, and political and economic priorities generally override environmental concerns.

·❧ 2 ❧·

The Natural, Demographic, and Economic Setting

IN THIS chapter we provide a brief overview of Cuba's principal natural, demographic, and economic features to place in context the environmental issues addressed in subsequent chapters. We begin with a description of the dominant natural features of the country, including climate and precipitation, and follow with a review of demographic trends in this century, particularly during the last fifty years. We then address the main characteristics of the Cuban economy, both before and after the 1959 revolution. Included is a discussion of the economic crisis that has gripped Cuba since the collapse of the socialist community in the early 1990s.

Physical Setting

Cuba, the largest of the island nations of the Caribbean, is actually an archipelago that includes the island of Cuba proper, the Isla de Pinos (called by the socialist government Isla de la Juventud), and several thousand small and large cays and inlets (see Map 2.1). The country has a total land area of 110,922 square kilometers, of which the main island accounts for 105,007 square kilometers (94.67 percent of the total), Isla de Pinos for 2,200 square kilometers (1.98 percent), and the cays and inlets for 3,715 square kilometers (3.35 percent). The two largest sub-archipelagos along the northern coast are Los Colorados (with about sixty cays and inlets) off Pinar del Río Province and the Sabana-Camagüey (with about four hundred cays and inlets) that runs along central Cuba. The most important along the southern coast are the Canarreos, which includes the Isla de Pinos and about three hundred cays, off the Zapata Swamp, and the Jardines and Jardinillos de la Reina in eastern Cuba. Because of its elongated and narrow shape, the aerial distance in the island of Cuba between the two further southeast (Punta Maisí) to northwest points (Cabo San Antonio) is 1,250 kilometers, and at its most narrow the island is only 32 kilometers.

Administrative Division, Principal Geographical Features, and Contemporary Land Use

For most of its modern history, and throughout the period lasting from 1902 (when the country formally gained its independence) until 1976, Cuba was divided into six provinces (from west to east: Pinar del Río, La Habana, Matanzas, Las Villas, Camagüey, and Oriente). The revised administrative division implemented at that time created eight new provinces. Pinar del Río remained roughly as it had been, the city of La Habana became a province in its own right (Ciudad de la Habana), and the former province of Las Villas was divided into three new provinces (Villa Clara, Cienfuegos, and Sancti Spíritus). The province of Camagüey was split into two (Ciego de Avila and Camagüey), and Oriente was divided into five (Las Tunas, Holguín, Granma, Santiago de Cuba, and Guantánamo). Isla de Pinos, renamed Isla de la Juventud, though not a province, was granted its own administrative identity (see Map 2.2).

Plains dominate the Cuban geography. The most important mountain ranges are found in the eastern part of the country, and secondary but much smaller mountain ranges are found in central and western Cuba (see Map 2.1). By world or even Caribbean standards, the elevation of most Cuban mountains is modest. Only in the Sierra Maestra in eastern Cuba do the peaks reach altitudes of 1,200 meters and higher above sea level, the Pico Turquino in Granma Province (formerly part of Oriente Province) being Cuba's highest at 1,974 meters. The tallest mountain in central Cuba is the Pico San Juan at 1,156 meters in the Sierra de Trinidad in Cienfuegos Province, and in western Cuba it is the Pan de Guajaibón in the Sierra del Rosario (in Pinar del Río Province) at 692 meters.

At the time of arrival of Spanish explorers, forests dominated Cuba's landscape, accounting for close to two-thirds of the total land area, with much of the remaining land being covered by natural meadows. The types and quality of Cuba's soils are varied, with ten major soil groupings and twenty-nine soil types having been identified. Many of Cuba's soils, particularly if intensively cultivated,

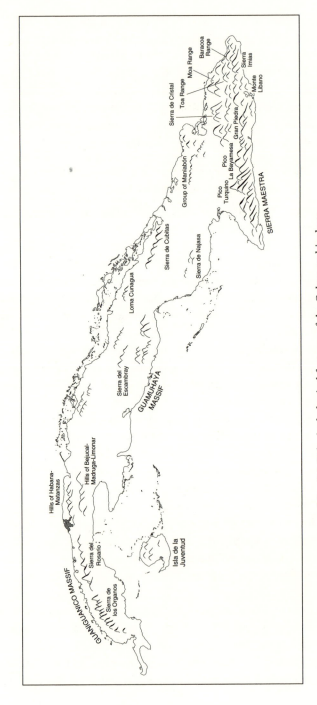

Map 2.1. Principal physical features of the Cuban archipelago

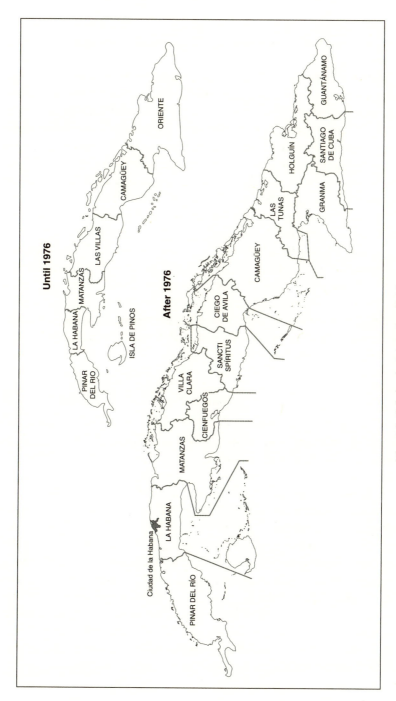

Map 2.2. Cuba's administrative divisions

are prone to erosion, due in particular to the country's wet, humid, hot climate and the intensity of its rainy season. Flooding and waterlogging are recurring problems.

Many of these soils are rich in various metals with economic value, including iron and nickel, of which the country has some of the major world reserves. Most iron and nickel reserves are found in northeastern Cuba. The country also has significant copper deposits in its southeastern and westernmost regions. Chrome, manganese, and zinc are among other metals found in commercial quantities. Hydrocarbon deposits, however, are quite limited and mostly found in locations along the north-central coast. Some geological markers suggest that important hydrocarbon deposits may be found offshore, but to date the results of explorations have been disappointing. Asphalt deposits are extensive in the marshy region of the Zapata Swamp.

In 1989, of the total land area of 11 million hectares, 6.8 million hectares, or 61.5 percent, were used for agriculture, whether under permanent or temporary crops. Of these, 4.4 million hectares were cultivated, whereas about 2.4 million hectares were devoted to pastures or kept fallow. Of the remaining 4.2 million hectares not dedicated to agricultural uses, 2.6 million hectares were covered with forests; human settlements occupied 694,000 hectares; and 606,000 hectares were regarded as "unusable." The balance of the land was covered by water bodies.

Water Bodies

Although the country has numerous rivers and streams, watersheds are limited and the course of water flows is generally short due to Cuba's narrow and elongated shape. Except during the rainy season, the beds of many small rivers and streams dry out. The vast majority of the rivers run from the central spine of the island to either the northern or southern coasts (see Map 2.3). The mean length of the country's major rivers is only 93 kilometers. The longest river, with a length of 370 kilometers, is the Cauto, which runs from the mountains of eastern Cuba to the southern coast. Some of the Cauto's tributaries are also among the longest Cuban rivers (of particular note are the Salado at 126 kilometers, the Contramaestre at 96 kilometers, the Bayamo at 89 kilometers, and the Cautillo at 84 kilometers). The rivers with the two next longest courses are the northward-flowing Sagua la Grande (163 kilometers) and the southward-flowing Zaza (155 kilometers), both in central Cuba. Arguably the country's best-known river, the Almendares, which runs through the western outskirts of the city of La Habana, is only 52 kilometers long.

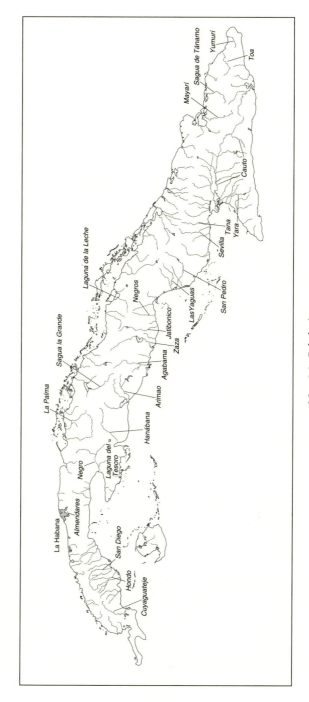

Map 2.3. Cuba's rivers

Naturally occurring lagoons and other bodies of still water are few and generally small. The two largest saltwater lagoons are the Laguna de la Leche (67 square kilometers) and Laguna Barbacoa (19 square kilometers). The two most important freshwater lakes are the Ariguanabo and Laguna del Tesoro (9 square kilometers each). Many man-made reservoirs were built, particularly in the 1960s and 1970s. At the beginning of the 1990s, Cuba had two hundred large dams and eight hundred minidams. Water bodies covered 330,000 hectares, or 3 percent of the country's land area.

Coastlines

Cuba's coastline measures 5,746 kilometers, 3,209 of which constitute the north coast, and the coastline of Isla de Pinos runs for 327 kilometers. The archipelago has hundreds of kilometers of natural, sandy beaches, many of which remain largely in their natural state. Cuba is also dotted with many natural harbors, some of which, due to the depths of their channels (the deepest being Matanzas and Guantánamo bays) and their bag or horseshoe shape, offer excellent protection. Among the largest are Nuevitas (with a length of 22.2 kilometers and a width of 25.9 kilometers), Nipe (25.9 and 16.8 kilometers), Puerto Padre (11.1 and 13.8 kilometers), Guantánamo (20.3 and 9.2 kilometers), and Cienfuegos (18.5 and 6.4 kilometers). La Habana Bay, although the most heavily used, is relatively small (3.7 and 4.6 kilometers) in comparison with others. Living coral reefs are abundant, and in some sections of the country extend for hundreds of kilometers. Only Australia's Great Barrier Reef is longer than the reef along Cuba's northern coast. About half of Cuba's coastline contains fossilized coral formations.

Climate and Precipitation

Cuba has a generally warm and humid climate. Annual mean temperatures range between 24 and 27 degrees centigrade, and annual fluctuations in monthly mean temperature in different parts of the country range between 4.8 and 6.8 degrees centigrade. Regional temperature fluctuations are mostly determined by altitude and by the moderating effects of the sea. Maximum temperatures recorded have reached 40 degrees centigrade, and the thermometer has dipped as low as 1– 2 degrees centigrade on the country's highest mountain peak.

The country has two well-established seasons. During the rainy season, which runs from May through November, Cuba receives abundant rainfall. The amount of annual and regional rainfall is profoundly influenced by hurricanes, with the hurricane season lasting from July through November. In the dry season temperatures moderate, particularly in western Cuba. The country receives alternate cold and warm fronts as dry and Arctic air masses reach Cuba and interact with warm southern winds. Because of these alternating patterns, some rainfall occurs in Cuba even during the dry season. Regional annual precipitation ranges between 400 and 5,000 millimeters. Eight distinctive ranges of precipitation have been identified. The largest volume of rain falls in the Moa–Toa–Baracoa region of eastern Cuba, and the least in the semidesert cactus-scrub belt of southern Guantánamo Province. Across the island, the level of precipitation tends to increase from the coast to inland locations, the amount of precipitation also being positively correlated with altitude. The drier areas tend to correspond with the southern side of mountain ranges, particularly in eastern Cuba. The northern side of mountain ranges tends to be the wettest (Borhidi 1991, 36–48).

Fauna and Flora

Cuba, like other large islands, has a large number of animal species, including a good many endemic (found only there) to the island. Some 12,400 animal species have been identified, of which 42 percent are endemic to Cuba (Silva Lee 1996, 58). Mammals, of which thirty-one endemic species are known, are not abundant and generally are of small size. The country has twenty-three species of bats—the most widespread mammal—six of rodents, and two of insectivorous animals. The survival of some of these species, such as the *almiquí*—an insect-eating mammal considered to be a living fossil—various types of *jutías* (a rodent), and the *manatí* (a water mammal also found in the southern United States and in other parts of the Caribbean, including La Hispaniola and Puerto Rico) is threatened, and some of these species are believed to be at the brink of extinction, if not already extinct. Many exotic species of mammals, domestic as well as wild, have been introduced in the country since pre-Columbian times. This has led to the extinction of some endemic species. Most recently, during the period of socialist rule, two dozen species of exotic mammals were introduced by an institution "supposedly dedicated to protect the island's flora and fauna . . . among them were 5 species of goats, 6 of antelopes and 3 different cervids, all liberated in special enclosures for acclimatization. Five species of monkeys were also imported (about 500 animals) and later disembarked on large keys of fragile and valuable nature" (Silva Lee 1996, 140), raising fears that other endemic species may meet a similar fate.

Cuba has about three hundred species of birds, seventy of which are endemic. The archipelago is also the wintering ground for numerous northern species and serves as a rest stop for flocks of other species in their yearly migration further south. Flocks of these birds, numbering in the millions, congregate in the Sabana-Camagüey sub-archipelago, an area targeted during the 1990s for major tourism developments. Cuba's fauna also includes some eighty species of reptiles, a majority of which are endemic to the island. The waters in and around the Cuban archipelago are home to some 720 species of fish, including thirty-five types of sharks.

The richness of Cuba's fauna is particularly evident with respect to crustaceans and mollusks, of which over 38,000 and 1,500 species, respectively, are known. The crustaceans include some economically valuable species, such as the spiny lobster and various kinds of shrimps, as well as several types of land crabs. More than 18,000 species of insects, ranging from butterflies to mosquitos, have been identified.

Because of the size of the main island and the location and relative isolation of the Cuban archipelago, scientists believe that Cuba probably has the highest degree of biodiversity in the Caribbean and is extraordinarily rich in marine resources. It is claimed that there are probably more endemic species in Cuba than in the entire Amazon region of Brazil (Cole and Domínguez 1995, 2).

The flora of Cuba is equally rich and varied. Approximately 6,700 species have been identified, with an endemism of 51.4 percent (Acosta Moreno 1995a, 25). A. Borhidi (1991), in his monumental study of Cuba's phytogeography and vegetation ecology, classifies, on the basis of physical and ecological characteristics, ten primary vegetation types (rainforests, seasonal evergreen forests or seasonal rainforests, semi-deciduous forests, tropical karstic forests, dry forests and scrubwoods, semidesert cactus scrubs, coniferous forests, savannas and grasslands, freshwater vegetation formations, and coastal vegetation), each of which is subdivided into numerous subtypes. This plant diversity has been under constant siege for centuries because of large-scale deforestation, agriculture, and the spread of invading exotic species.

Demographic Setting

In comparison to most Latin America and Caribbean countries, contemporary Cuba has attained a unique demographic profile. Cuba's rate of natural increase (the difference between the birth and death rates, expressed as events per a comparable base, usually 1,000 population) is expected to become negative by the year 2005 as the number of deaths exceeds the number of births (Acosta 1996). Negative natural increase and continued emigration (if it were to continue at the rates recorded during the first half of the 1990s) will make Cuba, together with several Western European nations, one of the few countries in the world experiencing net population size declines.

Population Growth

Cuba's potential declining population size represents a major historical reversal, since for most of this century Cuba's population grew rapidly. In 1899 Cuba had only 1.572 million people. The next three decades saw a major population increase that resulted primarily from high levels of immigration and fertility; in this interval the national population more than doubled, reaching 3.962 million in 1931. Growth rates during these three decades were close to or above 3 percent per annum. During the 1931–1943 intercensal period, the annual population growth rate declined to 1.6 percent, as many migrants left Cuba propelled by the adverse economic conditions associated with the Great Depression and as the birth rate began to decline. In the next intercensal period (1943–1953) the population growth rate rose to 2 percent per annum, partly as a result of a rapid mortality decline, with the national population reaching 5.829 million by 1953.

The antecedents for the current trend of slow demographic growth date back to the early part of the century, as Cuba was well into its demographic transition by the 1930s, when mortality and fertility rates had already started to decline (Díaz-Briquets and Pérez 1982; Hernández Castellón 1988). As can be seen in Table 2.1, which provides UN demographic estimates and projections, with a population of slightly less than six million in 1950, the country's population growth rate was 1.85 percent for the period 1950–1955, well below that for Latin America and the Caribbean as a whole, although it was slightly higher than in the smaller Caribbean island nations. Population growth accelerated during the 1960s and early 1970s as Cuba experienced a short-lived baby boom, total population size increasing by more than 50 percent between 1950 and 1975.

After 1975, however, population growth rates began to rapidly decline, except during the late 1980s and early 1990s when the baby boom cohorts of the 1960s and 1970s began to have their own children. Between the late 1950s and early 1990s Cuba also managed to sustain a gradual mortality decline. This can be assessed by examining the trends in life expectancy at birth and in infant mortality shown in Table 2.1, which are among the lowest in the Western Hemisphere.

Since the 1975–1980 quinquennium, the country's total fertility rate (or the number of children the average woman is likely to have over her lifetime) has been below the level required to assure the replacement of the population (at

Table 2.1. Demographic trends in Cuba, 1950–2010

Year	Population			Population density (per square kilometer)	Total fertility rate[a]	Infant mortality rate[a]	Life expectancy at birth[b]
	Size (in thousands)	Growth rate (%)	Median age				
1950	5,850	—	23.3	53	—	—	—
1955	6,417	1.85	23.2	58	4.10	81	59.4
1960	6,985	1.70	23.4	63	3.68	70	62.3
1965	7,754	2.09	23.7	70	4.67	59	65.3
1970	8,520	1.88	22.3	77	4.29	50	68.4
1975	9,306	1.76	22.7	84	3.55	38	70.9
1980	9,710	0.85	24.2	88	2.13	22	73.0
1985	1,0102	0.79	25.7	91	1.78	17	73.9
1990	10,958	0.96	27.8	96	1.78	13	74.6
1995	10,960	0.67	30.2	99	1.65	12	74.6
2000	11,182	0.40	33.0	101	1.55	11	75.3
2010	11,408	0.20	38.7	103	1.58	9	76.0

[a]Rates refer to the quinquennium preceding the reference period.
[b]Both sexes combined.
Source: United Nations 1995a, 602–603.

Cuba's current mortality, this level roughly corresponds to about 2.1 births per woman). Notice, too, that the fertility decline accelerated in the 1990s as women began to adjust their fertility decisions in light of the difficult economic conditions associated with the economic crisis known as the "special period in time of peace" (more on this later).[1]

The UN projections, however, do not fully capture the rapidity of the fertility decline in the 1990s. Other estimates suggest that by 1993 the total fertility rate was at 1.5 births per woman (Ministerio de Salud Pública 1994, 22; Population Reference Bureau 1996). By 1997, it is likely to have declined even further. Based on current trends, the population of Cuba may never reach 12 million people (Acosta 1996). Major fertility upturns are not to be expected, even if economic conditions improve sharply, since contraceptive use and abortion rates are high (Díaz-Briquets and Pérez 1982; Hernández Castellón 1988; Alvarez-Vázquez 1993; Noble and Potts 1996), and because Cuban women have consistently expressed a desire for small families (Acosta 1996). As a result of the declining fertility trend, Cuba's median age has increased substantially (reaching thirty years in 1995), as has the share of the elderly in the population. These trends, together with population stabilization, carry important long-term social, economic, and environmental implications.

Population Density

Following declining fertility and large-scale emigration over the last forty years, population density (persons per square kilometer) is unlikely to double from the level of the early 1950s given that population growth is expected to cease or even to become negative. As can be seen in Table 2.1, population density increased fairly rapidly from 1950 until about 1995. It is expected to stabilize after 1995 and will most likely remain at about one hundred inhabitants per square kilometer (barring unforeseen upturns in fertility or immigration), or close to 50 percent higher than when socialist development policies began to be implemented in the early 1960s. This means that the per capita availability of all fixed natural resources has declined by about half in the last half century. Agricultural land area per capita, for example, declined from 1.12 hectares per capita in 1950 to 0.6 hectares per capita in 1995. Cuba's population density continues to be lower, however, than that of other small island nations in the Caribbean but considerably higher than in other Latin American and Central American countries.[2]

Rural-Urban Population Shifts

This demographic transformation has been accompanied by the continued urbanization of the country. By 1996, Cuba was primarily an urban country, with 78 percent of its population residing in cities and towns.[3] The percentage of the population residing in urban areas, according to the Centro Latinoamericano de Demografía (CELADE 1995, 31–32), was just slightly above the figure for Latin America as a whole (74 percent) but below that of several countries that are even more urbanized than Cuba (e.g., Argentina, 88 percent urban; Chile, 84 percent; Uruguay, 90 percent; Venezuela, 86 percent). In 1996, 8.56 million Cubans lived in localities designated as urban, and 2.48 million in rural areas. Because of the overall changes in population growth rates, urban growth was accompanied by declining urban population growth rates and by a gradual but significant decline in the size of the rural population. Whereas rural population growth rates were consistently negative during the period in question, urban rates through the period were consistently about twice as high as national growth rates. Between 1970 and 1995, the size of the rural population declined by close to one million people, from 3.398 million to 2.481 million, or by 27 percent. These trends may have been reversed during the special period as the government, in an attempt to increase agricultural production, began to implement several initiatives to shift part of the urban population to the countryside. During the period of socialist rule, La Habana, the capital city, has continued to maintain its prominence as the country's major urban agglomeration, accounting today for approximately 20 percent of the country's total population, a percentage similar to that in 1953. Other major urban centers have grown considerably in population size (for further details, see Chapter 9).

Urban population growth together with agricultural intensification appear to have had a significant role in the environmental stresses experienced in Cuba over the last several decades. Water use rates increased not only to expand the farm land under irrigation but also to serve the needs of a burgeoning urban population. According to census data, the percentage of dwellings receiving water from wells, rivers, and springs declined, for instance, from 42 percent in 1953 to 8 percent in 1981; whereas the percentage of dwellings without sanitary facilities dropped from 23 percent to 9 percent during the same time period (Díaz-Briquets 1988, 57–58). By the end of 1990, 83 percent of urban residents and 56 percent of rural residents were served by aqueducts; the corresponding percentages for sewerage systems were 39 percent for urban residents and 3 percent for rural residents (COMARNA 1992, 3). These achievements were made possible partly through a vast expansion in the national water distribution and sewage collection networks. Prior to the revolution, water supply problems were already in evidence and intensified with the increased demand for water to serve the needs of a better serviced and rapidly expanding urban population. However, the new facilities have been improperly maintained, and the percentage of the population being served is actually much lower than suggested by the statistics (see Chapter 9).

In summary, to the extent that demographic factors contribute to environmental stresses, Cuba has achieved a relatively advantageous situation. Within the next few years, the size of its population will stabilize, if not begin to contract. Four out of every five Cubans live in urban areas. The environmental threats associated with rapid population growth in rural areas, such as population growth-induced subsistence agriculture, were of no consequence in Cuba, although they were of some significance in the country's urban areas. Environmental concerns at present are largely divorced from demographic trends and mostly arise from past economic development policies as well as from some of the economic and development policy choices being made under the pressures of the special period.

Economic Setting

Prior to the revolutionary takeover of 1 January 1959, the Cuban economy was predominantly capitalistic. With some notable exceptions (e.g., railroads), the means of production were owned by either domestic or foreign individuals and corporations. The sugar industry, which dominated economic activities and relations with the United States—Cuba's largest commercial partner and source of investment—was of paramount importance.

By 1963, in a matter of a few years, the revolutionary government led by Fidel Castro had nationalized most private property, announced its socialist ideology, and adopted central planning. The Soviet Union and its socialist allies had replaced the United States as Cuba's main economic partners. The Cuban government replicated in the island the bureaucracies that mismanaged the economies in the Soviet Union and Eastern Europe and was well on its way to adopting the Soviet development model, abandoning sugar production, and promising to turn Cuba into an industrial power. Although the dream of Cuba as an industrial giant was short-lived and sugar was again embraced as the engine of development, the basic tenets of the Soviet development model of gigantism and development of heavy industry prevailed, with adverse implications for the environment.

The Cuban Economy in the 1950s

The outbreak of World War II had a favorable effect on the production and price of sugar, Cuba's main product and export. The decline in trade caused by the war increased foreign exchange and promoted both the establishment of new industries and the expansion of existing ones to meet domestically the demand for commodities that were formerly imported.

In the 1950s government policies were deliberately geared toward the stimulation of the nonsugar industrial sector and the diversification of agricultural production. Government institutions were created to implement new directions in the economy.[4] The new government lending institutions, combined with the commercial banks, established a credit system to finance the development of agricultural, industrial, and social activities. Further encouragement to industrialization came from the Decree-law on Industrial Stimulation of 1953, which offered fiscal and import tariff incentives to new industries established in the country. The process of "Cubanization" of the economy, and, in particular, of the sugar industry, was accelerated during this period. While in 1939 Cuban nationals owned 54 sugar mills that accounted for 22 percent of total sugar production, by 1952 the number had increased to 113, accounting for 55 percent of sugar production.

Over the period 1950–1958, the Cuban economy grew at an average annual rate of 4.6 percent at current prices. Considering that population expanded at an average annual rate of under 2 percent and prices at an average annual rate of about 1 percent, the real per capita growth rate was closer to an anemic 1.5 percent per annum. An average of 28–29 percent of the island's gross national product (GNP) was generated by the sugar industry, and sugar and its by-products accounted for 84 percent of total exports, with exports of tobacco and minerals accounting for most of the rest. The United States was by far the largest investor in Cuba, and an average of nearly 69 percent of Cuba's foreign trade was with the United States (Mesa-Lago 1971, 278).

Cuban GNP statistics by broad economic sector for the period 1951–1958 prepared by the Banco Nacional de Cuba[5] show a positive growth rate, subject to substantial year-to-year fluctuations primarily resulting from ups and downs in the sugar sector and political instability. Between 1956 and 1957, for example, the product generated by the sugar sector grew by 47 percent, but it declined by 24 percent in 1958. In 1958 Cuban GNP was approximately $2.2 billion (the Cuban peso was at par with the U.S. dollar) and per capita GNP was $356; this per capita income level placed Cuba fourth in this category (after Argentina, Venezuela, and Uruguay) in the Latin American region (Illán 1964, 15).

One of pre-revolutionary Cuba's most serious socioeconomic problems was unemployment. In 1956–1957, 39 percent of the labor force was engaged in primary activities (agriculture, fishing, mining, cattle-ranching), 20 percent in secondary activities (industry, construction, electricity), 36 percent in tertiary activities (communications, commerce, services), and the rest had no specific occupation. The share of the labor force employed in primary activities (mainly agriculture) fell from 49 percent in 1919 to 39 percent in 1957, but increases in employment in industry and services sectors were not sufficient to absorb population growth and rural-to-urban migration. It has been estimated that on the eve of the revolution, on average of 16.4 percent of the labor force was unemployed and an additional 13.8 percent was underemployed. Compounding the problem, employment tended to be seasonal, rising during the four-month sugar harvest and falling sharply thereafter (Mesa-Lago 1971, 259).

Another serious socioeconomic problem was unequal distribution of income and wealth. Although reliable statistics are not available, fragmentary information and estimates suggest that income distribution in Cuba was extremely skewed. Urban–rural differentials were significant (Mesa-Lago 1981, 142). Based on statistics from the 1953 census, a foreign expert has estimated that the two lowest quintiles of the population (those receiving the lowest share of income) received a combined 6.2 percent of national income, whereas the top (highest income) quintile received 60 percent of the national income (Brundenius 1979, 43). Nevertheless, on average, at the end of the 1950s Cubans enjoyed quite high standards of living compared with other developing countries in Latin America. As sociologist Lowry Nelson has put it:

> Although by 1959 Cuba was still far from achieving its potential economic growth, by Latin American standards it was reasonably well

provided with food and the daily average consumption was exceeded only by Argentina and Uruguay. It had more motor vehicles in relation to population than any of its Caribbean neighbors except Venezuela and exceeded all of the Caribbean in telephones and newspapers per thousand population. It ranked near the top of all Latin America in the number of radios and had well established television networks. . . . Cuban workers enjoyed a number of benefits gained through their organizations, including the eight-hour day with time and a half for excess hours, along with many fringe benefits such as vacations with pay. Worst off were the agricultural workers, who suffered long periods of unemployment and underemployment. . . . Educational facilities were slowly improving, although they were still inadequate. The illiteracy rate of more than 20 percent in 1953 for the country as a whole was by no means the worst in Latin America; only about three other countries had a more favorable percentage. Nevertheless, it was the rural people who suffered the greatest deprivation in education. Illiteracy for all rural Cuba was more than 40 percent and in Oriente province rose to 50 percent. (Nelson 1972, 44–45)

In 1959 Cuba's economy was neither stagnant nor ready for a take-off into a higher development plane, as supporters or detractors of the Cuban revolution have argued. Significant positive steps had been taken by successive Cuban governments to create a set of institutions that could support economic growth and diversification, but serious structural problems remained.

The Economy of Socialist Cuba

Collectivization

Soon after taking power, the Cuban revolutionary government issued a series of law-decrees that brought about the confiscation of property and funds controlled by the deposed dictator Batista and his collaborators. In February 1959 the Ministerio de Recuperación de Bienes Malversados (Ministry for the Recovery of Misappropriated Assets) was created and given the mandate of nationalizing property belonging to individuals and companies alleged to have benefited from ties with the previous regime. Under these provisions, the state took control of several sugar mills, construction companies, agricultural enterprises, factories, hospitals, transportation companies, and so on.

This was followed by passage of the Agrarian Reform Law in May 1959, which confiscated land holdings beyond 30 *caballerías* (about 400 hectares) by any individual, with some exceptions for sugar and rice plantations and cattle ranches with productivity well above the national average. Land holdings beyond the upper limit were to be expropriated with compensation, divided, and distributed to landless peasants. In reality most of the large farms that were expropriated were not divided up, but instead were organized into state-controlled production cooperatives along the lines of the Soviet *kolkhozy*.

Using the pretext that business executives were mismanaging their firms as a form of economic sabotage or that labor-management conflicts existed, the government began in the second half of 1959 to take control of key industries, including those owned by foreign investors. It nationalized enterprises in the chemical, oil refining, textiles, and metal products industries, among others. In June 1960 the government "intervened" in three oil refineries owned by foreign oil companies (Esso, Texaco, and Shell) and, in response to a reduction by the United States of Cuba's sugar quota, also nationalized the largest U.S. investments in the island: thirty-six sugar mills, the telephone and electricity companies, and the oil refineries. By August 1960, the state had gained overwhelming control of the economy: 40 percent of the land, about 38 percent of the sugar industry, key public services (electricity, telephone), and a significant portion of the industrial sector, including almost 50 percent of the fourteen industrial enterprises with more than five hundred workers.

A law passed in October 1960 that permitted the nationalization of the remaining investments owned by U.S. citizens, those controlled by other foreign nationals, and key enterprises owned by Cuban citizens virtually sealed the fate of private enterprise in Cuba. Pursuant to this law, the Cuban government took control of the banking system, insurance companies, and 382 large corporations (sugar mills, factories, department stores, wholesalers, warehouses). In December 1962 large and medium-sized firms engaged in retail sales of clothing, footwear, and hardware were nationalized and placed under the control of the Domestic Trade Ministry. In 1963 the Second Agrarian Reform Law nationalized land holdings over 5 *caballerías* (67 hectares), almost doubling the state's share of agriculture. Finally, in March 1968, the government launched a "revolutionary offensive" to eliminate small enterprises, with the state nationalizing corner grocery stores, butcher shops, poultry and fish stores, vegetable and fruit stands, laundries, dry cleaners, barber shops, photo shops, lodging and boarding houses, shoe and auto repair shops, bars, restaurants, and snack shops, as well as stores engaged in the sale of garments, shoes, hats, furniture, books, flowers, hardware, and electrical appliances. The state also nationalized remaining small businesses manufacturing handicrafts, plastic, leather, rubber, wood, metal, and chemical products.

To summarize, from a situation in 1958 where private ownership of the means of production was the norm, by 1968 Cuba had shifted to one of overwhelming government control, with the government controlling 100 percent of industry, construction, transportation, retail trade, wholesale, foreign trade, and education; agriculture was the only area where a significant private-sector presence remained. Over the next two decades, the private sector's presence in agriculture eroded, so that by 1988, its presence in the formal agricultural sector had fallen to about 8 percent.

Leninist industrialization

In the early 1960s, as collectivization advanced, the Cuban government began to introduce elements of the socialist centrally planned model. In March 1960 the Junta Central de Planificación (JUCEPLAN, Central Planning Board) was created to coordinate government economic policies and to guide the private sector through indicative planning. In 1961, as the role of the government in the economy had risen, JUCEPLAN began to develop annual and longer-term plans, which took on the character of mandatory physical planning as practiced in the Soviet Union and socialist countries.

An economic plan for the four-year period 1962–1965, prepared with the assistance of foreign specialists, reflected the adoption by the Cuban regime of the Leninist industrialization model. According to a Cuban economist (Rodríguez Mesa 1980, 82–85), the essential elements of the Leninist "rapid industrialization" model being pursued by Cuba were: (1) investment of a large share of national income; (2) focus on investment in the industrial sector; and (3) emphasis on investment in heavy industry, especially in the metalworking, chemical, and energy industries.

In a challenging speech to Latin American nations at Punta del Este, Uruguay, in August 1961, then Minister of Industries Ernesto Guevara laid out the following vision of the Cuban economy:

> In the next four years, our overall economic growth rate will be 12 percent, that is, over 9.5 percent in per capita terms, transforming Cuba into the most highly industrialized country in Latin America in relation to its population, as evidenced by the following indicators: first place in Latin America in per capita production of steel, cement, electric energy, and—with the exception of Venezuela—oil refining. First place in Latin America in tractors, rayon, footwear, textiles, etc. Second place in the world in the production of metallic nickel (until now, Cuba had only produced concentrates) . . . 8 and one half to 9 million tons of sugar and beginning of the transformation of the sugar industry into a sucro-chemical industry. . . . To accomplish these objectives . . . we will make

investments in industrial plant for 1000 million pesos . . . establishing 205 industries, including a new nickel production plant . . . an oil refinery . . . the first steel mill . . . expansion of our plants producing steel pipes . . . tractors (5000 units per annum) . . . motorcycles (10,000 units per annum) . . . three new cement plants and expansion of existing ones . . . expansion of glass plants . . . a new bagasse boards factory . . . an ammonium nitrate plant . . . a plant capable of producing 81,000 metric tons of superphosphates . . . plants to produce ammonia and nitric acid . . . [and] 8 new textiles plants. (Guevara 1961, 431–32)

Rather than resulting in high and sustained economic growth, the implementation of the rapid industrialization model in Cuba brought an economic crisis in 1962–1963. Three years later, Guevara spoke about some of the problems with this strategy:

It is fair to acknowledge some of the errors we made. Fundamentally, these errors have to do with the technological and economic characteristics of some of the new industrial plants we have been installing. Influenced by the heavy unemployment that existed and by foreign trade pressures, we acquired a large number of plants to substitute imports and to create jobs for urban dwellers. With regard to many of these plants, we found out later that their technological efficiency, in international comparative terms, was low, and their net effect on import substitution was limited since the raw materials needed also had to be imported. (Guevara 1964, 622)

Although Guevara did not dwell on this point, it is clear that the industrial plants imported from the Soviet Union, Eastern Europe, and China embodied the same environmentally unfriendly technology that caused significant environmental damage in these countries. Moreover, environmental considerations were not taken into account in the siting of the plants, which were often located in highly populated urban areas.

Agriculture-led development

In 1963 the Cuban government shifted gears and redefined its development strategy to give agriculture—and sugar in particular—a central role. An ambitious plan for the sugar sector was drawn up to support production of ten million tons of sugar by 1970. Elements of the plan included significant increases in the area devoted to sugarcane cultivation, planting of improved varieties of sugarcane, increases in irrigation, mechanization of sugarcane harvesting, and modernization of sugar mills. Sugarcane lands that had been nationalized and turned into cooperatives were converted into large state farms to reap economies of scale in mechanization, irrigation, fertilizer application, and so on. Progressively, private farmers were co-opted to give up their land and to turn it over to the state to increase the size of state farms and other forms of collective farming.

Cuba failed to meet its sugar production target of 1970. More importantly, the concentration on a single economic objective brought about significant misallocation of resources and resulted in yet another economic downturn in the early 1970s.

After nearly a decade of ad hoc economic management, in the mid-1970s Cuba began to implement an economic development strategy inspired by the Soviet reform model. Central planning was reintroduced, and a number of market-oriented tools—costs, profits, interest, taxes, depreciation—began to be used again. Cuba made large investments in the sugar industry to support large, and sustained, sugar production levels. Throughout this period, the Soviet Union and the socialist countries of Eastern Europe were Cuba's principal economic partners, purchasing the bulk of Cuba's imports and providing most of the island's imports, including its industrial plant and equipment. The Soviet Union, in particular, was Cuba's main source of development financing, through price subsidies for Cuba's exports, grants, and loans. Cuba formalized its economic linkages with the socialist countries in 1972 when it joined the Council for Mutual Economic Assistance (CMEA or COMECON), the organization that coordinated trade and investment relations among socialist nations. Under these trade agreements, and with a substantial level of foreign subsidies, Cuba was able to diversify its industrial structure somewhat, although its primary economic role within CMEA continued to be that of a supplier of primary goods.

In the second half of the 1980s Cuba began to feel the impact of the political and economic changes that were occurring in the Soviet Union and Eastern Europe. Glasnost in the Soviet Union resulted in publicity about the large cost to the Soviet Union of supporting Cuba, the mismanagement of resources on the island, and Cuba's inability to meet export commitments to the Soviet Union.

The Crisis of the 1990s

The breakup in 1989–1990 of the economic relations that existed between Cuba and Eastern European countries, and subsequently the Soviet Union, caused a veritable economic depression on the island. The Cuban leadership refers to the present crisis as the "special period in time of peace," a term that conveys the urgency of the crisis, its national security implications, and the need to approach it with military-like resolve.

The strains on the external sector of the Cuban economy caused by political and economic changes in Eastern Europe and the Soviet Union are most evident in the performance of Cuba's merchandise trade accounts. In 1989 the last year in which economic relations with the socialist countries can be deemed as being "normal," Cuba's overall exports amounted to 5.4 billion pesos and imports to 8.1 billion pesos, for a trade deficit of nearly 2.8 billion pesos or 52 percent of the value of exports. The bulk of this trade deficit was incurred with the Soviet Union and financed by that nation through a combination of grants and credits.

By 1993, after the Soviet Union had changed its foreign economic relations to require that they be conducted on normal commercial terms (e.g., prices for exports and imports at world market prices in convertible currency, transactions financed through commercial credit institutions), Cuba's exports shrunk to 1.1 billion pesos (about one-fifth of their 1989 value) and imports to about 2 billion pesos (about one-fourth of their 1989 value). The reductions in imports cut across all categories of imported goods: oil, food, capital goods and spare parts, medicines and medications, raw materials for industry, transportation equipment, fertilizers and pesticides, consumer goods, and so on. In the absence of foreign subsidies agricultural and industrial production suffered major setbacks.

Cuba's economic crisis of the 1990s is perhaps the most serious that the island has suffered since independence. According to statistics published in a Cuban magazine in 1994, the nation's gross domestic product (GDP) in 1993 was about 10 billion pesos, a 48 percent decline from the 19.3 billion pesos recorded in 1989 (Terrero 1994, B40). This coincides with estimates in reductions in national income, as measured by the global social product, made by other analysts.[6]

A new statistical series on the GDP published recently by the Cuban National Bank in two reports (BNC 1994, 1995) suggests that the economic contraction during 1989–1993 was 35 percent (Table 2.2), compared to the roughly 50 percent decline suggested by the gross social product (GSP) series. The GDP series further suggests that the down slide in economic activity bottomed out in 1993 and positive economic growth was recorded both in 1994 (0.7 percent) and 1995 (2.5 percent). Irrespective of which of the two series more closely tracks the Cuban economic crisis of the 1990s, it is clear that the crisis has been broad and deep, affecting all sectors of the economy. The economic recession of the 1990s has lifted some of the taboos that prevented or discouraged certain forms of economic development (e.g., tourism) and reduced the already-meager level of resources that were available to prevent environmental disruption and reverse it.

The leadership has designated four sectors as key for the reactivation of the economy in the 1990s. In two of these sectors, tourism and mining, Cuba has considerable comparative advantage because of its natural beauty and many un-

Table 2.2. Gross domestic product, by economic activity
(in millions of pesos at 1981 prices)

	1989	1990	1991	1992	1993	1994	1995
Gross domestic product	19,585.8	19,008.3	16,975.8	15,019.9	12,776.7	12,868.3	13,184.5
Agriculture, hunting,							
forestry, and fishing	1,924.5	1,756.3	1,334.9	1,197.1	924.9	879.4	915.5
Mining and quarrying	123.0	91.6	81.6	105.7	96.4	97.5	152.1
Manufacturing industries	4,886.8	4,640.2	4199.7	3,506.5	3,103.6	3,340.6	3,555.2
Electricity, gas, and water	452.0	454.6	426.5	378.2	335.2	350.0	384.2
Construction	1,349.7	1,508.1	1,085.2	603.7	385.7	383.9	412.1
Trade, restaurants,							
and hotels	5,150.6	4,936.3	4,396.4	4,050.2	2,936.4	2,935.2	2,984.8
Transport, warehousing,							
and communications	1,352.6	1,202.3	1,058.9	911.6	733.3	708.7	748.4
Finance, real estate,							
and business services	584.9	603.2	639.2	543.9	513.4	492.4	483.8
Community, social, and							
personal services	3,761.7	3,815.7	3,753.4	3,713.0	3,747.8	3,680.6	3,548.4

Source: BNC 1994 and 1995.

spoiled beaches and its considerable mineral reserves. Another priority sector is agriculture, which, although continuing to be the country's most important, remains mired in grave problems. A priority role has also been assigned to the biotechnology industry in which major investments began to be made in the 1980s. Despite some successes in production of biotechnology products for the domestic market and for export, this sector accounts for only a small share of the national economy. A rapidly expanding tourist industry and expanding mining activities pose many environmental challenges; if improperly regulated, these sectors could lead to severe environmental degradation.

❧ 3 ❧

Law and Practice of
Environmental Protection

BEFORE THE 1959 revolution, Cuba, like most other Latin American countries, had a largely ineffectual legal framework of environmental protection. Although many laws had been enacted since colonial times, principally to preserve shrinking forest resources and watersheds, most were largely ignored and rarely enforced.

By Latin American standards, socialist Cuba has a well-developed legal framework of environmental protection. The 1976 Constitution, as amended in 1992, takes an expansive approach toward the environment, recognizing the duty of the state and of all citizens to protect the environment, and incorporating the modern concept of sustainable development. In 1981 Cuba enacted a comprehensive environmental law and issued several pieces of complementary legislation and implementing regulations. Cuba's Asamblea Nacional del Poder Popular (ANPP, National Assembly of People's Power, Cuba's highest legislative body) approved a more modern environmental law in July 1997 that superseded the 1981 statute. Cuba has also ratified most of the core international environmental conventions and treaties.

As is documented in the second part of this chapter and elsewhere in this book, the practice of environmental protection in Cuba is in sharp contrast to the positive picture painted by the legal framework. Because Cuban environmental laws are general, they are difficult to enforce. Moreover, environmental protection institutions are weak, and their ability to enforce laws and regulations is severely limited by their lack of authority to interfere in matters under the control of economic-sector ministries. To date, the role of environmental protection institutions has been that of motivators and coordinators rather than enforcers.

Law of Environmental Protection

During most of the time period covered by this book, Cuba's legal framework of environmental protection has consisted of (1) the 1976 Constitution, as amended; (2) Law No. 33 of 1981, a broad statute establishing the legal basis for the protection of the environment and the rational use of natural resources; (3) Decree-law No. 118, creating a national system of environmental protection pursuant to Law No. 33; (4) regulations that implemented Law No. 33 and complementary laws; and (5) international environmental agreements to which Cuba is a party. Law No. 81, a new broad environmental law, was adopted by the ANPP in July 1997. The impact of this new law on environmental protection in Cuba remains to be seen.

Cuba's Constitution

Latin American constitutions enacted prior to the UN Conference on the Human Environment in 1972 tended to have relatively few references to the environment (Brañes 1991). To the extent they did, references were specific to the protection of a specific natural resource, the protection of health from hazardous environmental effects and the consequences of urban development, or the conservation and rational use of natural resources. Cuba's 1940 Constitution, and the 1959 Fundamental Law that replaced it, are examples of the older vintage of constitutions with respect to the environment. The only vaguely related reference in these two charters to the protection of the environment and the conservation of natural resources is found in Article 88, second paragraph, in a section dealing with ownership issues. After setting out that the nation's subsoil belongs to the state and the exploitation of underground natural resources should be done pursuant to concessions granted by the state, the 1940 Constitution and 1959 Fundamental Law proclaimed: "The soil, the forests, and concessions for the exploitation of the subsoil, use of waters, means of transportation and provision of any other public services, will be exploited in such a way that they promote the well being of the population" (de la Cuesta 1974, 260, 351).

The 1976 Constitution, Cuba's first socialist charter, was approved by the First Congress of the Cuban Communist Party in December 1975 and subsequently by a popular referendum held in February 1976. Its adoption was proclaimed on 24 February 1976 (Domínguez 1978, 243, 257). The 1976 Constitution is an example of the more recent vintage of Latin American constitutions with respect to the environment. A very brief provision, Article 27, stated that "To ensure the well being of the citizenry, the State and society protect nature. It is incumbent upon the competent organs and every citizen to see to it that water and the atmosphere are kept clean, and that the soils, flora, and fauna are protected" (Constitución 1976, 16). There is no legislative history available on this article.

A commentary on the 1976 Constitution by a Cuban legal scholar sheds very little additional light: "There is no doubt that the [environmental] challenge faced by the Cuban State, as well as by the international community at large, calls for the development of a new body of law of considerable proportions, rather than on the codification of current legal norms and practices. It is not possible to define, in exhaustive terms, all the subject matters and issues that must be taken into account in restructuring legal norms, but we must be aware that something must be done if we do not want so see Earth turned into an uninhabitable planet" (Alvarez Tabío 1981, 125). This commentator refers to the large quantities of toxic emmissions produced by firms in the United States and rejects the thesis that environmental degradation is a by-product of economic growth and scientific-technical development irrespective of the social system. He argues that only in a socialist system are there the conditions to preserve the environment and concludes: "Developing laws and standards to preserve the environment is only possible in a society where there are no social groups with selfish interests, who are not concerned about protecting the environment or the health of the population and instead are only interested in obtaining high profits" (125).

1992 Amendments to the Constitution

In September 1992 the ANPP adopted significant changes to the 1976 Constitution. Among others, Article 27 was amended to read as follows:

> The State protects the environment and natural resources of the country. It recognizes their close link with the sustainable economic and social development required to make human life more rational and to ensure the survival, welfare, and security of present and future generations. It is incumbent upon the competent organs to implement this policy. . . . It is the duty of citizens to contribute to the protection of waters and atmosphere and to the protection of soils, flora, and fauna, as well as the rich potential of nature. (Constitución 1992, 5)

In addition Article 10 of the 1976 Constitution (renamed Article 11 in the revised Constitution) was amended to include the concept that the state exercises sovereignty over "the nation's environment and natural resources." Finally, Articles 105 and 106, dealing with the duties of the Provincial Assemblies of People's Power and the Municipal Assemblies of People's Power, respectively, assigned to these bodies jurisdiction and responsibility for overseeing environmental protection in every province and community within the nation.

A Cuban analyst has stated that the changes to Article 27 responded to a desire to comply with the spirit of the conclusions of the 1992 UN Conference on the Environment and Development (the so-called Earth Summit Declaration), which promoted a link between protection of the environment and sustainable socioeconomic development (Acosta Moreno 1995b, 65).

Law No. 33

In December 1980 the ANPP took up the discussion of a comprehensive environmental law. Prior to that time, Cuba had enacted sectoral legislation with environmental relevance (e.g., water laws, mining laws, forestry laws) that protected certain natural resources or sheltered the environment from the negative impact of disruptive activities. However, the legislation discussed by the ANPP in 1980 reflected a more modern approach that views the environment as a systematically organized whole and advances the notion that environmental problems must be tackled through a comprehensive, systemic approach. Commenting on this new legal philosophy, a Latin American scholar observed that "in recent years, environmental legislation has begun to change around the world as the result of a renewed vision of how the environment should be treated, which derives from a holistic and systemic concept that is transforming legislation and has given rise to laws that establish principles aimed at protecting the environment as a whole, taking all its components and interactive processes into account. However, these new laws have not completely replaced sectoral legislation with environmental relevance" (Brañes 1991, 14).

The law that emerged from these deliberations, Law No. 33, "On the Protection of the Environment and the Rational Use of Natural Resources," was approved by the ANPP in December 1980 and finalized in January 1981; it took effect on 12 February 1981, upon its publication in the *Gaceta Oficial* (Ley No. 33 1981). In his closing speech to the December 1980 ANPP session, President Castro described Law No. 33 as "very important, and perhaps [a law that] one day may be considered as historic" (Castro 1980).

The basis for Law No. 33 was a draft bill prepared by an interagency working group established by the ANPP president and consisting of representatives of the Cuban Academy of Sciences; the Ministries of Agriculture, Basic Industry, Sugar, Construction, Transportation, Fisheries, Public Health, and Culture; and the Secretariat of the Executive Committee of the Council of Ministers. Legal and environmental experts provided technical advice to the working group (Tiende 1980, 1). The establishment of the working group, and its charge to draft a broad environmental bill, resulted from ANPP discussions in July 1980 of a narrower draft bill, "On Rehabilitation of Soils and Conservation of Terrestrial Waters." During these discussions, it became apparent that the narrow formulation of the bill missed important environmental concerns such as conservation of soils and forest resources. Spurred by Raúl Castro, the ANPP suspended action on the draft bill and formed a working group to draft broader environmental legislation for consideration at a subsequent ANPP session (Resolución 1980).

According to the chairperson of the interagency working group (from the Academy of Sciences), the working group prepared five drafts of the bill from July to December 1980 (Síntesis 1980). In developing the proposal, the working group considered relevant environmental legislation in foreign countries (among them the German Democratic Republic, Bulgaria, Hungary, Poland, the Russian Soviet Federal Socialist Republic, Colombia, Mexico, Sweden, and Venezuela) as well as general guidelines for environmental legislation issued by the UN Environmental Program. The working group also consulted with experts from the German Democratic Republic and the Soviet Union. The chairperson of the working group stated that the approach of the law—a framework statute setting general principles, to be fleshed out by laws and regulations dealing with specific natural resources or environmental issues—followed closely the approach to environmental law in socialist countries (Tiende 1980, 2).

Law No. 33 consisted of a preamble, four major chapters, special provisions, transitory provisions, and final provisions. Chapter I (Articles 1–21) contained provisions of general application; a very lengthy Chapter II (Articles 22–116) addressed the protection and rational use of specific natural resources; Chapter III (Articles 117–126) dealt with management of the protection of the environment and the rational use of natural resources; and Chapter IV (Articles 127–130) addressed violations of norms and regulations regarding the protection of the environment and the rational use of natural resources.

Preamble

The preamble set out that the environment is a fundamental endowment of society, whose protection and conservation is of vital importance to the state. According to the preamble, there was a need to address environmental deterioration that had resulted from conditions inherited from the past as well as from the development of the Cuban economy since 1959, coupled with the lack of knowledge among the population about the proper use and conservation of nature. Thus, economic and social development plans of the state must take into account the protection of the environment and the rational use of natural resources; all state institutions, enterprises, cooperatives, mass organizations, and the citizenry at large should do the same, working to develop a culture that recognizes the importance of nature and the imperative to comply with measures to protect the environment and use resources rationally.

Demonstrating that environmental protection in Cuba is not free from ideological considerations, the preamble stated that "In socialist society, man is the most precious resource, and the social character of property facilitates the adoption of measures that guarantee the integral protection of the environment and the rational use of natural resources, unlike in capitalist regimes, where the interests of private property are at odds with those of society."

General provisions

The stated objective of the law was "to establish the basic principles for the conservation, protection, improvement, and transformation of the environment and the rational use of natural resources, in accord with the country's integral development policies, in order to use optimally the productive potential of the nation" (Article 1). Environment was defined as "the system of abiotic, biotic, and socioeconomic elements with which man interacts," and natural resources as "the abiotic and biotic natural elements that are available to man to meet his economic, social, and cultural needs" (Article 2).

The protection of the environment may be accomplished through a variety of means: conservation or planned transformation; systematic efforts to eliminate the causes and factors that give rise to contamination, damage, or deterioration; application of preventive measures that contribute to the elimination or diminution of contamination, damage, or deterioration; and, where appropriate, reclamation (Article 7). Environmental considerations must be taken into account in the evaluation of sites for investment projects and in territorial planning (Article 10). Measures to protect the environment approved by the competent state environmental organization must be an integral part of investment projects (Article 11).

State organs, institutions, and enterprises that generate waste products *(residuos)* must (1) provide for, and guarantee the proper functioning of, treatment and disposal systems and not introduce any modifications in such systems without previous authorization by competent authorities; and (2) conduct or promote scientific-technical research to use such wastes as raw materials in other economic activities (Article 15). State organs, institutions, and enterprises have the obligation to make available and update the information base that would permit the diagnosis of environmental impacts of their activities and to promote the adoption of measures that protect the environment and foster the rational use of natural resources (Article 20).

Cuba participates in international efforts to protect the environment through international agreements. Cuba also cooperates with other nations on environmental matters pursuant to the principle of proletarian internationalism (Article 21).

Protection of natural resources

Natural resources addressed in specific sections of Chapter II were (1) water (Articles 22–38); (2) soils (Articles 39–47); (3) mineral resources (Articles 48–60); (4) marine resources (Articles 61–75); (5) flora and fauna (Articles 76–86); (6) atmosphere (Articles 87–93); (7) agricultural resources (Articles 94–100); (8) human settlements (Articles 101–107); and (9) landscape and tourism resources (Articles 108–116). In addition to defining, in general terms, the specific natural resource, each of these subsections set out, in very general terms, norms for its conservation.

National System for the Protection of the Environment and the Rational Use of Natural Resources

Chapter III of Law No. 33 established the Sistema Nacional de Protección del Medio Ambiente y del Uso Racional de los Recursos Naturales (National System for the Protection of the Environment and the Rational Use of Natural Resources, or National System). The objective of the National System was "to focus national and global attention on the environment and on natural resources." State organizations and enterprises, cooperatives, political, social, and mass organizations, and the public at large would participate "harmoniously" in the National System (Article 117).

Implementation of the National System was the responsibility of national governmental agencies together with local organs of People's Power (Article 118). In addition to the National System, the law provided for the creation of subsystems with responsibility for the conservation and rational use of specific natural resources (Article 119). The organization, structure, and functions of the National System and its subsystems were to be determined by the Council of Ministers (Article 120).

The principal functions of the National System were (1) coordination and implementation of policies for the protection of the environment and the rational use of natural resources ; (2) development and implementation of such measures based on scientific information; (3) consideration of environmental factors within the process of socioeconomic development, with the aim of protecting the environment or mitigating its deterioration; and (4) development of contingency plans to address environmental catastrophes (Article 122).

The Council of Ministers was empowered to develop and issue regulations and complementary laws to protect the environment and promote the rational use of natural resources in each of the specific areas set out in Part II of Law No. 33, namely water, soils, mineral resources, marine resources, flora and fauna, atmosphere, agricultural resources, human settlements, and landscape and tourism resources (Article 124).

Within 180 days of publication of the law in the Gaceta Oficial, the Academy of Sciences was directed to present to the Council of Ministers a proposal regarding the organization, structure, and operation of the National System (Transitory Provision).

Penalties

Chapter IV dealt with sanctions that could be imposed against violators of laws and regulations promoting environmental protection and rational use of natural resources. Violations—either by commission or omission—were sanctionable by administrative fines and/or by seizure, elimination, shutdown, physical removal from the area, remediation of damage, and so on. The competent administrative authority was empowered to stop actions that violate the law and to order measures to remediate harm to the environment or to natural resources (Article 127).

Special and Final Provisions

The Cuban Academy of Sciences was given responsibility for defining all technical terms used in the law. Economic disputes arising from the implementation of the law would be decided by the National State Arbitration Organ. The Council of Ministers was empowered to issue directives to implement the statute.

Decree-law No. 118

As mentioned earlier, Article 117 of Law No. 33 established the National System, and Article 120 authorized the Council of Ministers to issue regulations to operationalize it. Moreover, one of the transitory provisions of Law No. 33 called for the Cuban Academy of Sciences to submit to the Council of Ministers, within

180 days of publication of Law No. 33 in the *Gaceta Oficial* (i.e., 180 days from 12 February 1981), a proposal regarding the organization, structure, and operation of the National System. In an upbeat presentation before the ANPP introducing the draft of Law No. 33, the president of the Cuban Academy of Sciences made explicit reference to the 180-day period for the preparation of the proposal on the National System and the high priority assigned to the development of implementing regulations dealing with specific natural resources such as water, soils, fisheries, and mineral resources. He predicted that "by the end of 1981, a significant portion of the legal framework necessary to implement this law [Law No. 33] will be in place" (Síntesis 1980).

It has not been possible to establish whether or not the Academy of Sciences complied with the directive to develop a proposal on the National System on a timely basis. What is clear, however, is that the Council of Ministers did not pass such a law setting the requisite implementation parameters until nearly ten years later. The enabling legislation, Decree-law No. 118, "On the Structure, Organization, and Operation of the National System for the Protection of the Environment and the Rational Use of Natural Resources," was approved by the Council of State on 18 January 1990 and published in the *Gaceta Oficial* of that same date (Decreto-ley No. 118 1990).

Decree-law No. 118 consisted of a preamble, seven chapters, special provisions, transitory provisions, and final provisions. Chapter II (Articles 4–6) defined the structure of the National System, Chapter III (Article 7) did the same for subsystems, and Chapter IV (Articles 8–9) set out duties and principal functions of the National System. Chapter V (Articles 10–14) addressed issues related to the operation of the National System and established a body with national responsibility for its implementation. Chapter VI (Articles 15–20) established provincial and municipal bodies. Chapter VII (Articles 21–34) set out specific duties and principal functions of each of the participants in the National System.

Structure of the National System

The National System was composed of the Comisión Nacional para la Protección del Medio Ambiente y el Uso Racional de los Recursos Naturales (COMARNA, National Commission for the Protection of the Environment and the Rational Use of Natural Resources), which directed the National System; central government agencies (state committees, ministries, institutes), which were involved in the protection of the environment and natural resources; subsystems responsible for the protection of specific natural resources; organizations that directed the subsystems; and provincial and municipal commissions charged with protecting the environment and promoting the rational use of natural resources (Article 4). In addition to central government agencies, cooperatives and the public at large also participated in the National System (Articles 5–6).

Decree-law No. 118 created eleven subsystems, one for each of the natural resources identified in Chapter II of Law No. 33. The subsystems, their specific responsibilities, and the central government agency responsible for their direction are as follows (Article 7):

Atmosphere: Protection and improvement of air quality; under the direction of the Ministry of Public Health.

Human settlements: Protection and improvement of health and environmental quality in cities and rural communities; under the direction of the Ministry of Public Health.

Soils: Protection and rational use of soils apt for agriculture and forestry; improve fertility of soils and prevent degradation; restore mining areas; under the direction of the Ministry of Agriculture.

Flora and fauna: Protection of endangered plant and wildlife species; optimal utilization of forest areas; protection and rational use of resources for bee's honey production; under the direction of the Ministry of Agriculture.

Agriculture: Protection of permanent crops, plant health, protection of quality of seeds; under the direction of the Ministry of Agriculture with assistance from the Ministry of the Sugar Industry.

Livestock: Animal health; rational exploitation of different kinds of livestock, including poultry; under the direction of the Ministry of Agriculture.

Minerals: Protection and rational use of commercial mineral resources, including those located offshore; under the direction of the Ministry of Basic Industry.

Fisheries: Protection and rational use of fisheries resources (in rivers, coastal zones, bays, beaches, territorial waters, economic zone), including their ecosystems; under the direction of the Ministry of Fisheries.

Maritime shipping: Protection and rational use of bays, ports, and territorial waters; protection against ocean pollution created by maritime shipping; under the direction of the Ministry of Transportation.

Water: Protection and rational use of terrestrial water, including underground resources, and protection of aquifers; under the direction of the National Institute of Hydraulic Resources.

Tourism: Protection and rational use of natural resources applicable to tourism, including landscapes; under the direction of the National Tourism Institute.

COMARNA

Decree-law No. 118 also created COMARNA, which reported to the Council of Ministers, to direct the implementation of the National System (Article 10). Representatives of more than two dozen central government agencies, provincial environmental protection commissions, and mass organizations were designated as members of COMARNA (Article 11). COMARNA was headed by a president and a vice president, both chosen and appointed by the Council of Ministers, and a secretary, chosen and appointed by the president of COMARNA (Articles 11 and 12). COMARNA had an Executive Secretariat, composed of the president, vice president, and secretary; the Executive Secretariat is advised on technical matters by a group of experts selected by the president of COMARNA (Article 14). COMARNA was directed to meet at least twice a year to coordinate the implementation of government policies regarding the protection of the environment and the rational use of natural resources (Article 8).

The National System was given the following primary functions: (1) each of its members, with guidance from COMARNA, was to coordinate and control the implementation of policies regarding the protection of the environment and the rational use of natural resources within its subject matter of competence; (2) the agencies directing the subsystems, COMARNA, and the State Committee on Standards were to develop and issue standards based on scientific information and analysis for the use and conservation of natural resources; (3) state agencies with national responsibilities, the agencies directing the subsystems, COMARNA, and other scientific organizations were to take into consideration environmental factors in socioeconomic development activities in order to protect the environment and mitigate environmental deterioration; and (4) all participants in the National System, in coordination with the High Command of the Civil Defense System, were to develop emergency plans to address catastrophic events that might harm the environment and assign resources for the implementation of such plans.

The duties and responsibilities of COMARNA were very broad, and included developing and proposing national environmental policies, enforcing environmental laws, proposing the expenditure of resources to protect the environment, overseeing environmental inputs into investment decisions, and developing materials to enhance environmental awareness by the population (Article 21). With regard to enforcement, COMARNA was responsible for, among other duties, conducting environmental assessments of projects that required such assessments; recommending to government authorities the suspension or permanent shutdown of activities that harm the environment or use natural resources irrationally; proposing emergency measures that redressed environmental damage or irrational use of natural resources and therefore avoiding the suspension or permanent shutdown of facilities; recommending to agency heads the imposition of fines against violators of environmental laws; and raising to higher levels within the government the failure of agencies in enforcing environmental laws (Article 22).

COMARNA was supported in carrying out its duties by the Executive Secretariat (Article 23). In emergency situations the Executive Secretariat could act on behalf of COMARNA, carrying out all of the duties of COMARNA set out in Article 22 (Article 24). Provincial and municipal commissions could do the same, bringing emergency situations to the attention of the executive committee of the appropriate local organ of People's Power (Article 31).

Provincial and municipal commissions

In addition to establishing COMARNA, Decree-law No. 118 also established a network of provincial and municipal Commissions for the Protection of the Environment and the Rational Use of Natural Resources, under the aegis of the executive committee of People's Power at the provincial and municipal levels (Article 15). Membership in the provincial and municipal commissions and in their management structure replicated the practices of COMARNA (Articles 16–18 and 19–20, respectively). The duties of provincial and municipal commissions were similar, within their area of competence, to those of COMARNA (Article 31).

Implementing Regulations and Complementary Laws

As noted earlier, Law No. 33 was a very broad statute. Its actual implementation was to be based on regulations issued by agencies of the Cuban government with primary responsibility for the use and conservation of specific natural resources or by decrees issued by the Council of Ministers. According to our research, during the time period when Law No. 33 was in effect, implementing instruments were issued only in the areas of work with radioactive materials, protection of water and soils, and use of natural resources in certain agricultural activities. Laws dealing specifically with air pollution, urban pollution, and tourism resources were not issued (Borges Hernández and Díaz Morejón 1997, 18). Other laws, decrees, and regulations were also issued by the Cuban government that, although not implementing Law No. 33 per se, complemented the environmental legal structure. We briefly describe these here.

Regulations

Among the decrees approved by the Council of Ministers and regulations issued by agencies of the central government to implement Law No. 33 are:

▶ Resolution No. 63/83 of the State Committee on Finance, issued in November 1983, implementing Article 44 of Law No. 33 with regard to financial responsibility for reclamation of land damaged by construction or mining activities (Resolución No. 63/83 1983).

▶ Decree No. 142, approved by the Council of Ministers in February 1988, governing work involving radioactive substances, including exposure limits to sources of radiation as well as safety measures in the workplace (Decreto No. 142 1988);

▶ A joint resolution issued by the Ministry of Public Health, the Institute of Hydraulic Resources, and the Ministry of Fisheries in June 1990, regulating commercial fishing and use of gasoline-powered boats in dams and reservoirs (Resolución Conjunta 1990);

▶ Several decrees issued by the Council of Ministers at the behest of the Ministry of Agriculture in 1992–1993, setting out rules, and assessing penalties for violations of rules, regarding: (1) registration and transfer of cattle (Decreto No. 174 1992); (2) quality of seeds (Decreto No. 175 1992); (3) the protection of apiculture and honey resources (Decreto No. 176 1992); and (4) the protection of soils (Decreto No. 179 1993).

▶ Decree-law No. 138, approved by the Council of State in July 1993, setting out regulations regarding rational use, conservation, improvement, and protection of terrestrial waters, including the rational use of hydraulic resources (Decreto-Ley No. 138 1993) and penalties for violation of regulations protecting hydraulic resources and promoting their rational use (Decreto No. 199 1995).

Complementary laws

Among the most significant instruments that complemented the legal environmental structure created by Law No. 33, Decree-law No. 118, and related instruments were amendments to the occupational safety and health code, the public health code of 1982, and aspects of the tax code of 1994 and of the foreign investment code of 1995.

In August 1977 the State Committee on Labor and Social Security (the agency that replaced the Ministry of Labor in Cuba's central administrative structure) amended provisions issued in 1963 dealing with occupational illnesses caused by toxic substances used in the work environment (Resolución No. 34 1977).

The public health code, adopted in April 1982 (Decreto-ley No. 54 1982), set out norms to be applied by the Ministry of Public Health in the areas of public safety and health and of epidemiological control. The code dealt specifically with the following topics: (1) prevention and control of diseases; (2) sanitation measures (air, soil, solid wastes, wastes entering water bodies, water quality, water and sewage services, towns and urban areas, introduction of foreign live species); (3) manufacture, labeling, and sale of medications; (4) food safety; (5) occupational safety and health; (6) school safety and health; (7) sanitary control measures (including the authority to close down buildings or workplaces that violate sanitary standards); and (8) penalties for violations of sanitary standards.

The very broad tax code adopted by the National Assembly in mid-1994 (Ley No. 73 1994) as part of Cuba's economic restructuring program inter alia levied a tax on the use of natural resources. For example, Chapter XI of the tax code proclaimed a tax to be levied on the use of natural resources (Article 50), whether by domestic or by foreign investors (Article 51). The modalities for the actual implementation of the tax (i.e., the basis on which to levy the tax, the form of the tax, and the payment schedule) were left to the discretion of the Minister of Finance and Prices; the law also foresaw the possibility of tax reductions or exemptions, subject to the discretion of the Minister of Science, Technology, and Environment (Article 52).

The new foreign investment law, adopted in September 1995, instructs the Ministry of Foreign Investment and Economic Cooperation to consult on investment proposals with the Ministerio de Ciencia, Tecnología y Medio Ambiente (CITMA, Ministry of Science, Technology, and the Environment); CITMA would evaluate the project from an environmental perspective and decide whether an environmental impact assessment was required. It is also charged with deciding whether to grant environmental licenses required by foreign investments and with establishing environmental monitoring and inspection mechanisms (Article 55). Finally, the new foreign investment code also establishes (Article 56.2) the principle that a natural or juridical person who is found to be responsible for damaging the environment is liable for restoring it to its earlier state as well as for redressing material losses incurred by other parties (Ley No. 77 1995).

International Agreements

There are over two hundred international agreements (treaties, conventions, protocols, and so on) dealing expressly with international environmental questions (Magraw 1991, 12). These agreements constitute a significant body of international public law (*Global Partnership* 1993, 216). In most countries formal ratification by the government of an international agreement is equivalent to adopting domestic legislation, since formally ratified international agreements and instruments carry the force of law.

Cuba has ratified, and therefore is a party to, over two dozen international environmental agreements, including most of those that the World Resources Institute defines as "critical international conventions and regional arrangements protecting the environment" (WRI 1992, 357–66). In the Cuban legal system, international agreements become part of national legislation upon ratification (Acosta Moreno 1995, 63). Thus, these agreements are part of Cuba's environmental legal structure. International environmental agreements to which Cuba is a party, their status, and Cuba's position are given in Table 3.1. For purposes of discussion, they are grouped according to subject matter: wildlife and habitat, oceans, atmosphere, hazardous substances, regional agreements, occupational safety and health, and miscellaneous arrangements.

Cuba has ratified, or is in the process of ratifying, ten international environmental agreements since 1991 (Acosta Moreno 1995b, 61). These actions are consistent with its efforts since 1990 to redefine its foreign policy, giving emphasis to "multilateralism against the unipolarism and North-centrism of the international system" (Suárez Salazar 1995, 92). More important than the act of ratification, however, is the extent to which countries effectively implement agreements to which they are parties and devote resources to their enforcement.

Wildlife and habitat

Cuba is a party to most of the major global conventions dealing with wildlife and habitat. Thus, Cuba has ratified the Antarctic Treaty, the Ramsar Convention on Wetlands, and the Heritage Convention. In 1991 Cuba ratified the Convention on International Trade in Endangered Species (CITES) and, in 1994, the Convention on Biological Diversity (Rio Treaty). Cuba is not a party to the Convention on the Conservation of Migratory Species of Wild Animals, which protects wild animal species that migrate across international borders.

Oceans

Cuba has ratified three major global conventions in this area: the Convention on the Prevention of Marine Pollution by Dumping of Wastes and Other Matter, the Ship Pollution Convention (MARPOL), and the UN Convention on the Law of the Sea. Cuba has also ratified the Brussels International Convention on Civil Liability for Oil Pollution Damage, the Treaty on the Prohibition of the Emplacement of Nuclear Weapons and Other Weapons of Mass Destruction on the Sea-Bed and the Ocean Floor and in the Subsoil Thereof, the Ottawa Convention on Future Multilateral Cooperation in the Northwest Atlantic Fisheries, the London Convention on Future Multilateral Cooperation in the Northeast Atlantic Fisheries, and the Rio de Janeiro International Convention for the Conservation of Atlantic Tunas.

Table 3.1. Principal international environmental agreements and instruments to which Cuba is a party

Agreement or instrument	Status	Cuba's position
Wildlife and habitat		
Antarctic Treaty (Washington)	Adopted 1959 Entered into force 1961	Ratified
Ramsar Convention on Wetlands of International Importance Especially as Waterfowl Habitat	Adopted 1971 Entered into force 1975	Ratified
UN Educational, Scientific, and Cultural Organization Convention Concerning the Protection of the World Cultural and Natural Heritage (World Heritage)	Adopted 1972 Entered into force 1975	Ratified
Washington Convention on International Trade in Endangered Species of Wild Fauna and Flora (CITES)	Adopted 1973 Entered into force 1975	Ratified
Convention on Biological Diversity	Adopted 1992 Entered into force 1993	Signatory
Oceans		
Rio de Janeiro International Convention for the Conservation of Atlantic Tunas	Adopted 1966 Entered into force 1969	Ratified
Brussels International Convention on Civil Liability for Oil Pollution Damage, as Amended by the 1976 Protocol	Adopted 1969 Entered into force 1975	Ratified with declarations
FAO Convention on the Conservation of the Living Resources of the South-East Atlantic (Rome)	Adopted 1969 Entered into force 1971	Ratified
Treaty on the Prohibition of the Emplacement of Nuclear Weapons and Other Weapons of Mass Destruction on the Sea-Bed and the Ocean Floor and in the Subsoil Thereof	Adopted 1971 Entered into force 1972	Ratified
International Maritime Organization Convention on the Prevention of Marine Pollution by Dumping of Wastes and Other Matter	Adopted 1972 Entered into force 1975	Ratified
Ottawa Convention on Future Multilateral Cooperation in the North-East Atlantic Fisheries	Adopted 1978 Entered into force 1979	Ratified

Table 3.1. *(continued)*

Agreement or instrument	Status	Cuba's position
International Convention for the Prevention of Pollution of Ships, 1973, as amended by the 1978 Protocol (MARPOL)	Adopted 1978 Entered into force 1983	Ratified
London Convention on Future Multilateral Cooperation in the North-West Atlantic Fisheries	Adopted 1980 Entered into force 1982	Ratified
UN Convention on the Law of the Sea, with Annexes	Adopted 1982 Entered into force 1984	Ratified
Atmosphere		
Vienna Convention for the Protection of the Ozone Layer	Adopted 1985 Entered into force 1988	Ratified
Montreal Protocol on Substances that Deplete the Ozone Layer	Adopted 1987 Entered into force 1994	Ratified
UN Framework Convention on Climate Change	Adopted 1992 Entered into force 1994	Ratified
Hazardous substances		
Vienna Convention on Civil Liability for Nuclear Damage	Adopted 1963 Entered into force 1977	Ratified
Convention on the Prohibition of the Development, Production, and Stockpiling of Bacteriological (Biological) and Toxin Weapons, and on Their Destruction	Adopted 1972 Entered into force 1975	Ratified
IAEA Convention on Early Notification of Nuclear Accident	Adopted 1986 Entered into force 1986	Signed
IAEA Convention on Assistance in the Case of a Nuclear Accident or Radiological Emergency	Adopted 1986 Entered into force 1987	Ratified
Basel Convention on the Control of Transboundary Movements of Hazardous Wastes and Their Disposal	Adopted 1989 Entered into force 1992	In process of ratification
Regional arrangements		
Convention for the Protection and Development of the Marine Environment of the Wider Caribbean Region, with Annexes	Adopted 1983 Entered into force 1986	Ratified
Protocol Concerning Co-operation in Combatting Oil Spills in the Wider Caribbean Region	Adopted 1983 Entered into force 1986	Ratified

Table 3.1. *(continued)*

Agreement or instrument	Status	Cuba's position
Protocol Concerning Specially Protected Areas and Wildlife to the Convention for the Protection and Development of the Marine Environment of the Wider Caribbean Region	Adopted 1990 Not in Force	Signed
Occupational safety and health conventions		
Occupational Safety and Health (ILO Convention No. 155)	Adopted 1981 Entered into force 1983	Ratified
White Lead (Painting) (ILO Convention No. 13)	Adopted 1921 Entered into force 1923	Ratified
Benzene (ILO Convention No. 136)	Adopted 1971 Entered into force 1973	Ratified
Working Environment (Air Pollution, Noise, Vibration) (ILO Convention No. 148)	Adopted 1977 Entered into force 1979	Ratified
Hygiene (Commerce and Offices) (ILO Convention No. 120)	Adopted 1964 Entered into force 1966	Ratified
Marking of Weight (Packages Transported by Vessels) (ILO Convention No. 27)	Adopted 1929 Entered into force 1932	Ratified
Protection Against Accidents (Dockers) (ILO Convention No. 32)	Adopted 1932 Entered into force 1934	Ratified
Occupational Safety and Health (Dock Workers) (ILO Convention No. 152)	Adopted 1979 Entered into force 1981	Ratified
Miscellaneous agreements		
Convention on the Prohibition of Military or Other Hostile Use of Environmental Modification Techniques (ENMOD)	Adopted 1977 Entered into force 1978	Ratified

Sources: The Environment 1995, 69–71; WRI 1992; Weiss, Magraw, and Szasz 1992; ILO 1988 and later issues.

Atmosphere

In 1992 Cuba ratified two of the major global environmental conventions dealing with the atmosphere: the Vienna Convention for the Protection of the Ozone Layer and the Montreal Protocol on Substances that Deplete the Ozone Layer. In 1994 it ratified the UN Framework Convention on Climate Change. Cuba is not a party to the 1963 Treaty Banning Nuclear Weapons Tests in the Atmosphere, in Outer Space, and Under Water (also known as the Limited Test

Ban Treaty), which prohibits atmospheric and underground nuclear weapons tests and other nuclear explosions and prohibits tests in any other environment if radioactive debris would be present outside the territory of the country conducting the test.

Hazardous substances

Cuba is a party to the major environmental global conventions dealing with hazardous substances: the Convention on the Prohibition of the Development, Production, and Stockpiling of Bacteriological (Biological) and Toxin Weapons, and on Their Destruction, the Convention on Early Notification of a Nuclear Accident, and the Convention on Assistance in the Case of a Nuclear Accident or Radiological Emergency; Cuba ratified in 1991 the latter two conventions, developed by the International Atomic Energy Agency (IAEA) in the aftermath of the Chernobyl nuclear power plant accident. Cuba reportedly is close to ratifying the Basel Convention on the Control of Transboundary Movements of Hazardous Wastes and Their Disposal.

In addition to the two nuclear-related conventions already noted, Cuba is a party to the Vienna Convention on Civil Liability for Nuclear Damages. Cuba is not a party, however, to one of the key agreements in this area, the Treaty on the Non-Proliferation of Nuclear Weapons (also known as the Non-Proliferation Treaty or NPT), which prohibits the transfer of nuclear weapons and technology to non-nuclear-weapon states and establishes a system of safeguards to prevent the diversion of nuclear energy from peaceful purposes to nuclear weapons.

Regional agreements

The regional environmental agreements to which Cuba is a party include the Cartagena Convention for the Protection and Development of the Marine Environment of the Wider Caribbean Region, the Cartagena Protocol Concerning Co-operation in Combating Oil Spills in the Wider Caribbean Region, and the Kingston Protocol Concerning Specially Protected Areas and Wildlife to the Convention for the Protection and Development of the Marine Environment of the Wider Caribbean Region. Cuba is also reportedly close to signing the Agreement on the Creation of the Inter-American Institute on Climate Change.

In 1995 Cuba announced that it would become a signatory to the Tlatelolco Agreement, the treaty that created a nuclear-weapon-free zone in Latin America and the Caribbean. Signatories to the treaty agree to abide by rules prohibiting the testing, use, manufacture, production, or acquisition, as well as the receipt, storage, installation, deployment, and any form of possession of any nuclear weapons.

Occupational safety and health conventions

Cuba has been a member since 1919 of the International Labor Organization (ILO), a specialized agency of the UN that seeks to improve working conditions, create employment, and promote human rights globally. Two of the chief functions of the ILO are to develop international labor standards and to supervise their observance (ILO 1991, 5).

Cuba has ratified eight of the twenty conventions developed by the ILO that deal with occupational safety and health (ILO 1995). These instruments ratified by Cuba include a general convention (Convention 155) calling on signatories to develop a coherent national policy on occupational safety and health and the working environment; three conventions dealing with protection of workers against specific risks (white lead; benzene; air pollution, noise, and vibration); and four conventions setting occupational safety and health standards in specific sectors of the economy (commerce and offices; dock workers).

Law No. 81

Law No. 81, the "Law on the Environment," was passed by the ANPP on 11 July 1997 (Ley No. 81 1997). Several years in the making, Law No. 81 superseded Law No. 33, the general environmental law that had been in effect since 1981, and Decree-law No. 118, the enabling legislation that had been in effect since 1990. According to a Cuban official, the then newly created CITMA began to draft a new environmental law in 1995 that "should reflect the most important natural and international advances [in the environmental protection] sphere as well as the institutional changes that have taken place in the country" (Rey Santos 1997, 42–43). Legislative action was expected as early as 1996, but the ANPP did not take up the bill until 1997.

A lengthy statute, Law No. 81 consists of a preamble, fourteen titles containing 163 articles, special provisions, transitory provisions, and final provisions. The preamble states that Law No. 33 was an early effort to create a legal framework for environmental protection. New socioeconomic realities, as well as domestic and international legal innovations regarding environmental protection, required a new legal instrument that would promote environmental protection and sustainable development, a concept accepted in the 1992 Earth Summit Declaration and already incorporated in the Cuban constitution through the 1992 amendments. The most novel aspects of the bill, according to CITMA Minister Rosa Elena Simeón, who presented the bill and led the discussion at the July 1997 ANPP session, are its emphasis on prevention, its efforts to promote participation by the community and by the people in general in environmental actions, and its emphasis on sustainable development (ANPP Ninth Regular Session 1997).

Compared to Law No. 33, Law No. 81 places a greater emphasis on the responsibilities of individuals (citizens or firms) in protecting the environment and on the ability of the state to impose sanctions on polluters. At the time the earlier statute was drafted, the bulk of the nation's productive resources were controlled by the state, and therefore the probability that individuals outside the control of the state would harm the environment was remote. This has changed in the 1990s as a result of structural changes in the economy that have promoted self-employment, agricultural cooperatives, and joint ventures with foreign investors. Along the same lines, Law No. 81 highlights the importance of environmental education as a way to prevent environmental disruption.

Law No. 81 identifies CITMA as the government agency responsible for developing and implementing environmental policy (see later discussion on implementation). The elaborate system of interagency bodies (with COMARNA at its head) created by Law No. 33 and Decree-law No. 118 seems to have been made moot by Law No. 81, although this is not explicitly stated in the law.

The new statute requires an environmental license for an activity "likely to have significant environmental effects" (Article 24); however, the term "significant environmental effects" is not defined in the legislation, granting CITMA a great deal of discretion on when to require licensing. The responsibility for seeking a license as well as any fees associated with it fall on the individual or organization initiating the activity. CITMA is empowered to suspend projects or activities that lack the appropriate environmental license.

In addition to an environmental license, the following new projects or activities are subject to an environmental impact assessment, to be conducted by CITMA: (1) dams, reservoirs, irrigation channels, drainage works, port dredging, and other projects that might bring about the reduction or elimination of natural water flows; (2) steel plants; (3) chemical or petrochemical plants; (4) facilities for handling, transporting, storing, treating, or disposing of hazardous substances; (5) mining activities; (6) electricity generation plants, transmission lines, and substations; (7) nuclear power generation stations and other nuclear reactors, including research reactors used for producing fissionable materials, and areas for disposing of nuclear wastes; (8) railways, roads, highways, *pedraplenes* (stone embankments), and gas and oil pipelines; (9) airports and ports; (10) oil refineries and storage facilities; (11) facilities to gasify and liquify hydrocarbons; (12) tourism facilities, especially those in coastal areas; (13) large housing developments; (14) free trade zones and industrial parks; (15) agricultural, forestry and aquaculture activities, particularly those introducing exotic foreign species or using domestic species difficult to reproduce or in danger of extinction; (16) changes in the use of soils that might lead to deterioration of their quality or affect their ecological balance; (17) facilities that transmit or collect urban sanitary sewage; (18) drilling of oil wells; (19) hospitals and other health facilities; (20) biotechnology facilities; (21) cemeteries and crematories; (22) sanitary landfills; (23) projects in protected areas; (24) sugar mills and plants producing sugar derivatives; (25) plants of the metalworking, paper and cellulose, beverages, dairy, meat processing, cement, and transportation equipment industries; and (26) any activity that takes place in ecologically fragile areas. Environmental impact assessments are also required for expansion, modification, or restart of any of the listed activities as well as for ongoing activities at the discretion of CITMA. The cost of conducting environmental impact assessments is to be borne by the entity that owns or manages the activity.

An innovative aspect of Law No. 81 is the adoption of economic tools—such as taxes, fees, and differential pricing—to promote environmental objectives. This is a departure from the command and control approach in Law No. 33 and its implementing laws. Among the economic measures contemplated by Law No. 81 are reducing and eliminating import tariffs on technology and equipment to control and treat pollutants as well as on raw materials and parts to build equipment for the same purpose, using accelerated depreciation for environmental projects and equipment, and tax holidays.

In the area of sanctions against polluters, Law No. 81 is more far-reaching than its predecessor. The thrust of Law No. 33 was on administrative sanctions imposed on polluters. Law No. 81 retains the concept of administrative sanctions but allows individuals who have been harmed by environmental disruption to seek remediation of the problem and compensation for damages incurred. In addition to civil prosecution, the law also foresees the possibility of criminal prosecution where the actions or omissions are deemed to be "hazardous to society" and are subject to sanctions under the penal code.

Practice of Environmental Protection

Before the mid-1970s, Cuba had no governmental institutions specifically charged with protecting the environment. As discussed earlier, COMARNA was established in 1977, and a network of provincial and municipal commissions in 1980. In 1990, Decree-law No. 118 formalized the role of COMARNA. As a result of the reorganization of central government functions in April 1994 (Decreto-ley No. 147 1994), COMARNA was absorbed by the newly created CITMA.

Institutions of Environmental Protection

In the early 1970s Cuba began a process of creating formal institutions charged with legal, political, and administrative duties. One of the milestones of this process was the First Congress of the Cuban Communist Party, held in December 1975.

Consistent with Marxist-Leninist precepts, the resolutions approved by the First Congress staked out an aggressive role for science and technology in the island's socioeconomic and cultural development. One of the specific objectives of a national scientific policy foreseen in the document was "the protection and improvement of the environment and the rational use of natural resources." The text of this section of the resolutions stated:

> In modern society, increasing attention is being focused on the protection and improvement of the environment and the rational use of natural resources. The unbridled exploitation of renewable and non-renewable resources must be avoided. The dumping of waste products that harm the environment and the indiscriminate use of substances that can harm the health of the population must be controlled. The harmonious solution of these problems can only be achieved within the socialist system.
>
> To achieve this objective, high priority should be given to scientific-technical research related to soil erosion and salinization; the pollution of underground, terrestrial, and ocean waters; air pollution; the inadequate exploitation of mineral resources; and harm being inflicted on the flora and fauna, especially native species.
>
> In carrying out research about our natural resources, the positive contribution that can be made by technological developments regarding the study of earth by aerial surveillance methods should be taken into account.
>
> In order to give special attention to these problems it is necessary to create a national institution with the required authority to recommend legislative measures and technology to protect and improve the environment and utilize rationally our natural resources. (Tésis 1976, 446)

COMARNA

In response to the preceding resolution, COMARNA was formally established in early 1977 pursuant to Law No. 1323 of 1976. For administrative purposes, COMARNA was placed under the guidance of the State Committee on Science and Technology (de la Osa 1977).

The tasks assigned to COMARNA at the time of its establishment were: (1) to develop the conceptual foundation of national scientific-technical policies to protect the environment and use resources rationally; (2) to analyze problems associated with the pollution of water and of agricultural areas; (3) to institute studies to support the establishment of medium- and long-term plans to protect the environment and conserve natural resources; and (4) to investigate the most pressing aspects of the problem of conserving sand in beaches (de la Osa 1977). COMARNA was also charged with maintaining close contact with central government agencies to analyze proposals for investments in new industrial plants in order to ensure that the environment was not adversely affected by such investments (de la Osa 1978).

The management council of COMARNA consisted of a president, a vice president, a scientific secretary, and the president of the scientific council. The scientific council consisted of representatives of the following central government agencies (de la Osa 1977): Cuban Academy of Sciences; National Institute of Forestry Resources; National Tourism Institute; National Institute of Physical Planning; National Institute of Sports, Physical Education, and Recreation; Ministry of Agriculture; Ministry of Higher Education; Ministry of the Interior; Ministry of Public Health; Ministry of the Chemical Industry; Ministry of Fisheries; Ministry of Education; Ministry of the Armed Forces; Ministry of the Sugar Industry; Ministry of Mining and Geology; Ministry of Construction; Ministry of Transportation; National Association of Small Farmers; Cuban Workers Central; and Committees to Defend the Revolution. Also participating in COMARNA were representatives of the Federation of Cuban Women and the presidents of COMARNA's provincial commissions (COMARNA 1992, 20).

COMARNA's work was carried out by several subcommissions responsible for the protection of: (1) the atmosphere, (2) water, (3) soils, (4) underground resources, (5) flora, (6) land-based fauna, (7) marine fauna, and (8) tourism resources. Other subcommissions dealt with environmental issues related to human settlements and the handling and use of toxic substances (de la Osa 1977). COMARNA established local offices throughout the country; provincial and municipal commissions began to operate in 1980 (Castro 1993, 44; COMARNA 1992, 21; Collis 1995, 1).

Spurred by the leadership of Guillermo García Frías, who at the time was a member of the political bureau of the Cuban Communist Party and a vice president of the Council of State and of the Council of Ministers, COMARNA seems to have been active in the late 1970s. In March 1978 a meeting of COMARNA's National Scientific Council in Camagüey criticized the lack of attention paid to the protection of the fauna and flora. The council designated a 5,000-hectare area southeast of the city of Camagüey as the "Najasa Protected Zone and Hunting Preserve" (Analizan 1978). In September 1978 the council met in Pinar del Río to discuss the creation of protected areas in that province, including in the peninsula of Guanahacabibes, a national park in the mountainous area known as "Mil Cumbres," and another national park in an area bordering the Cuyaguateje River (Presidió 1978). COMARNA was instrumental in focusing attention on the problem of erosion and loss of sand in beaches because of the excessive extraction of sand from cays and channels for use by the construction industry (Pagés 1981).

COMARNA was also involved in drafting Law No. 33 and other supporting laws and regulations. Among the latter are those creating a system of protected areas and hunting preserves throughout the island (including the aforementioned reserve areas as well as similar areas in the municipality of the Island of Youth, in the Zapata Swamp, and in the Gran Piedra zone within the Sierra Maestra) (Resolución No. 21/79 1979); regulating industrial pollution and toxic discharges; and protecting beaches and coastal zones (Oro 1992, 10). According to a former official of the Cuban Ministry of Basic Industries, "In 1987, at least 3,000 [environmental] activists were working in COMARNA and were writing numerous columns in state-controlled newspapers and journals, providing an intensive flow of information on subjects concerning pollution and environmental control" (Oro 1992, 11).

As discussed earlier, the mission and structure of COMARNA, and of the provincial and municipal commissions, were formalized by Decree-law No. 118 of 1990. A reading of this law does not give a hint that the structure of environmental organizations that are foreseen in this instrument had already been in existence for over thirteen years (in the case of COMARNA) and over ten years (in the case of the provincial and municipal commissions).

Implementation

According to COMARNA Vice President Helenio Ferrer, in an article published in 1993, Cuba's approach to environmental protection was "preventive" (Reed 1993, 33). The thrust of the Cuban approach to environmental protection, he states, was on safeguards that prevent environmental disruption from occurring: ecological studies before new industries are established; recommendations for location, raw materials, waste recycling, and disposal; studies of impact on the surrounding area. Indeed, examples of vigorous enforcement of environmental laws and the imposition of sanctions by the Cuban government are rare. Ferrer cites two examples: an oil spill that occurred in the Bay of Cienfuegos because of the carelessness of workers off-loading a tanker—the responsible parties were taken to court; and a sugar mill that was forced to pay for damages it caused to a neighboring fish-breeding farm because of wastes dumped into a reservoir (Reed 1993, 33).

In what may be a reference to the latter incident, the Cuban media reported in 1997 that the Fish Inspection Office of Granma Province sued the "Roberto Ramírez" Sugar Mill for dumping toxic materials into the Bay of Niquero. It was believed that the industrial wastes would do damage to the local fauna, including species in high demand by the tourism industry, such as lobster (Granma Province 1997). This is the only instance that we have been able to document where a legal process has been used by the government to sanction a state enterprise that is damaging the environment.

Earlier, in the 1980s, a case was reported in the press that illustrates the ineffectiveness of the regulatory mechanism. The fishing enterprise *Empresa Combinado Pesquero Industrial de Casilda,* through the State Dispute Resolution Organization (*Organo de Arbitraje Estatal*) of the city of Sancti Spíritus, sued Pulpa Cuba, a paper manufacturing plant, and the sugar mills "FNTA" and "Ramón Ponciano" for damages suffered because of the excessive discharge of pollutants into the Agabama and Manatí Rivers. The paper manufacturing plant agreed with the validity of the claim but alleged that it did not have the financial capacity to compensate the fishing enterprise. The fishermen were never compensated and, had they been compensated, payment would have come from the central budget rather than from the resources of the polluting enterprises (Dávalos 1984b,c).

In fact, there are several problems that limit the effectiveness of Cuban environmental laws and regulations: (1) they are hard to enforce because of their nonspecific nature; (2) the difficulty of enforcement is aggravated by unclear lines of authority and uncertainty about who is to be held responsible for environmental damage; (3) bureaucratic units are responsible for ensuring their own compliance with environmental laws, with conflicts of interest often arising because those responsible for attaining production goals are also responsible for overseeing and enforcing environmental regulations; and (4) economic needs often override environmental concerns (Espino 1992, 339; Collis 1995, 1).

Broad laws and regulations

Cuba's environmental laws and regulations are so broad that, in the judgement of specialists, they "could never be implemented legally" (Barba and Avella 1995, 5). For example, with regard to Article 29 of Law No. 33:

What represents residual substances? Industrial waste? Agricultural waste? Or household waste? If so, how should household waste be treated? Incineration? Landfill? Recycling? What constitutes surface and groundwater that needs to be protected? There is no provision in the law clarifying fresh water and brackish water or establishing the salinity of the water, as defined by the Total Dissolved Solids concentration. What are the levels of contamination established by the Law? Or more important, what are the contaminants? Are naturally occurring minerals such as zinc, copper, chromium, and lead classified as contaminants? It is known that lead, arsenic, and cadmium, to name a few, are hazardous to human health in large quantities, but those values need to be quantified and some parameters need to be established. (Barba and Avella 1995, 5)

Referring to a law passed in the early 1980s aimed at reducing industrial pollution and toxic discharges, former Cuban official José Oro states that it "contained a great deficiency, in that it did not clearly state the maximum permissible amounts of concentrations of these products [carbon dioxide, dust, soot, oil, chlorine, pesticides, mercury, arsenic, etc.]. The law stated that violators would be penalized with fines and in repeated or extremely dangerous operations, with the cancellation of operational licenses, severe bonds, prosecution, and eventual imprisonment. Unfortunately, the lack of specificity in this law hampered its effectiveness" (Oro 1992, 10).

Unclear lines of authority

Although COMARNA undoubtedly had the lead within the central government in developing environmental policies, its authority to enforce environmental laws and regulations, particularly in controversial matters, was severely constrained. COMARNA had to rely on the administrative office of each ministry to investigate allegations of violations of environmental laws and regulations (Oro 1992, 12). Moreover, "In addition to COMARNA's lack of real authority, its capability to enforce and penalize violators of environmental protection laws was dubious. In several interviews, the author [Collis] was told that most violations were corrected 'voluntarily' because Cuba is 'a solution oriented, not penalty oriented' country. No one interviewed could cite an example of violators being taken to court and fined. The majority of Cuban enterprises were (and still are) state operated, and a COMARNA suit against the state was inconceivable. In essence, COMARNA had no enforcement capability and its only recourse was to work through the existing structures and hope disputes could be solved voluntarily" (Collis 1995, 2).

Finally, areas such as research, development, and application of nuclear techniques were outside of the scope of COMARNA's mandate altogether. The

Secretariado Ejecutivo de Asuntos Nucleares (SEAN, Executive Secretariat of Nuclear Affairs) was responsible for research and development regarding nuclear energy as well as for control of potential hazards and risks (Oro 1992, 13).

Conflicts of interest

As noted earlier, COMARNA was a coordinating, rather than an enforcement, agency. The ability to enforce environmental laws and regulations, and the resources to do so, rested with other agencies of the Cuban state that were also responsible for economic production decisions: "When questioned in early 1994 about the apparent lack of enforcement capabilities, the National Commission of COMARNA pointed out that its staff of approximately 20 people was too small to effectively monitor compliance with regulations. Enforcement was left up to the Ministries which had 'more personnel and resources.' For example, the Ministry of Agriculture, which both supplied food to the population and promoted agricultural exports, was also responsible for enforcing environmental regulations governing cultivators. In effect, the Ministries were judge and jury of their own affairs" (Collis 1995, 2).

Cuban analysts believe that the mere creation of CITMA has eliminated this conflict. "With the creation of CITMA as the organization responsible for managing the environmental policy of the nation, a contradiction regarding direction of environmental policy was resolved. Prior to this time, Ministries responsible for environmental management of certain resources were at the same time responsible for their exploitation for productive purposes, making them both judge and party" (Borges Hernández and Díaz Morejón 1997, 14).

Economic needs versus environmental concerns

Environmental impact assessments are systematic examinations of environmental consequences of projects, policies, and programs. They are conducted to provide policymakers with an assessment of the implications of alternative courses of action *before* a decision is made (Avella 1995, 396). Law No. 33 arguably required that an environmental impact assessment be conducted for each investment project carried out in Cuba since 1981 (this can be inferred from Articles 10 and 11 of Law No. 33). Despite this laudable requirement, as we show throughout this book, when environmental concerns clashed with the achievement of the state's economic objectives, environmental concerns were set aside.

One documented example of economic needs prevailing over environmental concerns involved the development of tourism resources in Cayo Coco, a cay located along Cuba's northern coast (see Chapter 10). Other instances where economic needs have taken precedence over environmental protection—

allegedly in violation of environmental laws and regulations—are plans for extracting peat from the Zapata Swamp to be used as fuel and plans for constructing a giant hydroelectric dam system on the Toa and Duaba Rivers (near the city of Baracoa) even though these rivers are located within national parks (Oro 1992, 12).

There is also evidence that environmental impact assessments are not thorough and that planners consider them as afterthoughts rather than integral parts of the planning process. For example, project documentation regarding treatment and disposal of liquid wastes has tended to be very general, at best making reference to the general treatment that would be used but without detailing design specifications, expected efficiency levels, expected quality of effluents after treatment, and other key parameters "whose estimates or measurement are necessary to conduct a serious evaluation" (Terry Berro 1997, 43). It was common practice on the part of the ministries undertaking investments to consider project completion as "their problem," with environmental measures and mitigation relegated to secondary roles (43).

Conducting environmental impact assessments has been hampered by lack of trained personnel—particularly outside of La Habana—and lack of technical and financial resources. In particular, monitoring systems that obtain the necessary information to evaluate the environmental effects of a project before and after its completion require technologies that are not always available in Cuba. Moreover, these technologies tend to be expensive and difficult for Cuba to acquire during the current economic crisis (Terry Berro 1997, 44).

An official of CITMA's Environmental Management and Inspection Center has very effectively summarized the reluctance of some investors to engage fully in the environmental evaluation process: "The need to conduct environmental assessments as instruments for planning and decision making, although recognized as such by many, is still perceived as being a 'waste of time and money' by others, who do not see as beneficial in the long term a decision to abandon a project early on if all its alternatives become unacceptable from an environmental standpoint, avoid expensive corrective actions once a project has been carried out, or prevent the destruction or deterioration of environmentally sensitive areas with unique value for the quality of life and the enjoyment of current and future generations" (Terry Berro 1997, 45).

In summary, even though Cuban law since 1981 has required that environmental impact assessments be conducted for all investment projects, it is not clear that this has occurred or that they have been given proper consideration by policymakers (Atienza Ambou 1996). This issue has become more critical during the special period (see Chapter 10), when foreign investments in environmentally sensitive industries, such as tourism and mining, are being carried out.

Governmental Structures

In 1994, in the midst of an aggressive campaign to cut government expenditures to reduce a huge budget deficit, the Cuban government undertook a major reorganization of its central administration, shifting and consolidating responsibilities among agencies. Pursuant to the central administration structure, Cuba created six new ministries and eliminated thirteen institutes and state committees. One of the new agencies is CITMA, which absorbed COMARNA as well as the Cuban Academy of Sciences, the Atomic Energy Commission, and the Executive Secretariat of Nuclear Affairs. CITMA has the following functions:

▶ Direct and control the implementation of policies integrated with the sustainable development of the country aimed at protecting the environment and promoting the rational use of natural resources;

▶ Propose and establish national strategies to protect specific natural resources and biodiversity;

▶ Develop and control the implementation of programs to enhance environmental control, the appropriate management of agricultural and industrial wastes and the introduction of clean technologies;

▶ Oversee the implementation by organizations of regulations to protect, conserve, and use natural resources rationally;

▶ Settle disputes between organizations and other entities related to environmental protection and the rational use of natural resources, implementing decisions or raising them to higher authorities for action;

▶ Approve environmental impact assessments;

▶ Direct, evaluate, and control monitoring activities related to climate, chemical composition, and pollution of the atmosphere;

▶ Direct, evaluate, and control monitoring activities related to radioactivity, and conduct studies of earthquakes, climate change, and radioactivity, as well as of other natural or man-made disasters. (Ministerio de Ciencia 1997a)

The rationale for creating CITMA was "to resolve a contradiction in the country's former environmental management structure in which some ministries were responsible for protecting natural resources at the same time that they were responsible for their exploitation for productive purposes" and to "contribute to making the management of the nation's environment more effective" (Ministerio de Ciencia 1995, 73).

Although the reorganization of government functions occurred in April 1994, it was not until January 1995 that the structure of CITMA was announced.

CITMA brought together under a single ministry—led by a minister, two vice ministers, four agency heads, and eight directors—the activities formerly carried out by four independent bureaucracies. According to one observer, the new ministry "sought to address the previous structure's conceptual problems, including COMARNA's obsolescence and the need for information sharing between the scientific institutes and policy makers.... The new Ministry's structure is geared toward policy formulation, and features four agencies which provide information needed to formulate policy, then implement those policies once they are defined" (Collis 1995, 2). In the specific area of environmental protection, CITMA's principal functions include: (1) broadening and strengthening the system of inspections to ensure compliance with environmental law; (2) redesigning environmental impact analyses and ensuring the implementation of recommendations that emerge from such analyses; (3) updating and completing national environmental legislation; and (4) achieving a closer relationship between environmental management and national scientific developments (Ministerio de Ciencia 1995, 73–75).

Environmental Policy Directorate

The Dirección de Política Ambiental (Environmental Policy Directorate) has overall responsibility for the design of national environmental policies and, with the cooperation of entities responsible for the development of economic policies, of economic mechanisms to guarantee the protection of the environment, consistent with Cuba's current economic situation. Implementation of such policies is the responsibility of the Agencia de Medio Ambiente (Environmental Agency). The Environmental Agency has a great deal of administrative and financial autonomy from CITMA; it is composed of three centers, five research institutes, and three other institutions and brings together more than 2,000 workers with scientific background.

Environmental Management and Inspection Center

The Centro de Gestión e Inspección Ambiental (Environmental Management and Inspection Center) has direct responsibility for implementing the nation's environmental policies, including measures to promote the rational use of resources, to protect fragile ecosystems, and to reduce pollution. It also approves and controls the carrying out of environmental impact assessments and the implementation of recommended measures. It is supported by the five research institutes of the Environmental Agency: (1) Instituto de Meteorología (Meteorology Institute), which is responsible for protecting the atmosphere and monitoring climate change issues; (2) Instituto de Oceanología (Oceanology

Institute), which is responsible for protecting coastal areas; (3) Instituto de Ecología y Sistemática (Ecology Institute), which is responsible for preserving biodiversity; (4) Instituto de Geografía (Geography Institute), which is responsible for physical planning; and (5) Instituto de Geofísica y Astronomía, (Geophysics and Astronomy Institute), which is responsible for conserving soil and water resources.

Information and Environmental Education Center

The Centro de Información, Divulgación y Educación Ambiental (Information and Environmental Education Center) is responsible for gathering and diffusing data and specialized information on the environment and development and for environmental education. In its responsibilities related to carrying a national environmental education program, the center relies heavily on three institutions associated with CITMA: (1) the Museo Nacional de Historia Natural (National Natural History Museum); (2) the Acuario Nacional (National Aquarium); and (3) the Zoológico Nacional (National Zoological Park).

National Protected Areas Center

The Centro Nacional de Areas Protegidas (National Protected Areas Center) is responsible for the management of the National System of Protected Areas, guaranteeing its optimal operation.

Under the new administrative arrangements, according to David Collis (1995, 3), enforcement of environmental regulations will be carried out by special provincial delegations that allegedly will serve as "independent overseers." What degree of independence these overseers will have from the state, how their independence will be protected, what power they will have in conflicts with other state agencies, and how such conflicts will be resolved are all matters that remain to be clarified. Collis (1995, 3) sees the raising of environmental issues to the ministerial level and the establishment of the "independent overseers" as positive steps for the protection of the environment. However, he cautions, it is still too early to tell whether the "independent" provincial delegations will in fact carry out their enforcement roles.

Summary and Implications

On paper, Cuba has a well-developed legal framework of environmental protection. Cuba's socialist constitution, as amended in 1992, recognizes the duty of the state and of all citizens to protect the environment and incorporates the concept of sustainable development. Cuba first adopted a broad and general environmental law in 1981. In July 1997 the Cuban legislature enacted a new environmental law that superseded the earlier statute.

Cuba has ratified over two dozen international agreements, conventions, or protocols dealing with wildlife and habitat, oceans, atmosphere, hazardous substances, regional concerns, occupational safety and health, and miscellaneous issues. These international instruments constitute a significant body of international public environmental law. In the Cuban legal system, international legal instruments become part of the national legal structure upon ratification.

In practice, environmental protection in socialist Cuba is weak. Cuban environmental laws are general and difficult to enforce. Following the practice in other (former) socialist countries, the Cuban economy is controlled by sectoral ministries that have both promotion and regulatory functions. When conflicts arise over environmental matters, the feeble environmental protection institutions are routinely overruled by sectoral ministries that are able to appeal to the imperative of promoting economic objectives.

↝ 4 ↜

Agriculture and the
Environment

THE HISTORICAL development of land use patterns in Cuba is closely related to the variance in soil quality and evolving institutional factors that have determined tenure arrangements, including ownership and size of holdings. These land use patterns, in turn, are important determinants of the country's changing environmental situation, including a long-term trend of soil degradation. A World Bank publication, in a recent examination of soil conservation issues in Central America and the Caribbean, offers a summary of some of the factors behind soil degradation that provides a useful framework for assessing Cuba's situation (Lutz et al. 1994).[1] After observing that soil degradation has on-farm as well as off-farm consequences and that it often occurs as a result of cultivation practices, Stefano Pagiola goes on to note how cultivation practices can damage the soils and how, in some instances, they can have a cumulative impact:

> Cultivation practices can expose soil to water and wind erosion; repeated tillage can weaken soil structure; crop production can remove nutrients; and use of machinery can lead to soil compaction. Reductions in soil depth through erosion are the best-known form of degradation, but far from the only one. In many cases, different forms of soil degradation are correlated. For example, soil compaction can result in increased runoff and higher rates of erosion; conversely, erosion can

carry away nutrients and weaken the soil's physical structure. An important characteristic of such damage is that it is usually cumulative; its effects in any one year can be minor or insignificant but become important as they accumulate over time. Whatever its forms, soil degradation is usually reflected in lower yields or, if compensating measures are taken, in higher costs for a given yield.

A soil's vulnerability to degradation depends on how easily it is damaged and on how significant that damage is for crop production. The erodibility of the soils varies, for example, according to their structure and their chemical and physical composition. The effects of environmental conditions must also be considered. The predominant soils in the humid tropics (oxisols, ultisols, and alfisols), for example, tend to be less erodible, under equivalent conditions, than temperate region soils. However, erosivity is generally much higher in these regions.

In turn, the effects of degradation on productivity depend on complex soil characteristics and on crop requirements. The impact of erosion, for example, depends on the distribution of plant nutrients in the soil profile, on the crop's rooting depth, on plant-available water resources, and on physical and chemical properties of subsoil horizons. Where soils are deep and subsoil characteristics are favorable, even substantial rates of soil loss may have little effect on productivity. (Pagiola 1994, 22)

Soil degradation may also have important off-farm consequences, such as when it damages reservoirs and waterways. Damage in one location, furthermore, may accentuate soil degradation elsewhere. For example, by contributing to rapid water run-off in one location, it may lead to soil erosion downstream. Conversely, in some cases, soil degradation may be harmful where it occurs, but it might have beneficial consequences in other locations, such as when fertile topsoil eroded from one location ends up enriching other farm land (Pagiola 1994, 22).

In this chapter we examine patterns and long-term trends of land use in Cuba by briefly reviewing and discussing some of the principal physical features of its soils, including agricultural quality, and how they have influenced land use patterns before and after the 1959 revolution. We also consider the determinants of soil degradation, including factors that have contributed to it in Cuba, and conclude with a brief review of some of the measures instituted since the early 1990s primarily to arrest a dramatic decline in agricultural output but also to address the worrisome soil degradation problem (Rosset and Benjamin 1993; Weinberg 1994).

Cuba's Soils

Cuba's soils are heterogeneous and include some of the world's richest as well as some poorly endowed for agriculture. Until recently, detailed analyses of Cuba's soils were lacking. For many years, the classic reference source on Cuba's soils was the reconnaissance survey conducted by Hugh H. Bennett and Robert V. Allison in 1927 (Bennett and Allison 1928). This survey, organized by the Tropical Plant Research Foundation with financial support from the Cuban Sugar Club, provided a tentative soil classification that was to have been followed by a detailed soil survey. The detailed survey was never conducted, however, due to funding problems arising from the world depression (Foreign Policy Association 1935, 460). In their taxonomy, Bennett and Allison divided Cuba's soils into 18 major families, which were further divided into 104 subtypes. This survey served as the primary source of information on Cuba's soils until a new taxonomy was developed with Soviet technical input during the 1970s. The latter, based on an analysis of the geological processes leading to the evolution of soils, identifies 10 major soil groupings and 29 soil types (Academia de Ciencias 1989, IX.1.1).

For our purposes, it is sufficient to rely on a less detailed scheme, based on the Bennett and Allison taxonomy, that has been the basis for several analyses of Cuba's soils. For example, Leví Marrero (1950, 95–103) identified five types of soils (clay, sabana, sandy, calcareous, and various). Similarly the World Bank in its 1950 study of the Cuban economy (World Bank 1951, 85–86) divided Cuba's soils into (1) several types of clay soils (according to their fertility), (2) sandy soils, and (3) thin and rocky mountainous soils.

Among Cuba's most fertile soils are the deep red Matanzas clays. Despite their high clay content, these soils are rich in iron oxide and can be readily cultivated even after heavy rains because of their good drainage quality. Matanzas soils are mostly found in central Cuba. This family also includes other fertile soils such as the "Habana." At the other extreme are the reddish-brown and yellowish clay soils dominant in eastern Cuba. In contrast to the Matanzas clay, these soils are plastic in the extreme. Other clay-rich soils have intermediate physical characteristics, some containing considerable amounts of organic matter and others having a high salt content. Many of these soils, particularly those with impermeable clay subsoils, drain rainwater poorly and are thus difficult to cultivate during the rainy season.

Cuba's abundant sandy soils, of which there are many types, are mostly found in the flatlands (sabanas) of central Cuba as well as in the western part of the island, including the Isla de la Juventud. Although sandy soils drain well, their fertility is below that of clay soils because they are deficient in humus and lime.

Over 8 percent of Cuba's territory (including many of the small islands in the archipelago) corresponds to water-soaked clays (Ministerio de Ciencia 1995, 5), often with a high content of peat, muck, and marl. Finally, the mountain regions at both extremes of the island have rocky and thin soils poorly suited for agriculture other than in alluvial accumulations in intermountain valleys.

A major problem with many of Cuba's soils (approximately 4 million hectares, or 37 percent of the country's territory) is that they drain poorly. A further 1 million hectares (or about 11 percent of the national territory), particularly in eastern and central Cuba, suffer from salinization, although natural salt contents are not extreme, except in coastal areas where seawater intrusions are common. Salinization has been on the increase, however, in poorly drained irrigated areas.

An Agricultural Production Potential Soil Typology

A more recent typology takes into account the physical constraints of Cuba's soils and divides agricultural soils into six basic types according to their actual or potential fertility (Atienza Ambou et al. 1992, 8). This typology has been derived from the detailed studies conducted in Cuba since 1959 under the aegis of the Academia de Ciencias de Cuba (ACC, Cuban Academy of Sciences) and other research institutes with Soviet technical support (Núñez Jiménez 1972, 349–50). It takes into account soil productivity and characteristics such as depth, texture, permeability, incline, erosion potential, natural drainage conditions, salinity, and flooding capacity. According to this typology, agricultural soils account for 6.6 million hectares, or 60.6 percent of the country's total land area of slightly over 11 million hectares.

The most fertile soils, agricultural types I and II (clay and clay loam rich in humus), are deep and permeable and well suited for most tropical crops. They account for 12 percent (approximately 800,000 hectares) of the country's agricultural land area. Next in order of potential agricultural fertility is soil type III. These lightly waterlogged clay soils represent 35 percent (2.3 million hectares) of the country's agricultural land area. Although type III soils lack permeability and tend to retain too much water, with appropriate drainage they can become highly productive. The productivity of the remaining agricultural soils (53 percent of the agricultural land area) is low, with some, particularly soil type VI, generally regarded as poorly suited for agricultural pursuits. Soil types IV–VI are prone to salinization and/or easily eroded.[2]

In general, the Cuban provinces with the most productive soils (in terms of overall physical characteristics and geographical features) are La Habana and Ciego de Avila, and the least productive are, at both extremes of the island, Pinar del Río, Granma, Holguín, Santiago de Cuba, and Guantánamo. The provinces with soils of intermediate production capacity are Camagüey, Las Tunas, Sancti Spíritus, Cienfuegos, Matanzas, and Villa Clara (Academia de Ciencias 1989, IX.1.1).

Historical Land Use Patterns

The presence of Cuba's most fertile soils in the central regions of the country is an important variable in explaining the evolution of the colonial settlement pattern, even though Cuba's first villages founded by the Spanish settlers were established in the eastern part of the country. With a major expansion of the sugar industry in the early nineteenth century, sugar plantations began to radiate from La Habana in all directions (Moreno Fraginals 1964, 63–70), with land settlements proceeding eastward. By 1827, the province of Matanzas, famous for its deep red clay soils, produced a quarter of all the sugar exported by Cuba (Nelson 1950, 119). Sugar production was concentrated in a band running from the province of La Habana in the west to the city of Cienfuegos in the center of the island. Throughout the nineteenth century, according to Manuel Moreno Fraginals (1964, 67), 90 percent of all Cuban sugar was produced in the central and western regions of the country.

Settlements east of Sancti Spíritus were few and mainly found in coastal or inland towns dating to the early days of discovery, such as Santiago de Cuba, Puerto Príncipe, and Trinidad, even though Puerto Príncipe, first established on the seacoast but transferred in 1528 to a broad inland sabana area, was Cuba's second largest city at the time of the 1774 census. Although small amounts of sugar were produced in Puerto Principe, this region (together with Bayamo) was indirectly linked to the sugar industry as a supplier of crucial production inputs in the form of beasts of burden and salted beef to feed the slave labor force (Moreno Fraginals 1964, 68–69). So was Trinidad, which for a time was also an important sugar-producing region. To this day, Puerto Príncipe (known today as Camagüey) continues to be Cuba's center of livestock production. In western-most Cuba, agricultural production, aside from the famed tobacco leaves, was largely limited to subsistence agriculture. After the 1791 slave revolt in Haiti, coffee production became a dominant economic activity in the mountain areas of eastern Cuba. Small amounts of sugarcane were planted in the Santiago de Cuba region, and, as noted, Bayamo was an important cattle region. Mining activities rounded the economy of eastern Cuba. This changed with the expansion of sugar production into eastern Cuba during the first quarter of the twentieth century when the country's center of sugar production markedly shifted toward the provinces of Camagüey and Oriente.

With the eastward expansion of sugar production and the destruction of the forests during the first quarter of the twentieth century (see chapter 6), the fertility of many of Cuba's soils began to decline. Erosion and other physical processes degraded the soils as slash-and-burn agricultural practices intensified. Forest land cleared and planted with sugarcane would lose its fertility within five years, resulting in low yields over time and its eventual abandonment and conversion to pasture (Sáez 1997). By the mid-1940s, this process had pretty much run its course, although logging and mining interests, as well as subsistence farmers, still continued to push into the remaining forested areas. The soils in these areas, primarily in mountainous regions, were poorly suited for agriculture but quite vulnerable to erosion due to their textures, permeability, and inclines. As discussed in the next section, Cuba was a largely settled country by the mid-1940s, with about one-fifth of the total land area under cultivation and another two-fifths in pastures.

Land Use Trends

It is difficult to derive a reliable time series of estimates of land use according to categories such as land under cultivation, in pastures, forested, permanently flooded, or covered by water bodies. A cursory examination of the evidence reveals major stumbling blocks in terms of inexact and/or changing definitions and incomplete categorization. Even the historical estimates of total land surface vary, apparently because some sources consider the island of Cuba exclusively, whereas others also refer to the Isla de Piros and adjacent cays in the Cuban archipelago. Furthermore, we have been unable to locate historical estimates of land area in pastures or covered by water bodies, and for earlier periods only have broad estimates of forested area we have derived on the basis of uncertain historical data.

A reasonably accurate comparison of land use patterns in the mid-1940s and the late 1980s can be made by contrasting data from the 1946 agricultural census (with a 1945 reference period) and statistics for 1989 provided in the *Anuario Estadístico de Cuba*. Although the definitions used in both sources differ, they are sufficiently close to assess broad changes over time in the principal land uses. The main comparability problem is that land use statistics in the 1946 agricultural census refer only to land in farms, whereas the 1989 statistics refer to the entire national territory, whether in farms or not.

According to the 1946 agricultural census, 79.3 percent of the total land area was in farms in 1945. The census identified almost 160,000 farms, with an average size of 56.7 hectares. About a fifth of the farm area (21.7 percent) was cultivated, twice that amount was in pastures (42.9 percent), 14 percent was wooded, and smaller percentages were covered by *marabú*[3] or in fallow. Close to one-fifth (18.2 percent) of farm land was accounted for by "roads, ditches, buildings, and other unproductive land" (World Bank 1951, 87). In 1945, about 6.2 million hectares, or 53.8 percent of the country's total land surface, was used for agricultural purposes. The amount of land allocated to agricultural pursuits at the time was about 400,000 hectares below the 6.6 million hectares maximum amount of potential agricultural soils (see the earlier section of this chapter that describes an agricultural production potential typology of Cuban soils).

By 1989, nearly 62 percent of the nation's land was devoted to agriculture and 38 percent to nonagricultural uses. Out of the nearly 6.8 million hectares devoted to agriculture, about 4.4 million hectares (40.1 percent of total land) were cultivated (including about 1.1 million hectares of cultivated pastures) and 2.4 million hectares (21.4 percent) were dedicated to natural pastures or were in fallow. Nonagricultural land was distributed among forests (23.7 percent), settlements (6.3 percent), not usable (5.5 percent), or covered by water (3 percent). The total amount of agricultural land in 1989 (whether cultivated or not) exceeded by nearly 200,000 hectares the maximum estimated potential agricultural soils. The total amount of land being used for agriculture (61.5 percent) in 1989 exceeded by nearly one percentage point the amount of soils considered suitable for agriculture (60.6 percent). More noteworthy is that the total land area cultivated had increased by nearly 23 percent between 1945 and 1989, whereas pastures (excluding cultivated pastures, which in 1989 accounted for 9.8 percent of the total land area) had declined by half (from 34 percent in 1945 to 17 percent in 1989).

The most comparable data for 1945 and 1989 are associated with agricultural land, namely cultivated, pasture, and fallow land. The amounts of land assigned to these uses can be extracted directly from the sources. More problematic is the information for nonagricultural land, since the 1946 agricultural census includes a large residual land use category labeled "not in farms," which included 20.7 percent of Cuba's total land area, as well as a category labeled "other uses," which accounted for a further 14.4 percent of the total land area and includes, as noted earlier, "roads, buildings, and unproductive land."

Despite these comparability problems, several conclusions can be drawn from the data. The first is that the total amount of agricultural land between 1945 and 1989 increased by close to 8 percent. This trend is consistent with information on agricultural development policies of the socialist government that placed a premium on increasing the country's agricultural land area and is also suggested by a relative decline in the total land area classified as "settlements and other uses" between 1945 and 1989. Fidel Castro's remark that "not an inch of land should be left unused" (Deere et al. 1994, 211) was premised on the contention that in capitalist Cuba agricultural production had been limited because the country's sugar mills maintained a large amount of land in reserve to take advantage of upswings in world market demand (and prices) for sugar. Thus, "it was the level of demand and not the limitations imposed by productive capacity that determined the actual volume of sugar production . . . the possibilities of expanding the area under crops on a significant scale were inextricably linked even in the case of sugar states to the diversification of farm output. . . . In Cuba there was neither a land nor a labor shortage" (Bianchi 1964, 86).

The increase in cultivated land and the decline in pastures were not as sharp when considering cultivated and natural pastures together, the total land in pasture declining by only 7 percent during the 1945–1989 interval. Still, the amount of cultivated land (other than cultivated pastures) increased by nearly a third between 1945 and 1989. The number of hectares lying fallow increased substantially, but from a much lower base. Substantial increases in the land area cultivated were largely achieved by bringing under the plow formerly nonagricultural land either because it was being held in reserve (e.g., by the large sugar mills), or because it was considered agriculturally marginal. As Aída Atienza Ambou and her associates note (1992, 10), "by the 1970s, the country's best agricultural soils were cultivated, thus soils of inferior quality requiring major improvements and considerable expenditures had to be used to increase the cultivated land area." As noted earlier, only 12 percent of Cuba's soils are well suited for agriculture; a further 35 percent are potentially productive if appropriately drained; and the agricultural potential of the remaining 53 percent is poor.

Between 1945 and 1989, the cultivated land area (the sum of land in permanent and temporary crops) increased by 1,321 hectares, or by 66.6 percent, from about 2 million to 3.3 million hectares. Of the total amount of additional land brought under the plow, 877,000 hectares, or 66.4 percent, were assigned to sugarcane. In relation to land under permanent crops, the increase attributed to sugarcane was even more pronounced, amounting to 72.8 percent. Although the amount of land planted with citrus increased eight-fold between 1945 and 1989, by the latter date citrus plantations only occupied 150,000 hectares, or 2.4 percent of all agricultural land, as compared to the 2 million hectares planted with sugarcane (31.5 percent of all agricultural land). Other notable changes were a considerable expansion in the amount of land devoted to coffee, fruit trees, and rice, and a contraction in the land area dedicated to tobacco and the residual category of "other temporary crops."

A more detailed assessment of changes in land use by type of crop suggests that the overall acreage planted in tubers declined between 1945 and 1989, even though potato plantings nearly doubled. Land devoted to the production of vegetables, other than tomatoes, appears to have increased considerably, with major increases in the amount of land planted with onions and peppers. A striking difference in the planting trends for rice and corn occurred. Whereas the amount of land devoted to rice production increased three times between 1945 and 1989, the acreage devoted to corn declined by half. These divergent trends are consistent with Cuba's policy to increase domestic rice production (even before the 1959 revolution) and with the country's dependence on Soviet supplies of feed grains for livestock and poultry production. Millet, an important crop in 1945, is not even listed separately in 1989. Finally, the number of hectares planted with beans between 1945 and 1989 remained essentially unchanged.

Interpretation uncertainties apply to land in settlements/other uses, although it is unlikely that the land area occupied by settlements could have declined in view of continued urbanization, including the establishment of over three hundred new urbanized rural communities since 1959, accompanied by an approximate doubling in population size between 1945 and 1989. These trends suggest, therefore, that the main decline must have taken place among those lands classified as in "other uses" in 1945.

The major unknown factor relates to the changes in forested land area. Although it is claimed that the forested land area increased from 14 percent to 18 percent between the late 1950s and the late 1980s (see Chapter 6), there are no firm estimates regarding the extent of forests outside of farms for 1945, even though nonfarm land accounted for 21 percent of the country's total land surface. Estimates for the pre-socialist period are also silent regarding tree stands in places other than those classified as forests (e.g., small tree stands, tree fences).

A major conclusion is that the socialist government expanded the amount of cultivated land by bringing into production farm land that could be used only if it was improved considerably. The critical improvement necessary for making much of this marginal land (35 percent of the national territory) productive was proper drainage. Other alternatives were irrigation and the liberal application of chemical inputs. The latter approaches were apparently preferred by the socialist authorities who in their agricultural development policies neglected the country's drainage infrastructure, a situation made even more acute by the decision to expand several-fold the land surface irrigated (see Chapter 5). Not until the mid-1980s, when it became obvious that the country faced a drainage crisis of major proportions, did the authorities turn their attention to this issue. It was announced then that the 1986–1990 and later development plans would assign considerable resources to the country's drainage infrastructure. These plans failed to materialize with the end of Soviet subsidies and Cuba's economic crisis.

The Socialist Transformation of the Cuban Countryside

In Cuba, as elsewhere, poorly designed and implemented agricultural development policies have been a major contributor to the degradation of the country's soils. These policies date back to colonial times when the indiscriminate deforestation of Cuba began in earnest. Soil degradation continued apace with the agricultural practices that came to prevail in Cuba during the first half of the twentieth century as more and more virgin lands were brought under sugarcane and livestock production. There is mounting evidence that the pace of soil deterioration intensified during the second half of the century, principally because of the widespread adoption of modern agricultural practices, particularly from the 1960s to the late 1980s. As in the Soviet Union and other socialist economies, the collectivization of the rural sector was regarded as necessary for achieving the goals of a centrally planned command economy. Collectivization would make possible economies of scale and bring the advantages of mechanization and modern scientific agriculture. Collectivization was also justified ideologically, aside from its alleged intrinsic economic merits, since it was intended to ensure the political loyalty of the peasantry (Wunderlich 1995, 31, 54; see also Sáez 1997a).

The appearance of modern agricultural practices in Cuba, however, antecedes the socialist revolution. By the 1950s Cuba's agriculture was already changing appreciably, with the use of modern agricultural inputs (e.g., tractors, fertilizers) increasing rapidly (Grupo Cubano 1963, 1009–1011, 1130–32; Grupo Cubano 1965, 68, 70, 74). In 1945 rural Cuba was divided into 159,958 farms, 84.6 percent of which had 50 hectares or less. The largest amount of farm land was occupied by farms of between 100 and 500 hectares; 10,433 farms in this size class accounted for 24.1 percent of all farm land. Next were farms with 5,000 hectares or more: 114 of these farms occupied 20.1 percent of all Cuban farm land. Farms with 500 hectares or more controlled 47 percent of the national farm land. The high degree of land concentration was explained first and foremost by sugar production: in mid-1959, "the 28 largest sugar-cane producers owned over 1,400,000 hectares and rented 617,300 hectares, thus controlling over 20 percent of the land in farms and almost one-fifth of the Cuban territory" (Bianchi 1964, 76). Another factor explaining land concentration was the existence of large livestock farms (the forty largest livestock farms controlled 10 percent of all farm land) and large-scale rice production farms.

With the enactment of two agrarian reform laws, the revolutionary government completely transformed Cuba's land ownership patterns and set in motion a process that in a relatively brief time span would also radically change the country's farm-size distribution. The first Agrarian Reform Law, enacted on 17 May 1959, was designed to undo the historical pattern of land concentration by doing away with latifundia and distributing farm land to the peasantry. The law set a maximum size of individual landholding of 30 *caballerías* (some 400 hectares), except for certain farms with above-average productivity, and expropriated land exceeding this limit. Most of the expropriated land was not distributed to peasants, but rather organized into state-run agricultural cooperatives and subsequently into state farms. On 3 October 1963, with the enactment of the second Agrarian Reform Law, the bulk of the remaining farm land in private hands was taken over by the state, and all landholdings above 5 *caballerías* (approximately 67 hectares) were nationalized. As a result of these two laws, 70 percent of Cuba's agricultural sector ended up in state hands.

The drive to consolidate land ownership in the hands of the state did not end with the second Agrarian Reform Law. For the next twenty-seven years, the Cuban government pursued policies intended to voluntarily bring small, privately controlled land into the state sector. These policies relied as much on economic incentives (e.g., providing modern housing to peasants willing to cede their land to state farms and move into towns) as on disincentives (e.g., giving private farmers limited access to farm inputs). The rationale behind the drive to reduce—if not to eventually eliminate—small private farm holdings was to impose the "modern" large-scale agricultural model. Small farms were not only a more primitive form of agricultural organization, but their existence interfered with the implementation of central planning. With collectivization, the government would be able to acquire "the necessary direct control to carry out a consistent policy of economic development" (Pryor 1992, 48).[4] As a result of these policies, by 1986, "state farms averaged 14,260 hectares, or around seven times larger than average latifundios before the revolution, and employed an average of 1,372 workers. The sugar industry had 146 state enterprises, with an average size of 10,000 hectares per farm, while there were 130 cattle enterprises, with an average size of 21,000 hectares" (Sáez 1997).

In 1987 Cuba was among the socialist countries with the highest concentration of agricultural land in state farms (73 percent, with a further 12.5 percent in collective farms). This percentage was surpassed only by Sao Tomé (96.2 percent) and Bulgaria (90 percent) and was closely approximated by the former Soviet Union (67.8 percent). Whereas in Cuba private farms accounted for only 14.5 percent of total agricultural land, in other socialist countries this figure ranged from zero (Kampuchea, North Korea, and Mongolia) to 78 percent in Poland, 82.3 percent in Yugoslavia, and over 90 percent in several African socialist countries (Pryor 1992, 99–101). Frederic Pryor also found that, as a general rule, "the average state and collective farms are much larger than average size farms in advanced capitalists market economies" (142). By 1992, according to Héctor Sáez (1997a), the land consolidation process had advanced even further, with 80 percent of the arable land being in large-scale state farms, the remaining 20 percent of the agricultural land being almost equally divided between small private farmers and production cooperatives.

From State Farms to Production Cooperatives

Three decades of increasing rural collectivization came to an end in September 1993 when the socialist government divided large-scale state farms into smaller, and potentially more productive, units that were expected to perform as efficiently as private farms and cooperatives. According to many observers, this policy shift, ushered by a dramatic decline in agricultural productivity, amounts to no less than a third agrarian reform. It followed a period that began in 1986 with the decision to do away with peasant markets and other decentralization measures and to introduce the "Plan Alimentario" (Food Plan) whose objectives were to raise agricultural production by increasing yields and factor efficiency and to save foreign exchange by increasing self-sufficiency and reducing food imports (Roca 1994, 96–97). With the collapse of the Soviet bloc and the initiation of the special period, the Food Plan became an even greater national priority. In due course, however, it became increasingly obvious that it could not live up to the central planners' expectations and that its potential success had been further compromised by severe shortages of imported agricultural inputs.

Cuba's new cooperatives, the Unidades Básicas de Producción Cooperativa (UBPC, Basic Units of Cooperative Production), were created by breaking up state farms. Although UBPC members do not own the land, the government hopes to increase agricultural production by linking rural workers to the land by granting them land in usufruct for an indefinite period of time, allowing them to own what they produce and permitting them to sell their output. This entails turning many management decisions over to the members of the UBPCs (although decision making continues to be tightly regulated by the state), promoting the workers' self-sufficiency, and tying worker earnings to productivity (Alvarez and Messina 1996, 176–77). However, UBPCs have to sell their output to the state and purchase agricultural inputs from it. Cooperative members have also been given access to individual family plots whose output may be consumed or commercialized in farmer markets. Variants of these privileges have also been extended to small private farmers.

An important result of establishing the UBPCs and other agrarian tenure adjustments is a considerable decline in the average size of farm units. As summarized by José Alvarez and William Messina, Jr. (1996, 177), "Through August 2, 1994 the total number of sugarcane and non-sugarcane UBPCs established amounted to 2,643, with a total area of 221,300 *caballerías* (7.4 million acres, or three million hectares) and more than 257,000 members. This represented approximately 50 percent of the total area in state hands, with 93.5 percent of state cane area going to cane UBPCs and 29 percent of state non-cane area allocated to non-cane UBPCs. Average UBPC size is 84 *caballerías* (2,800 acres, or 1,133 hectares), with 37 members per UBPC, or 1.2 workers per *caballería* (33.3 acres, or 13.5 hectares) of total area." This means that between 1986 and 1994 average farm size in the state sector declined (by comparing state farms to UBPCs) from 14,260 to 1,126 hectares, or by a factor of eleven. Alvarez and Messina further observe that the shift to UBPCs is part of a broader land-tenancy realignment drive that has also included granting usufruct to farm land to families willing to relocate to the countryside. They cite official figures indicating that as early 1995, "about 6,000 families had received about 12,000 hectares of land in usufruct for tobacco production in the province of Pinar del Rio; more than 430 urban families, especially in the province of Santiago de Cuba, had moved to the mountains after receiving land for coffee production, while 2,600 individuals were in the process of obtaining such approval; and 369 workers and their families had received 19,870 hectares and livestock for dairy production in the province of Ciego de Avila" (Alvarez and Messina 1996, 4).

Aside from major economic ramifications, the significant shift in land tenancy arrangements and average land-holding size could have major future environmental implications. Plagued by a shortage of imported agricultural inputs, Cuban farming in the 1990s increasingly depends on traditional practices, as well as on selected modern techniques that are more benign to the environment (such as drip irrigation), than on former production-enhancement methods (such as large-scale irrigation projects) that caused considerable environmental damage. The post-Soviet agricultural model is less dependent on imported agricultural inputs and seeks to reintroduce formerly frowned-upon peasant agricultural practices that had been abandoned for their alleged backwardness: the large-scale use of beasts of burden, increased reliance on organic fertilizers and biological pest controls, the abandonment of marginal soils that had been brought under cultivation and made productive only by the intensive use of agricultural inputs, and a shift of urban labor to the countryside. Even more important, by devolving control of the land to farmers closely tied to it and whose economic well-being will depend on the stewardship of the natural resource base, there is the expectation that the trend of soil deterioration, if not arrested, may at least be slowed down. It is well established that the productivity of small, privately held farms in Cuba has consistently surpassed that of state farms and cooperatives (Alvarez and Puertas 1994), and, as noted by Sáez (1997a,b), small, privately held farms have also relied on far more environmentally benign agricultural practices. As Gene Wunderlich has observed, "large scale farming operations may be less sensitive to microenvironments; widespread ownership, to the extent that it encourages greater variety and management intensity in production, may induce more environmentally friendly practices" (Wunderlich 1995, 9).

It is too early to tell how successful the new agricultural policies will be. Food supply problems continue to plague the Cuban economy. Minor improvements in food availability reported in the mid-1990s seem to be attributable to the policy decision to increase self-sufficiency and the changing production incentive structure rather than to yield increases.[5] Cuban peasants, always sensitive to market signals, have rediscovered that it makes economic sense to plant, tend the crops, and bring their products to market.

Soil Degradation and the Capital-Intensive Agricultural Model

On the basis of field research and a review of the literature, several sources (Sáez 1997; Borges Hernández and Díaz Morejón 1997; Ministerio de Ciencia 1997a) have identified the main reasons behind the deterioration of Cuba's soils. Most are related to selected geophysical features of the country that have been aggravated and/or caused by past agricultural practices. They are erosion; soil compaction resulting from excessive use of heavy farm equipment; salinization and inadequate drainage, both partly associated with irrigation; acidification of the soils due to fertilizer overuse; and the contamination of potable water sources.

Erosion

With respect to erosion, Sáez (1997a) summarizes data that suggest that soil erosion is widespread. Although its precise extent is unknown, some estimates suggest that up to 70 percent of the national territory may be eroded to one degree or another, with the proportion of agricultural land thus affected reaching 90 percent.[6] He cites 1992 Ministry of Agriculture data indicating that of the 4.3 million hectares affected by erosion, 1.566 million hectares have either "high" or "very high" levels of erosion (see Table 4.1).[7] Although erosion is more apparent in the country's mountainous regions, it is also considerable in some of Cuba's most fertile and relatively flat provinces, such as La Habana and Matanzas, where high and very high levels of erosion affect, respectively, 21.4 percent and 9.8 percent of the land. The extent of erosion is extreme in Santiago de Cuba and Guantánamo Provinces, where 68.8 percent and 72.6 percent, respectively, of the land is highly or very highly eroded. Erosion is also a very grave problem in the provinces of Granma (50.7 percent of the land with high or very high levels of erosion), Pinar del Río (45.4 percent), Holguín (42.5 percent), and Cienfuegos (37.2 percent). Only in the province of Las Tunas (0.8 percent) and in the Isla de la Juventud (2.3 percent) is erosion of minor consequence.

A longitudinal study of twenty years of data collected in thirty-two hydrometric stations found that the rate of erosion is lowest in coastal areas, where it does not generally exceed 10–20 tons per square kilometer a year, increasing to as high as 300 tons in a station located at the Sierra del Rosario in western Cuba. Although altitude is the major determinant of the rate of rain-induced erosion, the erosion rate was considerably greater in western Cuba, where the mountain chains are much lower, than in the higher mountain ranges of eastern Cuba. The reason is that the extent of human intervention is much greater in the former than in the latter. The study also found much greater erosion rates for annual and seasonal crops, such as corn, tobacco, yuca, and boniato, than for pastures and permanent and semipermanent crops. Whereas in flat regions erosion rates for

Table 4.1. Level of erosion by province, 1992 (in thousands of hectares)

Province		Level of erosion				Percentage high and very high
	Total	Very high	High	Medium	Low	
Pinar del Río	390.5	114.5	62.9	74.0	139.1	45.4
La Habana	233.0	18.0	31.9	57.3	126.4	21.4
Matanzas	197.6	11.0	8.4	47.3	131.2	9.8
Villa Clara	434.5	39.1	40.0	78.0	277.4	18.0
Cienfuegos	185.3	26.0	43.0	35.3	81.0	37.2
Sancti Spíritus	466.0	10.0	96.0	99.0	261.0	22.8
Ciego de Avila	85.5	8.5	7.5	15.0	54.2	18.7
Camagüey	358.0	12.0	23.0	20.0	303.0	9.8
Las Tunas	111.2	0.5	0.4	14.2	96.1	.8
Holguín	361.5	62.5	91.0	32.0	156.0	42.5
Granma	325.4	72.0	93.0	63.4	97.4	50.7
Santiago de Cuba	495.2	191.0	149.5	39.0	115.7	68.8
Guantánamo	485.2	207.1	145.2	43.5	89.7	72.6
Isla de la Juventud	100.3	2.0	0.3	30.2	67.8	2.3
Total	4,229.3	774.2	791.5	667.9	1,995.7	37.0

Source: José M. Febles, "Apuntes sobre la necesidad de un cambio en el manejo y conservación de los suelos de Cuba." Conferencias y Mesas Redondas. II Encuentro Nacional de Agricultura Orgánica (May 1995): 13, in Sáez 1997, tab. 3.

annual and seasonal crops increased by approximately 40 tons per square kilometer per year for every increase of 100 millimeters in water run-off, the increase was limited to 10 tons for permanent and semipermanent crops. A major conclusion of the study is that "although erosion is a natural process, a high percentage of its current volume is due to human activity" (Pérez Zorrilla and Ya Karasik 1989, 73).

The extent of erosion by type of crop varies considerably, but regardless of crop, it is pervasive, affecting at least about half of all cropland. It is particularly widespread but perhaps not as severe in lands planted with perennials such as coffee, cocoa, and citrus—with over 90 percent of these lands affected—but it is also highly prevalent in pastures and in lands planted with sugarcane. It is also very prevalent in tobacco plantations.

These statistics suggest that soil conservation measures adopted by the revolutionary authorities (see, among many others, Ministerio de Ciencias 1995) were ineffective, and furthermore that the capital-intensive agriculture pursued from the early 1960s to the late 1980s did extensive damage to Cuba's soils. Although the statistics reviewed here warrant concern, care must be exercised regarding the economic and environmental consequences of the global erosion estimates. We have already noted that the empirical basis of erosion estimates is often weak but, more importantly, that the environmental and economic consequences of erosion are generally localized and context specific. As the World Bank has noted:

> Erosion rates, even when they are significant, may have very little effect on productivity under certain conditions. . . . Although erosion rates [may be] extremely high, the consequent effect on productivity is minor because soils [in a region of Costa Rica] are very deep (up to 1 meter in places) and have high organic matter throughout the soil profile. Moreover, the subsoil that underlies these soils is itself productive, although less so than the topsoil. Conversely, areas with shallow soils or unfavorable subsoils . . . can be very sensitive to even limited rates of erosion. The same is true of other forms of soil degradation. The impact of nutrient loss on productivity, for example, depends on the initial stock of nutrients and on their rate of regeneration. (Lutz et al. 1994, 5)

A final judgement must await the cumulative results of carefully conduced local studies based on the evaluation of the physical and chemical characteristics of soils throughout Cuba. Nevertheless, the data we have reviewed are alarming, particularly in light of the experiences of other countries that implemented the socialist agricultural development model and whose experience has been open to independent and verifiable scrutiny.

Policies Responsible for the Soil Degradation Trend

The historical record permits identifying specific policies responsible for the intensification of soil degradation in Cuba. They are mostly related to an implicit belief in man's ability to "conquer nature" by the systematic and widespread application of "scientific methods" (such as large-scale modern farming). Already discussed in the introduction was the fact that the culprits generally blamed for environmental deterioration in developing countries, namely population pressure, inequitable land tenure systems, unequal access to resources, and extreme rural poverty, have little explanatory value under the egalitarian conditions prevailing in rural Cuba during the first thirty years of socialist rule, when most of the land was under the direct control of the state. Other factors associated with the country's development model during this period appear to have greater explanatory power, in particular the reliance on capital-intensive agriculture and a marked tendency to uncritically implement central directives across agroeconomic zones. As Sáez (1994, 14) has noted, soil erosion in Cuba "runs parallel to the introduction and use of mechanized agriculture and irrigation," and directly follows from the disregard for soil conservation measures by the socialist leadership.

Capital-intensive agriculture, in turn, was related to the blanket application of centrally dictated production guidelines (including crop selection and large-scale irrigation) in very large-scale farms, which disregarded differences in the physical, hydrological, and environmental characteristics of soils throughout Cuba. It also resulted from the ingrained tendency to reward political loyalty and administrative docility over technical competence.[8] The views of a Cuban agricultural technician, as summarized by an observer, best convey what this entailed and what it eventually came to mean from an environmental perspective: "According to the bitter comments of an agronomic engineer 'in order to be an agricultural technician the only thing that was needed was to know how to read, to read what was established, and to implement it exactly.' In fact, noncompliance with norms carried more severe punishments (for the individual) than obtaining poor results. That is, 'if you lost the crop, but were able to show that the norms had been followed, you were not penalized. But you were punished, and harshly, if you had not done, or done something different from what had been established'" (Díaz 1995, 15).[9] In these respects Cuba did not differ from other countries that embraced a capital-intensive agricultural model under socialism. According to Frederic Pryor:

> The Stalinist conception of the state farm sector featured an extreme degree of organizational centralization, in which the subordi-

nate units had little decision-making autonomy and existed primarily to implement production orders determined from higher administrative agencies and incorporated in the plans sent down from the state planning commission to the ministry to the farms. In theory the collective farms had greater decision-making autonomy and also greater participatory management than state farms. Nevertheless, in most centrally planned Marxist regimes, all farm managers have had to follow a production plan in the same manner. Moreover, election of farm managers in the collective farms was pro forma in many Marxist regimes until the late 1980s because higher political authorities also intervened in production decisions of the farm. . . . This lack of decision-making autonomy for both the state and collective farms led to several kinds of principal-agent problems between the superior administrative agencies and the farm managers, reinforced in the case of state farms by the separation of ownership and control. (Pryor 1992, 163–64)

The first such problem mentioned by Pryor is loss of soil fertility: "The treatment of soil was costly to monitor by central authorities. Farm workers, however, had little incentive to preserve the fertility of the soil. The farm managers did have such an incentive, but only for the duration they would remain in this position" (Pryor 1992, 164).

These attitudes, and specifically the imposition across the board of directives devoid of sensitivity to specific environmental circumstances, led to soil degradation through the adoption of localized but cumulatively damaging agricultural practices. Another major factor behind the deterioration of the soil was the introduction and widespread adoption of technologies that when improperly applied and monitored could have adverse consequences. For example, following the Bulgarian and Soviet vertical integration models, large state farms and agroindustrial enterprises (Complejos Agro Industriales, CAI) were forbidden to intercrop beans with sugarcane in order to facilitate mechanization (Deere et al. 1994, 225). The practice of intercropping beans and sugarcane had been followed for many decades by Cuban farmers to preserve the fertility of the soil. Another instance was removing crops (e.g., beans in the Velasco region of the former province of Oriente) that contributed to soil preservation by promoting the growth of nitrogen-fixing (Rhizobium) bacteria (García 1992, 297) when crop diversification plans were introduced. Later attempts to reverse the resulting damage failed since these soils never regained their former productivity.

The failure to practice crop rotation was particularly damaging, despite evidence that in most "centrally controlled economies . . . strong emphasis [was placed] on one aspect of environmental control, namely careful crop rotation" (Wunderlich 1995, 59).[10] Failure to rotate crops in Cuba had already been noted by Rene Dumont in the 1960s (Dumont 1970a, 160–61). Héctor Sáez has summarized some of the reasons and consequences of these policies within the context of the socialist agricultural development model and the imposition of central planning by noting:

> First, the integration of peasants into regional production schemes ignored the intimate peasant knowledge of micro-ecological conditions. Thus, ecological conditions at the micro level were not taken into account in this centralized allocation of resources. Some general results of this kind of top-down resource allocation is that crops were often planted in ill-suited fields while hundreds of hectares were cleared at the same time, thus limiting and even destroying ecological balances such as plague controls, organic replenishment and moisture retention of soils, and others. Second, these policies reduced the scope of individual units' decision making, thus limiting their ability to take advantage and adjust to local ecological conditions and their fluctuations. Third, they spurred a process of de-campesinamiento (de-peasantization), which reduced the development of peasant knowledge and its transmission to a younger generation of agriculturalists. (Sáez 1994, 7)

Possibly the three most damaging large-scale interventions responsible for the deterioration of Cuba's soils were, first and foremost, extensive irrigation in the absence of, or unaccompanied by, proper drainage practices; second, the widespread use of heavy equipment in agriculture, leading to soil compaction; and third, excessive reliance on chemical inputs, such as herbicides, pesticides, and fertilizers, which have altered in many instances the natural balance and contributed to the acidification of the soils and contaminated potable water supplies.

Many of these problems resulted largely from policies implemented since the early 1960s, since pre-socialist agriculture was not characterized by extensive irrigation, widespread mechanization, or the heavy use of chemical inputs. The opposite was in fact the case, with pre-socialist agriculture being faulted by what was considered to be a relatively poor record of modernization. Andrés Bianchi, for example, in an influential assessment of the state of Cuban agriculture before the revolution, concluded that low pre-revolutionary farm output "could have been increased by introducing improvements in technology and organization [and that] in general, neither the use of fertilizers nor the irrigation of croplands was common practice. In 1945 only 4 per cent of all farms had land under irrigation; they represented just 3 per cent of the total cultivated area. Irrigation increased, however, during the 1950s, especially for rice, and by 1959 it was estimated that 10 per cent of the cultivated land was under irrigation. . . . The area fertilized and the number of farms using fertilizers were somewhat higher, but still l0–12 per cent of all holdings fertilized 7.4 per cent of the total agricultural land in 1945" (Bianchi 1964, 90).

Extensive Irrigation and Inadequate Drainage

One of revolutionary Cuba's agricultural development priorities was the creation of an extensive irrigation infrastructure (see Chapter 5). By the end of 1989, 101 dams with a reservoir capacity of 6.770 billion cubic meters had been completed and 32 others with reservoir capacity of 2.3 billion cubic meters were under construction.[11] In addition, 1,000 microdams with a further capacity of 900 million cubic meters of reservoir capacity had been built.[12] The irrigation infrastructure also included over 500 kilometers of master distribution channels, fifty electric pumping stations, irrigation channels serving over 1 million hectares of agricultural land, more than 1,300 kilometers of antiflooding works, and other facilities. These facilities were organized in thirty-five hydraulic enterprises that were assigned operational responsibilities. According to official data, in 1989 approximately 900,000 hectares, or 13.3 percent of the agricultural land area, had been prepared for irrigation. About 43 percent of irrigated land (388,000 hectares) was devoted to sugarcane cultivation, 18 percent (164,000 hectares) to rice cultivation, 12 percent (105,500 hectares) to citrus, and the remainder to pastures and animal feed, bananas, vegetables, potatoes, and tobacco.

Cuba is among the leaders in irrigation in Latin America and the Caribbean. The estimated 900,000 hectares of irrigated land in 1990 placed Cuba sixth in this category in the region, behind countries like Mexico (5.2 million hectares under irrigation), Brazil (2.7 million hectares), Argentina (1.7 million hectares), and Chile and Peru (each 1.3 million hectares), which are severalfold larger than Cuba, and far ahead of Central American and Caribbean countries. Moreover, the irrigated area in Cuba over the period 1965–1990—nearly a doubling of the area—is the fastest growing in the region for countries with a sizable amount of land under irrigation in 1965, except for Brazil.

In 1989, 84 percent of land prepared for irrigation in Cuba was actually irrigated; for sugarcane and pasture land, the share of land prepared for irrigation that was actually irrigated was 70 percent and 68 percent, respectively. Oddly, the area under citrus cultivation in 1989 that was irrigated exceeded land prepared for irrigation by 56 percent, raising questions about the efficacy of such practice. Atienza Ambou and her colleagues report that 18 percent of sugarcane land prepared for irrigation in 1990 in fact was not irrigated for the following reasons: water was not accessible (38 percent); no pumping stations were available (13 percent); irrigation systems were not completed (12 percent); or poor maintenance, reconditioning activities, and salinization or waterlogging (37 percent) (Atienza Ambou et al. 1995, 14). Water-use practices in agricultural enterprises were wasteful, leading the government to begin charging them for water in 1989, even before the economic crisis began (Atienza Ambou et al. 1995, 6).

Atienza Ambou and her colleagues also note that the anticipated efficiency of Cuba's irrigation plans largely did not materialize, often falling short of expected norms by 40–50 percent, primarily because the techniques used were not the most appropriate (Atienza Ambou et al. 1995, 14). In addition, crop response was generally poor because irrigation was used in ill-suited soils whose productivity could not be raised in the absence of previous improvements, poor soil preparation, and inadequate water-flow controls. Although it has been estimated that 73 percent of Cuba's agricultural areas could respond well to irrigation, 60 percent of this land must be improved first for irrigation to be effective (4). But the most serious constraint limiting the efficiency of irrigation has been the insufficient attention given to the drainage of irrigation water, since excess humidity harms root development, thus depriving plants of oxygen and reducing crop yields by 20–40 percent. For decades, drainage efforts were exclusively oriented to preventing floods and removing surface waters, with hardly any attention to waterlogging (14–15).

Poorly managed irrigation has also aggravated, and in some instances produced, soil salinization, a problem in many agricultural regions of the country. According to Sáez, salinization affects about 15 percent, or 1 million hectares, of Cuba's agricultural land, and acidity is prevalent in over 2 million hectares (1994, 15, 16). As shown in Table 4.2, the provinces most affected by salinization are Granma and Camagüey, but the problem is also of concern in the provinces of Holguín, Villa Clara, and Ciego de Avila, the latter being one of the provinces with some of the most productive soils. As early as 1984 a Cuban journalist had pointed out, when reporting on the results of salinization studies conducted in Cuba, that "Five years of study have led to the conclusion that salinization of irrigated soil is caused by a rise in the groundwater level. This points to man as the one to blame for indiscriminate treatment of the soil" (Bendoyro 1984, 8). Soil acidity is a particularly serious problem in the provinces of Pinar del Río, Villa Clara, Cienfuegos, and Sancti Spíritus, as indicated in Table 4.3.

Cuba is not unique in the serious problems it confronts arising from inadequate irrigation practices. The World Bank has found that the primary cause for the loss of agricultural land in irrigated soils is salinization. Irrigation, the World Bank concluded,

> has been a powerful force in fostering development in many countries. But when it is pursued injudiciously, it can become the progenitor of agricultural devastation, embodied in the form of irrigation-induced salinity. Irrigation-induced salinity has begun to cause drastic reductions in agricultural productivity in many parts of the world. . . . For many years, the rate of growth [of irrigation] has been 5 million hectares per year, but recently it has fallen to 2 million hectares per year. However, this growth is counteracted by 2–3 million hectares going out of production each year due to salinity problems—this means that cultivated land is being lost at a similar rate to new land being brought under cultivation. Indeed, the same countries who have invested heavily in irrigation infrastructure are now having to deal with serious salinity problems. (Umali 1993, xiii, 3)

The World Bank has provided a review of the factors contributing to salinization

Table 4.2. Salinization in land controlled by the Ministry of Agriculture, by province and level, 1992 (in thousands of hectares)

Province	Very high	High	Medium	Low	Total	Percentage very high and high
Pinar del Río	10.5	2.5	6.2	34.0	54.0	24.0
La Habana	0.2	2.6	2.7	9.7	15.2	18.4
Matanzas	–	1.2	11.8	4.2	17.2	7.0
Villa Clara	16.1	2.4	5.2	37.7	61.4	30.1
Cienfuegos	–	0.4	0.2	0.6	1.2	33.3
Sancti Spíritus	2.7	17.6	21.9	46.2	88.4	23.0
Ciego de Avila	11.3	16.2	12.1	22.3	61.9	44.4
Camagüey	10.7	41.1	32.5	61.9	146.2	35.4
Las Tunas	2.2	3.3	4.8	20.1	30.4	18.0
Holguín	8.6	18.8	12.5	54.0	93.9	29.2
Granma	20.6	54.0	108.6	4.0	185.2	40.3
Santiago de Cuba	0.2	0.3	1.2	3.3	5.0	10.0
Guantánamo	1.4	6.7	4.1	8.4	20.6	96.4
Total	84.5	165.5	223.8	306.8	780.6	32.0

Source: José M. Febles, "Apuntes sobre la necesidad de un cambio en el manejo y conservación de los suelos de Cuba." Conferencias y Mesas Redondas. II Encuentro Nacional de Agricultura Orgánica (May 1995): 14, in Sáez 1997, tab. 5.

(Umali 1993, 29-41). These factors include poor farm water-use efficiency related to inadequate management by farmers and irrigation authorities; government policies that underprice the true cost of providing irrigation water, leading to its wasteful use; poor construction and maintenance of irrigation facilities (such as canals); and, perhaps most crucial, poor drainage practices. The World Bank goes on to note that political factors often contribute to the neglect of the drainage infrastructure. These include a tendency to poorly assess the benefits and costs of irrigation facilities during project design; the proclivity of decision makers to ignore waterlogging and salinity issues; the high cost of developing drainage infrastructure and the fact that its benefits are not readily visible or politically rewarding; and the reality that drainage facilities are in many instances unpopular with farmers since they require that substantial amounts of land be devoted to it, as much as 15 percent of the available farm land if open drains are used.

These four factors behind the spread of salinization appear to fit well the Cuban experience. Cuban central planners gave little attention to the need for drainage facilities as complements to the ambitious irrigation projects begun during the 1960s and 1970s until the 1980 decade when it became obvious that

Table 4.3. Soil acidification in land controlled by the Ministry of Agriculture, by province, 1992 (in thousands of hectares)

Province	Amount of land affected (in thousands of hectares)
Pinar del Río	384.7
La Habana	57.7
Matanzas	61.7
Villa Clara	124.8
Cienfuegos	111.4
Sancti Spíritus	120.8
Ciego de Avila	13.4
Camagüey	97.9
Holguín	24.2
Las Tunas	14.8
Granma	40.2
Santiago de Cuba	9.4
Guantánamo	5.4
Isla de la Juventud	67.1
Total	1,133.5

Source: José M. Febles, "Apuntes sobre la necesidad de un cambio en el manejo y conservación de los suelos de Cuba." Conferencias y Mesas Redondas. II Encuentro Nacional de Agricultura Orgánica (May 1995): 14, in Sáez 1997, tab. 6.

there were problems. Furthermore, it appears that in the early planning phases, no attempt was made to include in the economic calculus the benefits of irrigation net the costs of providing adequate drainage facilities. This is perhaps explainable by the third factor cited by the World Bank regarding the limited political gains that arise from benefits not readily visible but entailing high investment costs. The final factor is also important, since the development of an extensive drainage infrastructure carried as well the loss of agricultural farm land, an outcome that ran counter to the policy priority of increasing the country's agricultural land, a problem that was compounded by the amount of land being lost to water reservoirs. The potential loss of land to water reservoirs would have amounted to 235,000 hectares if the hydraulic plans proposed in 1976 had been fully implemented (Atienza Ambou et al. 1995, 5).

Mechanization and Soil Compaction

Although during the 1950s the extent of mechanization increased markedly—the number of tractors rising from 1,888 in 1945–1946 to 19,700 in 1957–1958 (Grupo Cubano 1965, 68)—this increase was dwarfed by the growth in the number of tractors and other agricultural equipment following the revolution. By 1989, according to statistics in the Anuario Estadístico, Cuba had close to 79,000 tractors, including 1,918 bulldozers, 7,216 tread tractors, and nearly 8,000 heavy tractors and special tractors, as well as over 4,000 sugarcane combines, 626 rice-harvesting combines, and over 2,100 planters. By 1989, many agricultural operations for most crops were fully or highly mechanized (see Table 4.4): soil preparation, for example, was 100 percent mechanized in sugarcane, rice, several vegetables, beans, plantains, and other crops (Atienza Ambou, et. al 1992, 11). Before the revolution, in contrast, agricultural mechanization was largely limited to preparing the soil for sugarcane planting (Grupo Cubano 1963, 1009–1011). Most of the pre-revolutionary tractors were, furthermore, of the light variety whose use only marginally damages the soil.

According to data collected by the UN Food and Agriculture Organization, Cuba's stock of agricultural tractors in 1990 was 77,800 units, a number slightly lower than reported in the *Anuario Estadístico*. As with irrigation, Cuba was among the leaders in Latin America and the Caribbean in the use of tractors in agriculture, surpassed only by much-larger Brazil, Argentina, and Mexico. Cuba's stock of tractors in 1990 was more than twice that of Chile, an international agricultural export powerhouse. Some knowledgeable observers have suggested that the ability of some of Cuba's soils to retain their productivity for many years despite inadequate soil preservation practices resulted from the continuous planting of sugarcane as a perennial crop and exclusive reliance on light agricultural equipment (Soto Hernández 1994, 50). Contrast that situation with the one prevailing at the height of the implementation of the capital-intensive agricultural model under socialism when agricultural soils on the average were tilled from ten to twelve times a year; by 1992, because of fuel shortages, the average had declined to three to five times a year (Gersper et al. 1993, 17). Much of the tilling was done with heavy agricultural machinery, weighing between 2 and 7 tons (Sáez 1997a). Sáez cites data from COMARNA indicating that 1.6 million hectares, or 24.2 percent of the country's agricultural soils, are affected by compaction. Other estimates suggest that soil compaction may affect as many as 2 million hectares (Borges Hernández and Díaz Morejón, 1997, 15).

Excessive Reliance on Chemical Inputs

As already noted, before the revolution, fertilizers and other chemical agricultural inputs were increasingly being used, a substantial rise in their application occurring between 1945 and 1958. Particularly dynamic was the growth of domestic fertilizer production after 1951; by 1958 Cuba had sixteen or seventeen plants producing chemical fertilizers, some financed by the Cuban Agricultural and Industrial Development Bank (Banco de Fomento Agrícola e Industrial de Cuba, or BANFAIC). This institution was also sponsoring several studies to determine which fertilizers were most adequate for Cuba's different crops (Grupo Cubano 1963, 1130–32). In 1945, 1,787 tons of fertilizers (including potassium

Table 4.4 Percentage of mechanized operations in selected crops, 1989

Mechanized operation	Crop							Pastures and Hay
	Sugarcane	Rice	Tomato	Potato	Bean	Plantain	Citrus	
Soil preparation	100	100	100	100	100	100	100	100
Seeding	4	100	55	100	80	35	—	60
Cover seeds	100	100	55	100	80	35	—	60
Fertilization	100	100	100	100	100	90	90	100
Application of herbicides	80	80	–	100	100	100	100	95
Application of pesticides	–	–	100	100	—	—	—	—
Weeding	62	100	90	90	80	80	85	90
Cutting	70	100	—	—	—	—	—	100
Collecting	70	100	—	—	60	—	—	100
Lifting	100	100	—	—	60	—	40	100

Source: Atienza Ambou et al. 1992, 11.

muriate, ammonium sulfate, nitrogenous solutions, triple sulfates, and simple super-phosphate) were applied to 144,983 hectares, or 7.4 percent of the then total agricultural land area; by 1958, some 500,000 hectares, or 20 percent of the agricultural land, was fertilized. Close to half of the fertilizers were applied to lands planted with sugarcane, with four other crops (tobacco, rice, vegetables, and potatoes) consuming most of the remainder (Grupo Cubano 1965, 70–71).

The centrally directed, capital-intensive agricultural model promoted by the socialist authorities led to a massive increase in the use of fertilizers, pesticides, and other chemical inputs. An economic geography of Cuba published in the early 1970s by a Soviet scientist states: "Cuba occupies first place among the countries of Latin America in the use of artificial fertilizers per hectare of culti-vated land. In this it is ahead of a number of European countries. Thus in 1969 there were used 1.5 million tons of chemical fertilizers" (Valev 1972, 20).

Cuba's heavy use of pesticides cannot be attributed to the aggressive mar-keting by agrochemical multinationals. Rather, it reflects the eager adoption by Cuba of the capital-intensive agricultural model advocated by and practiced in the former Soviet Union. When pressed by a U.S. expert on the adverse environ-mental effects of heavy pesticide use, a Cuban agricultural official rejected the argument, stating that "pesticides could not be all that bad since the Soviet Union produced them" (Levins 1993a, 28).

In 1965 Cuba imported fertilizers valued at over 29 million pesos and her-bicides/pesticides valued at 5 million pesos. In terms of quantity, Cuba imported roughly 500,000 tons of fertilizers, about 500 tons of herbicides, and about 5,300 tons of pesticides. By 1985, imports of fertilizers had grown over fourfold, to 136 million pesos, and of herbicides/pesticides over twelvefold, to nearly 65 million pesos. In that year over 1.3 million tons of fertilizers, 17,500 tons of herbicides, and over 14,000 tons of pesticides were imported. In 1985 Cuba was using 192 kilograms of fertilizers per hectare of cultivated land (Díaz 1995, 7).[13] According to a 1986 study, sugarcane plantations consumed 677 kilograms of fertilizers per hectare; rice, 404 kilograms per hectare; vegetables, tubers, and plantains (cul-tivos menores) 1,793 kilograms per hectare; and, during their early growth, citrus fruit, 539.7 kilograms per hectare, and coffee and cacao, 707 kilograms per hectare (Bucek 1986, 16–17).

Official statistics are available on the application of fertilizers, herbicides, irrigation, and mechanical services to sugarcane production. According to these statistics, in 1989 balanced fertilizers and nitrogen fertilizers were applied to over 1.1 million hectares of sugarcane plantings; the intensity of balanced fertilizer application was 460 kilograms per hectare, and for nitrogen fertilizers it was 218 kilograms per hectare, for a combined fertilizer intensity of 678 kilograms per hectare. This application rate was slightly lower than the 699 kilograms per hectare recorded in 1985. In 1989 pesticides were applied to nearly 2.2 million hectares of sugarcane plantings. Application of fertilizers was increasingly carried out using either air dusting or mechanical equipment.

In brief, during the first three decades of socialist rule, the application of fertilizers increased tenfold, and of pesticides, fourfold. By 1989, Cuba was consuming close to 34,000 tons of imported and domestically produced pesticides and herbicides. Herbicides were being applied in the late 1980s to nearly 2 million hectares, or about one-third of the country's total agricultural land (Sáez 1997a). Many of these chemicals, particularly the organoclorine (e.g., dieldrin)—banned in the United States and other countries—and organomercurial compounds, are highly toxic and constitute a threat to human health (Murray 1994, 41–43; Sáez 1994, 11, 16; Weir and Schapiro 1981; WRI 1994, 111–15). In some instances, application of these products has led to changes in the chemical composition of the soil and caused the acidification of many of Cuba's soils (Sáez 1997a). Sáez cites figures that indicate that soil acidity in several provinces affects close to 90 percent of the land area.

In Central America, where organochlorides were extensively used to control pests in cotton fields, surface and ground waters were contaminated: "Rivers running through the flat coastal plains carried chemical residues to the coastal estuaries, where much of the aquatic life of the region could be found among its mangroves and lagoons. Indigenous communities in Nicaragua's cotton-growing region saw some of their primary food sources dwindle as shellfish and crayfish disappeared from the rivers and estuaries during the era of cotton expansion. Nearby, Salvadoran shrimp harvests dropped by 50 percent as cotton plantations expanded on the watershed for the fishing grounds" (Murray 1994, 42). Moreover, organochloride contamination extended to the deep wells of the cotton region of Nicaragua, as chemicals found their way into aquifers hundreds of feet below the surface. These chemicals were still percolating down through soil and rock strata into community water supplies years after their use ended (Murray 1994, 43).

Although it is difficult to judge the extent of contamination associated with the abuse of chemical inputs in Cuban agriculture, the consensus is that it is extensive. According to José Oro, a Soviet-trained geologist and former high-level official who defected to the United States in the 1990s, chemical pollution affects 1 million acres (405,000 hectares), 20 percent of which are severely polluted, primarily with phosphorus, nitrates, and chemical pesticides (Remos 1996, 11A). Among the most affected regions are the central zone of the Habana-Matanzas plain, one of the country's most fertile agricultural areas, and the rice-producing regions of Yara and Manzanillo (rice is a crop that is extensively treated with pesticides). In the Los Palacios region, contamination of the rice crop has been blamed on the excessive use of chemical inputs (Cabrera Trimiño 1994, 10). Oro believes that in some areas, chemical pollution levels are twenty-five to thirty times higher than in Okeelanta and Talisman, two sugar-producing regions in the state of Florida that have gained notoriety for their high pollution levels (11A).

The head of Cuba's Plant Protection Research Institute in the Agriculture Ministry takes a far more benign view of the pesticide contamination problem in Cuban agriculture. He states that although DDT was used in Cuban rice paddies for over twenty-five years with an average dose of 8 kilograms active ingredient per hectare per annum, residue levels in three important rice-producing areas (Vado del Yeso, Florida, and Los Palacios) were practically the same over the period 1976–1983 (Dierksmeir 1996, 386). Low pesticide contamination is attributed to Cuba's high soil dissipation rates, caused by climatic conditions (high relative humidity and high soil temperatures) coupled with the development of soil microflora that degrade pesticides efficiently when used over time in the same area.

A related potentially serious problem affecting Cuban agriculture and caused by the excessive application of chemical pesticides is that of secondary pests and their adverse effect on agricultural productivity (WRI 1994, 113). In Caribbean countries promoting nontraditional agricultural exports and in which pesticide use has been extensive, such as the Dominican Republic and Guatemala, serious environmental problems resulted when some crops had to be abandoned because of worsening pest problems and the appearance of secondary pests (Murray and Hoppin 1992). It is important to note that in these countries pesticide use—and chemical inputs in agriculture generally—has been far less prevalent than in Cuba. Numerous direct and indirect references in the Cuban agricultural literature suggest that a similar, if not worse, problem has occurred there.

Although directly comparable statistics are not available, it is possible to establish relative orders of magnitude regarding the application of fertilizers in Cuba, the Dominican Republic, and Guatemala in 1989–1990 by using data provided by the UN Food and Agriculture Organization (1994, 51, 83, 106, 119).[14] Whereas Cuba used 661,000 metric tons of nitrogen, phosphate, and potash-based fertilizers in 1989–1990, total consumption in the Dominican Republic reached only 73,000 metric tons, and in Guatemala 125,000 metric tons. At the time, 198 metric tons of fertilizer per 1,000 hectares of agricultural land were being used in Cuba as compared to 50 metric tons in the Dominican Republic.

These figures suggest that Cuba relied on chemical agricultural inputs severalfold more than other countries in which the adverse environmental consequences of pesticide and other chemical inputs have been independently documented.

Animal and Other Agricultural Waste By-products

Another environmental problem Cuba had to face—as do many other countries, including the United States—is how to dispose of the waste created by raising animals in commercial facilities where thousands of animals are concentrated in limited spaces. As in agriculture, the revolutionary government gave priority in cattle and poultry production to methods that heavily depended on economies of scale and required importing grain to feed cattle and poultry. As noted, large-scale cattle ranches were part of this strategy, as were poultry farms capable of handling between 50,000 and 250,000 birds and the commercial production of hogs. Many of these achievements were made possible by the large-scale availability of low-cost feed grains imported from the Soviet Union. Increased imports of feed grains led to a drastic decline in domestic feed grain production, as suggested by the data on number of hectares planted in corn and millet in 1945 and 1989.

By 1989, when the capital-intensive model of beef and poultry production was dominant, Cuba had to contend with how to dispose annually of 9 million tons of solid waste (4.6 million tons of cattle waste, 2.1 million tons of hog waste, and 0.64 million tons of poultry waste) and 27 million cubic meters of liquid waste. In addition, 2.5 million tons of solid waste were generated by the sugarcane industry, plus a further 1.7 million tons of solid waste of other agricultural by-products (Atienza Ambou et al. 1992, 12–13). The collapse of Soviet imports after 1989 was followed by a collapse of beef and poultry production, which in turn has significantly reduced the production of many animal waste products.

Environmental Legacy of Socialist Agricultural Practices in the Cauto River Basin

Many of the issues discussed in this chapter have come to a head in one particular region of Cuba, the Cauto River basin. The magnitude of the problems in this area is so severe that they can no longer be ignored. The basin is regarded as one of the most environmentally compromised regions of Cuba. This region covers about 8 percent of the country's territory and accounts for 9.3 percent of its agricultural land. Approximately 25 percent of the rice and 10 percent of the sugar produced in Cuba originates here. According to Armando H. Portela, a former senior researcher at Cuba's Academy of Sciences and the holder of a doctorate in geography from the Institute of Geography of the Soviet Academy of Sciences, "the depletion of the environmental quality of the Cauto River basin over the past three or four decades is a serious national problem. The damage covers a broad range of ecological issues. One-third of the basin suffers from severe erosion; salt water intrusions have spoiled most of the groundwater reservoirs; the natural runoff has been reduced by 60 percent in recent decades; the forested areas have been nearly annihilated" (Portela 1997).

In addition, Portela reports that a survey found that the water in the basin is heavily polluted by discharges from 652 sources, including industries, urban sewage systems, and cattle farms. There are grounds to believe that many of the environmental problems of the Cauto River basin are related to poor management of hydraulic development projects, a situation aggravated by extensive deforestation.

Concern over the conditions of the basin—described in *Granma,* the official Cuban newspaper, as critical—led CITMA to initiate a two-year study in 1997 to determine the range and scope of remediation measures needed to arrest its environmental decline (Riera 1998). Some measures being evaluated include building water treatment plants, tightening regulations for use of agricultural chemical inputs, and improving the management of irrigation waters. The situation is so alarming that in some regions of the Cauto River basin, agricultural yields have become so meager that fields have been left unplanted. The damage in coastal areas is also extensive.

Summary and Implications

The data reviewed in this chapter suggest that Cuban agriculture will face problems in years to come because of soil deterioration and environmental damage (e.g., erosion, water contamination); this environmental legacy is the result of implementing a capital-intensive agricultural development model not viable without significant foreign subsidies and unsustainable from an environmental perspective. The extent of the environmental damage can be partly assessed by examining the estimates compiled by Sáez (1997a) in Table 4.5, although it must be kept in mind that the bases for such global estimates are not always entirely clear and that proper corrective measures must depend on detailed and careful field investigations. Dealing with these realities, and the need to reactivate agricultural production, has entailed implementing or accelerating initiatives, some already under way in the 1980s (Levins 1990), to modify agricultural production methods. Some of these initiatives have involved returning to practices that were neglected if not discouraged for nearly three decades, and others are based on applying the latest agricultural technologies.

The inability to import farm machinery, fuel, and chemicals since 1990 has forced Cuba to embark on a substitution program (see Chapter 10) that has heavily depended on the extensive use of integrated pest-management methods, including biological controls; the reintroduction and wide-scale application of traditional farming methods, including beasts of burden; reduction in tillage; rehabilitation of the soils; increased reliance on biofertilizers; use of green manures and crop rotations; intercropping; waste recycling; and extensive biotechnology experimentation (Carney 1993; Díaz 1995; Gersper et al. 1993).

It is still too early to conclude how effective this program will prove to be; some observers find that it has been very successful in some respects while deficient in others (for reviews, see Carney 1993; and Díaz 1995). A broad consensus is that its chances for success are enhanced by the availability of a cadre of scientists and research centers that Cuba developed over the last three decades. Whatever its achievements, this policy shift augurs well for Cuba's environment, since it is part of a broader soil management program that, although forced on the country by extraneous circumstances, is consistent with the preservation of the natural environment.

Grounds for optimism must be tempered, however, for several reasons. The first is that many of the traditional and modern technological fixes available to address erosion and other forms of soil deterioration are very expensive and well beyond Cuba's reach as long as the country continues to be submerged in an economic crisis. Note was already made of the high costs attendant to desalinization efforts. The World Bank has also identified the high costs associated with using more water-efficient irrigation technologies such as sprinkler and drip systems and has recommend that these technologies be used exclusively with high-value crops (Umali 1993, 51–52). Yet, in Cuba, drip irrigation techniques have been

Table 4.5. Agricultural land affected by degradation problems

Type of problem	Hectares affected
Erosion	4,229,500
Compaction	1,600,000
Salinization	780,600
Drainage	2,700,000
Acidification	1,133,500

Source: Sáez 1997, tab. 7.

promoted to increase yields in the production of plantains, hardly a high-value crop—a decision that, while consistent with the need to attain food self-sufficiency, calls into question economic rationality given the country's current resource scarcity. Equally uncertain in many cases are the potential economic benefits of many known soil recovery approaches (Pagiola 1994).

The second reason is that many of the potential solutions being advocated in the agricultural sector carry potential but unknown environmental threats (e.g., biopesticides), are being ignored (e.g., waste products generated during the production of torula yeast) or cannot be properly addressed under current circumstances. For example, a drive to increase the production of torula yeast, a sugarcane derivative used as a substitute for imported animal feed, may aggravate stream and soil pollution through the discharge in the soil of the wastewater generated as part of its production. This discharge was already regarded in 1991 as one of the country's primary sources of pollution (Sánchez Hernández 1992). Resources may not be available to build the required treatment plants to minimize the environmental damage arising from increased torula yeast production. For these and other reasons, it may prove impossible to reverse in the near future many of the adverse environmental effects of three decades of misguided agricultural development policies.

⊷ 5 ⊶

Water and the Environment

IN THIS chapter we discuss selected relationships between water availability and use, development strategies, and environmental issues in socialist Cuba. We examine Cuba's endowment of water resources and selected policies regarding their use, giving explicit attention to the degree to which the Cuban socialist government has recognized the extent and significance of water-related development-environment interactions. The narrative includes a discussion of some water-related problems that predate the revolution and were a source of concern by the 1950s but that may have been aggravated since then, as well as new issues that have emerged largely as a consequence of policies associated with the socialist economic development model.

To set the context, we begin with an overview of the hydrological cycle and how natural and human interventions can disrupt its natural operation and include a brief discussion of how an integrated water management approach can provide a framework for minimizing potential environmental damage from the improper use of water. We then assesses the determinants of water availability in Cuba. This is followed by a discussion of selected agricultural and water policies instituted by the Cuban authorities and an evaluation of their environmental and economic consequences.

The Hydrological Cycle and Integrated Water Management

Mohan Munasinghe (1992, 34–40) provides a useful summary of how the natural hydrological process is affected by human intervention. Two premises

111

underlie his analysis. First, the satisfaction of humanity's water needs demands that the natural environment be altered. Second, once water is used, it must be returned to the natural ecosystem. How water is returned to the environment has important consequences; it could contribute to the lasting use of water and other renewable resources or to the degradation of the natural resource base. Mishandling of water supplies can also have detrimental consequences for a myriad of human activities.

The hydrological cycle encompasses "a number of atmospheric, surface, and underground sub-cycles of different magnitude where water moves from and between the air, the vegetation, the soil, the solid rock, and the rivers, lakes, and seas" (Munasinghe 1992, 35). The cycle is interactive, with every phase affected by the others. Precipitation rates are partly determined by topography, winds, and evapotranspiration (the process whereby water is returned to the atmosphere). Evapotranspiration depends on the vegetation's capacity to retain water, the ability of water to infiltrate different types of soils, the effect that the capillary action of the soil has on evaporation, and the amount of water that plants transpire through their leaves.

Water that filters underground accumulates and may result in a water table that rises to the surface as the amount of water percolating downward increases. Large water deposits in sandy and/or porous rock create aquifers. Water moves within the aquifers depending on the permeability of the rock and other factors. Lakes, springs, and rivers are formed in places where the water table meets the land surface. The way in which water flows and how it is stored depends on topographical and morphological features, including surface vegetation and other landscape cover.

There are many points at which humans can intervene in the hydrological cycle. The care with which these interventions are managed, and the efforts made to ensure the renewability and preservation of water resources, to a great extent determines the long-term viability of the intervention and the sustainability of the economic uses of water and other natural resources. Policies that permit the short-term exploitation of water resources at high social, economic, and/or environmental costs are ill-advised. These policies could bring about long-term, or even irreversible, environmental damage.

Determinants of Water Availability and Quality in Cuba

Cuba has a hydraulic potential[1] of 38.4 cubic kilometers (Atienza Ambou et al. 1995, 5).[2] Since much of this water cannot be recovered (e.g., rainfall in coastal areas) or is lost to evaporation, the usable hydraulic potential is estimated at 23 cubic kilometers, or 60 percent of hydraulic potential.[3] The amount of usable water is determined in part by the country's hydrology and in part by the availability of surface and underground waters. In Cuba 25 percent of this potential can be obtained from underground water sources and the remainder from surface sources.

Hydrology

As a subtropical nation, Cuba receives an abundant amount of rainfall. Cuba's mean annual precipitation, according to long-term observations, amounts to 1,410 millimeters (Egorov and Luege 1967, 15). Regional fluctuations in mean precipitation are substantial. Mean annual rainfall ranges from 3,000 millimeters in the mountainous region of northeast Cuba to under 600 millimeters in the semidesert southeastern coastal region between the Sierra Maestra Mountains and the Caribbean Sea, since the mountains prevent the rains from reaching the coast. Mean annual precipitation ranges from 1,100 millimeters along some of the coasts to 2,000 millimeters in mountainous regions (Figueras 1994, 21). As a general rule, rainfall tends to be relatively more abundant in the western and central plain regions than in the eastern plains and is most abundant in the mountain ranges.

Major seasonal and cyclical rainfall fluctuations give way to periods of overabundance of water followed by periods of water shortage. With marked wet and dry seasons, the monthly rainfall ratio (the ratio of the wettest to the driest month recorded in millimeters of precipitation) may be as low as 8:1. In some years the monthly rainfall ratio may be as high as 15:1, and in extreme cases (as in 1989), higher than 20:1. On average, rainfall between November and April, the dry season months, ranges from 32 to 99 millimeters per month, and during the rainy season it fluctuates between 200 and 260 millimeters per month (Egorov and Luege 1967, 15). The frequency and intensity of hurricanes has a major bearing on seasonal and secular rainfall fluctuations.

Surface Waters

Cuba's elongated and narrow shape, insular character, and extensive coastline (2,306 miles or 3,735 kilometers) accentuate the cycles of water overabundance and water scarcity. Most rivers flow from the center of the country to either the northern or southern coasts (see Map 2.3), and since most rivers have a short course, rainwater reaches the seas within at most a few hours (Report of Cuba 1976, 59). Due to the short course of rivers, and the rapidity with which rainfall reaches the sea, much rainfall cannot be stored superficially or as underground waters, thus reducing the country's hydraulic potential. Watersheds are geographically limited as well. Indonesia, for example, shares many of Cuba's geographical features; Java's watersheds are described as extremely shallow, with

most of its rivers less than 50 kilometers long. These characteristics, together
with deforestation and rural development, are responsible for the rapid runoff
and increased river flow variability observed in several of Indonesia's islands
(World Bank 1990, 79).

Eighty percent of the total flow of Cuba's rivers occurs during the rainy season (Osterling 1985, 67). In the dry season, many of the rivers and hundreds of
streams completely lose water flow. In some parts of the country, the limited permeability of the soils contributes to rapid runoff. In low-lying coastal areas, often
at or below sea level, flooding is a persistent problem. In some regions, flooding
is a serious concern during periods of intense rain and during the hurricane season, another factor that contributes to the reduction of the country's hydraulic
potential as rainwaters go directly to the sea. More than 35 percent of Cuba's total
land area is reported to be affected by excess humidity, whether on a permanent
(18 percent) or a temporary (17 percent) basis, including 27 percent of the total
agricultural land area (1.8 million hectares of 6.6 million hectares). It was estimated in 1979 that if the proper drainage infrastructure were established, flooding could be prevented in 1.2 million hectares. Of these, 658,000 hectares were
prone to flooding when affected by hurricanes, and a further 545,000 were subject to periodic flooding. Of the total agricultural land area subjected to flooding,
433,000 hectares (36 percent) were used for pastures, 447,000 were planted with
sugarcane (37.2 percent), 125,000 with rice (10.4 percent), and the remainder
with other crops (Informe acerca 1979, 3).

Two other factors contribute to rapid rainwater loss: high temperatures and
prevailing wind patterns. A high evapotranspiration rate is associated with the
country's tropical climate. The mean annual temperature is 25.5 degrees Celsius
(Osterling 1985, 69). Monthly mean temperatures only fluctuate within a narrow
band (from 22.5 degrees Celsius in the coldest month, January, to 27.8 degrees
Celsius in the hottest month, August). Mean annual evaporation experimentally
measured from a free water surface amounts to 1,995 millimeters, ranging from
over 2,000 millimeters in the rainy season to about a 1,000 millimeters in the dry
season (Report of Cuba 1976, 47; González and Gagua 1979, 29). The pattern of
persistent winds prevailing in most of Cuba further contributes to high evaporation rates. Less is known about regional patterns of evapotranspiration rates,
but some of the highest are believed to occur in the Cauto River watershed due
to the combined effects of high temperatures and persistent winds. These high
water losses promote the salinization of the soil by drawing to the surface underground waters that leave behind mineral deposits as they evaporate into the
atmosphere (Egorov and Luege 1967, 15).

A major man-made contributor to rainwater loss is the deforestation to
which Cuba has been subjected since colonial times. The capacity of tropical soils

to retain water is impaired when the natural forest cover is lost. Forests also contribute to the reduction of evapotranspiration rates. One of the most direct consequences of deforestation is an intensification of the runoff rate, with a detrimental impact on soil conservation. By the late 1950s, forested areas were mostly confined to the least inaccessible and inhospitable mountainous and swampy coastal regions. The last three decades have seen considerable reforestation efforts, but how much has been accomplished is open to debate (see Chapter 6; Espino 1992, 331).

Underground Waters

The abundant annual rainfall contributes to the replenishment of the country's aquifers. Water capacity in Cuba's aquifers varies greatly as does the quality of their waters (Egorov and Luege 1967; Academia de Ciencias de Cuba 1970, 13–13, 22–23; Cereijo 1992). Overall, however, the country has abundant underground water resources, especially in the karstic soils found under more than 60 percent of Cuba's land surface (Dos requisitos 1982, 79).[4] Geographically, the country's richest aquifers are concentrated in two regions. The first includes the aquifers running throughout the karstic soils of western Cuba, from Pinar del Río Province to much of Matanzas Province (except for the Zapata Swamp), and into northern Villa Clara Province. The second corresponds to the concentration of karstic soils extending through most of Sancti Spíritus and Ciego de Avila Provinces and parts of Camagüey and Las Tunas Provinces. This region contains the La Rocha Plains aquifer, Cuba's richest, which is capable of producing between 1,500 and 2,500 gallons of water per minute with 10- and 12-inch wells (Cereijo 1992, 80).

The regions most poorly endowed with underground water resources are northern Pinar del Río Province in westernmost Cuba; parts of Sancti Spíritus Province and the bulk of Camagüey Province in central Cuba; and parts of Granma Province and the Provinces of Holguín, Santiago de Cuba, and Guantánamo in the eastern part of the country. Cuba's mountain ranges are primarily located in regions with relatively poor endowment of underground water; the country's geographical pattern of alternating mountain ranges and plains contributes to the replenishment of the aquifers in the plains (see Map 2.1), since they receive part of the rainwater flow from the precipitation in the mountain regions (Egorov and Luege 1967, 14).

Water-rich, karstic aquifers can easily become contaminated, especially those that lack surface areas capable of containing the filtration of pollutants (González Báez and Jiménez Hechevarría 1988a, 6). Contaminants spread swiftly in karstic aquifers, since they allow underground waters to be rapidly displaced (González Báez and Jiménez Hechevarría 1988b, 24). This feature of karstic soils is of considerable concern in Cuba, since many of the country's principal aquifers are located in karstic soils along coastal areas, which are prone to salt-water infiltration.

Mineral Content of Cuba's Underground Waters

There is enormous variation in the mineral content of Cuba's underground waters. Waters are classified according to their mineral content expressed in milligrams per liter of water (Egorov and Luege 1967, 9). Waters with 1,000 milligrams per liter or less of minerals (total dissolved solids) are adequate for all human and agricultural uses;[5] those with mineral content up to 3,000 milligrams per liter may also be used for human consumption and for watering livestock; those with mineral content of 3,000–10,000 milligrams per liter are not suitable for human consumption, though they may be used for livestock; waters with mineral content of 10,000–50,000 milligrams per liter are unfit for human (other than for medicinal purposes) or agricultural use. In Cuba the supply of water with low mineral content is deemed to be abundant in the karstic aquifers but less so in regions with soils with limited permeability. In the richest Cuban Miocene age aquifers, calcium-bicarbonate type waters typically have a total dissolved solids range of 300–700 milligrams per liter. Total dissolved solids readings above 1,000–2,000 milligrams per liter usually indicate saltwater intrusions.

Coastal regions are prone to have waters with high total dissolved solids. In coastal regions of southwestern Cuba, the Zapata Swamp, and the northern coasts of Sancti Spíritus and Ciego de Avila Provinces, the mineral content of the aquifers tends to be high. In some of these regions deposits of water with low mineral content suitable for human and agricultural use are found above stores of marine water (Egorov and Luege 1967, 9). This makes the danger of excessive salinization of coastal region aquifers high, a danger made more immediate by flood-prone soils that are often at or below sea level. The aquifers in the Cauto River watershed, in some areas with a limited capacity to hold underground water stores, tend to have a heavy mineral load. In this area the range of variation in total dissolved solids in underground waters is considerable, from as low as 300 milligrams per liter to as high as 5,700 milligrams per liter (Egorov and Luege 1967, 72–75).

Cuba's Water Use Levels

From an international perspective, Cuba's renewable water resources of 3,340 cubic meters per capita in 1975 were about half the world average of 7,690 cubic meters per capita and well below the average for North and Central America (16,260 cubic meters) and South America (34,960 cubic meters). Cuba's relatively limited freshwater availability per capita is similar to that of other major Caribbean islands (2,790 cubic meters in the Dominican Republic; 1,690 cubic meters in Haiti, and 3,290 cubic meters in Jamaica) (WRI 1992, 328).

Where Cuba differed from the other large Caribbean island nations, at least up to the mid-1970s, was in the rate at which freshwater resources were withdrawn. In Cuba, as in most island-nations not receiving freshwater flows from neighboring countries, annual withdrawal of freshwater resources is calculated as a percentage using as a base internal renewable resources, exclusive of evaporative losses (WRI 1992, 334). In 1975 Cuba was withdrawing an estimated 8.1 cubic kilometers of water, the withdrawal volume increasing to 10 cubic kilometers by the early 1990s, or about 50 percent of the country's usable water potential (Figueras 1994, 22).[6] In comparison, 15 percent of freshwater resources were being withdrawn in the Dominican Republic (in 1987), 4 percent in Jamaica (1975), and less than 1 percent in Haiti (1987). These differences in water use rates can be attributed largely to the emphasis given to the use of hydraulic resources in socialist Cuba as compared to the other countries. Only Barbados (in 1962) withdrew water at a higher rate than Cuba, 51 percent.

The per capita volume of water withdrawal in Cuba in 1975 (868 cubic meters) was nearly twice as high as that of the Dominican Republic (453 cubic meters), and from five to twenty times as high as in the other island nations (46 and 157 cubic meters in Haiti and Jamaica, respectively). At the time, 89 percent of the freshwater withdrawn in Cuba was dedicated to agricultural uses, 9 percent to domestic use, and the balance to industrial use. Despite these efforts, it was already apparent by 1982 that important economic regions of the country such as the provinces of La Habana, Ciudad La Habana, and parts of Ciego de Avila were at or nearly at a point at which water demand exceeded supply. In three rice cultivation regions in the provinces of Pinar del Río, Camagüey, and Granma, underground water demand exceeded the amount of available water (Administración 1982, 35). In the early 1990s Ciudad de La Habana Province consumed 105 percent of its water resources and those of La Habana Province, with neighboring provinces supplying the capital city's shortfall. At the other extreme, the eastern province of Guantánamo consumed only 6 percent of its water resources (Celeiro Chapis 1995, 89).

Water Resources Management in Cuba

Active involvement in the management of water resources by Cuba's socialist government began in 1962 with the creation of the Instituto Nacional de Recursos Hidráulicos (INRH, National Institute of Hydraulic Resources) (Administración 1988, 31). The management of water resources went into high gear in the late 1960s and early 1970s with the implementation of an ambitious agricultural development program aimed at increasing output of sugarcane and other water-intensive agricultural products (MacEwan 1981, 98–99) and with the decision to increase the supply of water for human and industrial use. In 1976 Soviet and Cuban technicians conducted a general hydraulic assessment by reviewing the principal characteristics of the country's watersheds and the topographical and geological features of areas identified as potential sites for water reservoirs. Based on this study, it was determined that it would be possible to build up to 500 reservoirs with a flood area of 235,000 hectares and a total capacity of 14,400 cubic meters of water (Atienza Ambou et al. 1995, 5). In 1977 Cuba established the Instituto de Hidroeconomía (Institute of Hydroeconomy), within the Ministry of Construction to coordinate the development and administration of water projects involving other government agencies representing large water users, such as the Ministry of the Sugar Industry, the Ministry of Agriculture, and local organs of People's Power (with regard to household consumption of water).

Dam Construction Program

As part of this ambitious hydraulic development project, an extensive dam construction program has been carried out. Increasing water availability in man-made reservoirs was consistent with the mobilization of resources to meet production targets and minimize water shortages during the dry season. The dam construction strategy was also consistent with Fidel Castro's personalistic style of government and the weight of his views on the country's development policies. The influence of Castro on the country's water development policies—as in so many other national issues—should not be underestimated, since in 1962 he directed "that not a single drop of water be lost, that not a drop of water reach the sea . . . that not a single stream or river not be dammed" (El paisaje 1982, 52). He further stated in 1963 that "we will build hydraulic systems until the day when not a single drop of water reaches the sea. The sea will not be able to claim a single drop of rainwater falling on the land of our country. . . . The day must come when we do not waste a single drop of water. . . . The difference is 100 percent of our land being irrigated and no irrigation. And we will be totally safe. To be sure, a hurricane might come . . . but we are going to have very serious and well-grounded hydraulic studies of all approaches to conserve water, including lagoons, dams, reservoirs, etc., so that we can have an agriculture based on water and technology" (Castro 1992, 79–80).

Castro's wishes began to be translated into reality with the initiation in 1962 of an extensive Soviet water-related technical assistance program in which Bulgaria also collaborated. In the fifteen-year period between 1976 and 1990, it was projected that Cuba would invest between 10.4 and 16.8 billion pesos in hydraulic works (Report of Cuba 1976, 57). By 1992, Cuba had 200 dams and close to 800 microdams (COMARNA 1992, 16).[7] The dam and water reservoir construction program increased the stored water capacity from 48 million cubic meters in 1959 to 7,000 million cubic meters in 1987, or by a factor of nearly 150 (Editorial 1988).

A more detailed assessment of the dam construction program can be made by examining data for 1989 shown in Table 5.1. The first column of this table shows the number of dams with a capacity of over 5 million cubic meters completed and under construction in 1989. The number of dams per province appears to be determined at least in part by water demand (17 dams in La Habana Province to supply the capital city) and by the relative availability of underground waters. The number of dams is greater in the relatively water-poor and more rugged provinces of Pinar del Río and Granma than in underground-water-rich Ciego de Avila Province. Also of interest is the combined average reservoir capacity of dams completed and under construction by province. Reservoir capacity tends to be much higher in mountainous provinces with more favorable catchment basins. The smaller average reservoir capacities are found in the western provinces of Pinar del Río, La Habana, and Matanzas (with average reservoir capacities of 39.6, 31.7, and 39.0 million cubic meters, respectively). With the exception of Pinar del Río, these provinces have relatively few deep basins capable of holding large quantities of water. The largest water-holding capacities, in contrast, are found in the rugged provinces of Sancti Spíritus (216 million cubic meters), Villa Clara (131 million cubic meters), and Santiago de Cuba (111 million cubic meters).

Central Planning and Water-Pricing Issues

Socialist Cuba's seemingly inexhaustible demand for water was—and to some extent continues to be—driven, in part, by its central planning system. Under a central planning system, resources are allocated on the basis of a central physical plan rather than on their price (cost). That is, central planners determine a priori the volume of water that will be required to meet demand (from agriculture, household use, industrial plants, and so on) and command other branches of the economy to provide the required volumes, without explicitly taking into account the costs of obtaining and delivering the water to the user. The central planning system is also largely oblivious to environmental externalities. State enterprises are not held accountable for environmental costs or handle them through soft budget constraints cushioned by the central government (Dávalos 1984a,b,c,d).

At least through the early 1980s, water was a free commodity in Cuba

Table 5.1. Number and capacity of dams[a] (capacity in millions of cubic meters)

	Before 1959		*As of 31 December 1989*					
			Total		*Completed*		*Under construction*	
Province	*Number*	*Capacity*	*Number*	*Capacity*	*Number*	*Capacity*	*Number*	*Capacity*
Total	13	47.8	133	8,962.5	101	6,699.0	32	2,263.5
Pinar del Río			21	832.7	17	744.3	4	88.4
La Habana			17	538.4	16	458.4	1	80.0
Ciudad de la Habana			4	95.8	4	95.8		
Matanzas			3	116.9	1	38.5	2	78.4
Villa Clara	2	6.0	8	1047.6	4	779.4	4	268.2
Cienfuegos			6	3,20.4	4	257.5	2	62.9
Sancti Spíritus			6	1,296.7	5	1,238.7	1	58.0
Ciego de Avila			2	64.8	1	31.5	1	33.3
Camagüey	8	28.6	13	1,035.8	11	904.5	2	131.3
Las Tunas			5	204.9	4	184.0	1	20.9
Holguín	1	5.7	12	919.0	9	457.5	3	461.5
Granma			10	996.2	7	486.2	3	510.0
Santiago de Cuba	2	7.5	8	890.9	4	544.5	4	366.4
Guantánamo			6	344.3	4	300.8	2	43.5
Isla de la Juventud			12	258.1	10	177.4	2	80.7

[a]Dams with capacities of over 5 million cubic meters.
Source: Comité Estatal 1989, 174.

(Report of Cuba 1976, 56). Writing in 1982, the president of the Institute of Hydroeconomy (Dorticós 1982, 13, 18) argued for the institution of a system during the 1981–1985 five-year plan that would set a price for water and charge state enterprises that used water for what they consumed. Such a system would promote conservation and more efficient use of the resource, he argued. He further recommended that water providers and users be given the ability to enter into contractual relationships, subject to the commercial code and the national system of commercial arbitration, to ensure that users actually paid for the water they consumed.

The 1981–1985 five-year plan had as an objective the "adoption of measures to attain and guarantee the more rational use and conservation of water resources, whether surface or underground" (Lineamientos 1981, 73), but there is no evidence that a water-pricing system was introduced. This period, however, was characterized by low efficiency in the sector's investment coefficient, with poor coordination between the hydraulic and other socioeconomic sectors; the dispersion of investment funds among too many hydraulic works; limited construction capacity because of inadequate equipment and poor maintenance; inattention to the disruptions caused by construction projects; and inadequate feasibility studies, including poor technical and economic assessments, that interfered with sectoral coordination during the implementation of economic plans (Atienza Ambou et al. 1995, 8). The plan for the period 1986–1990 was even more cryptic than its predecessor on water pricing, calling instead for "preserving and controlling the quality of surface and underground waters, achieving their rational use, and increasing their conservation and recycling [*reutilización*]" (Lineamientos 1986, 50).

Growing recognition since the mid-1970s of the need to use water resources more efficiently—later made even more pressing by the financial crises provoked by the special period and the emphasis it gives to national food self-sufficiency—eventually gave impetus to what appears to be a major drive to further increase the country's water use capacity, as well as to use this resource more efficiently. In 1986 efforts began to be made to reinvigorate the hydraulic sector. Construction brigades in charge of dams and irrigation canals were reconstituted and provided with new equipment, the National Institute of Hydraulic Resources was reorganized, and the organizational and administrative capabilities of hydraulic complexes were strengthened (Atienza Ambou et al. 1995, 8).[8] By the early 1990s, the country had been divided into thirty-three hydraulic complexes, more than half of which had been created since 1990. To accelerate the rate of construction of dams and other hydraulic works, thirty-nine construction brigades were in operation in 1990, twenty-eight building dams and eleven building irrigation canals. Goals for 1995 were to complete 1,534 additional microdams (with a capacity of 1.4 billion cubic meters) to gain the capacity to extract an additional 5 billion cubic meters of underground water and to channel into water distribution systems 700 million cubic meters of surface runoff (Atienza Ambou et al. 1995, 8–9).

Water-pricing policies were instituted in 1989 when enterprises producing rice, sugarcane, and mixed crops ("*cultivos varios*") began to be charged for "excessive" water consumption. Water flows began to be regulated by the installation of measuring meters, most of which were in place by mid-1990, with the expectation that end users would use the resource more economically. Gaining efficiency in water use was a pressing issue, since it was projected that by 1995 demand for water would reach 15.5 billion cubic meters. To satisfy this growing demand, aside from using water more efficiently, it would be necessary to increase the surface water storage capacity by nearly 5.5 billion cubic meters and underground water withdrawals by 120 million cubic meters. These targets could not be attained, however, because of the special period austerity (Atienza Ambou et al. 1995, 6–7).

Agricultural Development and Water Demand

As discussed in Chapter 4, in the late 1960s and early 1970s Cuba embarked on an ambitious irrigation program that has relied heavily on (1) developing a network of water storage areas through a major dam reservoir construction program, and (2) increasing the extraction rate of underground water. The water storage areas were also designed to contribute to the artificial recharge of aquifers since there was concern about the high rates at which water was being withdrawn from underground stores. The irrigation strategy was intended to increase production of the country's export mainstay—sugarcane—and also to boost the cultivation of other products, prominent among them citrus and rice. The planned major expansions in the output of these three crops, all prodigious users of freshwater, could only be sustained with additional water supplies, particularly during the dry season.[9]

Expanding the agricultural land surface under irrigation was a development priority during the 1970s, but even more so during the 1980s.[10] Doing so demanded the investment of hundreds of millions of pesos in equipment, machinery, and wells. The 1986–1990 Cuba-USSR cooperation agreement alone assigned 40 million rubles for the expansion of the national irrigation system.

Urban Water Demand

Since the 1970s, the demand for water has mushroomed with the growth of the country's system of urban aqueducts; an estimated 74.1 percent of dwellings had access to running water in 1980 compared to 54.5 percent in 1953 (Rodríguez and Carriazo Moreno 1987, 141). Despite these accomplishments, the water needs of a considerable share of the urban population remained unsatisfied by the late 1980s. According to data provided by Atienza Ambou and associates (1995, 11–12), at this time 1.3 million urban residents did not have access to running water, and the dwellings of 5 million residents were not connected to sewerage systems. Furthermore, less than 8 percent of the people served by aqueducts had continued water access, others receiving water only for periods of time ranging between 8 and 15 hours a day. Among major cities, Santiago de Cuba had the worst situation, with water availability limited to only 8 hours a day. Why these problems remain with the urban water distribution and sewerage systems has been attributed to the relatively low priority assigned to the sector and inadequate management of investments, poor maintenance and operation of the existing infrastructure, inadequate resources, and organizational and labor problems.[11]

Medium- and long-term urban water goals include providing no less than 87 percent of the urban population with aqueduct services and 51 percent with sewerage facilities, while assigning first priority to provincial capitals and urban areas with population of 20,000 or more; increasing the number of hours of service, mainly by reducing the amount of water wasted and promoting the more efficient use of available water resources; treating all aqueduct waters and ensuring that high quality waters reach users; and promoting in the state sector the efficient use of water by improving existing systems, as well as charging for water use (Atienza Ambou et al. 1995, 12). These goals were to be achieved by maximizing the use of the country's pipe construction capacity, including expanding production of conventional pipes and acquiring the ability to domestically build plastic pipes; by enlarging existing aqueduct construction brigades; and by guaranteeing the rational use of available resources, in part by introducing more modern technologies for the development, maintenance, and reconstruction of water distribution and sewerage systems. Atienza Ambou and her colleagues (1995, 13) stress that regardless of strategy, "the priority should be proper maintenance, reconstruction, and completion of the existing infrastructure, improving the operation and control of the systems available, and every other measure that would imply saving water without compromising current levels of supply."

Water Demand in Other Sectors

Also taxing water resources were the establishment or expansion of mining operations (e.g., nickel) and industrial plants (e.g., cement) that are heavy water users and contaminate the environment. These development initiatives have further strained Cuba's freshwater resource availability with potential long-term economic and ecological consequences (see Chapter 7).

Environmental Impact: An Assessment

The enormous gross gain in reservoir capacity achieved by socialist Cuba must be evaluated against the potential contamination of a considerable volume of water in aquifers. Salinization of significant stores of underground water sources is known to have occurred partly because of overextraction of water from aquifers for irrigation and other uses, and partly because by retaining water in large surface stores, water flows are not being allowed to reach and recharge coastal aquifers, resulting in the infiltration of seawater.

The Cuban technical literature suggests that there is increasing concern in the country about the adverse environmental consequences of damming the country's rivers and streams. A major worry is that the low volume of freshwater flowing below the dams alters nature's ecological balance by drastically reducing the oxygen level of the water and the amount of freshwater reaching the coasts.

Fourteen rivers in the southern plain of Pinar del Río Province had their sources of water affected by the construction of eighteen dams, with adverse impact on the fauna and flora of the region. Such activity, coupled with excessive withdrawal of underground waters and loss of forest cover contributed to desertification of the plain from the municipality of Sandino to Candelaria. Desertification is reportedly advancing in an eastward direction by 11 kilometers annually. The municipality of Sandino has been particularly affected, with sand overtaking fields that were formerly used for tobacco production and many native plants disappearing due to the excessive acidity of the soil (Pedraza Linares 1997). Equally troublesome are seawater intrusions associated with reduced freshwater flows. In some cases seawaters have reached the walls of the dams. As a result, river beds and adjacent fields have become contaminated by salts (Tapanes 1981, 39).

Historically, water had been extracted from some of Cuba's major aquifers at a higher rate than they were recharged, but in recent years the pace of water extraction appears to have intensified. Reports of saltwater intrusions along coastal areas suggest that overexploitation of underground freshwater stores has occurred in areas where there has been overpumping (Cereijo 1992, 75). As the World Bank has observed,

> Excessive water withdrawal contributes to [other] environmental problems. In addition to displacing people and flooding farmland, damming rivers for reservoirs alters the mix of fresh and salt water in estuaries, influences coastal stability by affecting sedimentation, and transforms fisheries by changing spawning grounds and river hydrology. When groundwater is drawn off at a rate faster than the rate of natural recharge, the water table falls ... The costs are often substantial and go beyond the additional costs of pumping from greater depths and replacing shallow wells with deep tube wells. Coastal aquifers can become saline, and land subsidence can compact underground aquifers and permanently reduce their capacity to recharge themselves. (World Bank 1992, 49–50)

In Cuba's case, because of its narrow and elongated shape, the diversion of surface waters into man-made reservoirs is believed to have interfered with the natural recharge of aquifers along the coast line, a major reason why saltwater intrusions have become such a major concern. The reasoning behind the dam construction program was that by impeding the rapid flow of rainwater to the seas, the stores of usable surface water would be increased. Although in general terms this appeared to be a reasonable approach, as early as the 1950s hydrologists warned advocates of this strategy that it carried some major potential environmental dangers in a country with a geography such as Cuba's. By not allowing rainwaters to flow freely to the ocean, there was a considerable likelihood that coastal aquifers would be infiltrated by seawater, since the amount of freshwater reaching coastal areas could be insufficient for underground recharge. A sponge offers a useful analogy. If loaded, a sponge cannot soak more water, but if dry, it is capable of soaking water until it reaches its maximum absorptive capacity.

A study commissioned by one of Cuba's development banks during the 1950s warned about the dangers of salinization attendant to an ambitious dam development program then being considered.[12] The study also noted that given Cuba's geography, and in particular the limited availability of locations capable of holding large stores of water, the flood area required would be extensive, including some of the country's most fertile soils. The alternative strategy recommended at the time as more economically and environmentally effective was to carefully use and manage the country's abundant underground water resources.

According to the hydraulic plans drawn up in 1976 by Soviet and Cuban technicians, 235,000 hectares of land were to be flooded under the dam construction program being proposed, or 2 percent of the country's total land area (Atienza Ambou et al. 1995, 5). The percentage of agricultural land that was to be lost to water reservoirs is probably much higher, since some of Cuba's most important agricultural areas (La Habana and Matanzas Provinces, for example) are flat or dominated by low rolling hills, which would limit the depth of the reservoirs. The loss of agricultural land should be a concern, given that Cuba is a relatively densely populated country. One way to overcome these limitations (extensive use of farm land and limited volumes of stored surface water) that has been considered and used to some degree is to build long-distance water distribution canals. Building these canals entails additional expenses and, given Cuba's climate, major water losses through evaporation and perhaps the intensification of the salinization problem.

The coastal salinization problem was further compounded by the mismanagement of underground water resources, in particular by overpumping. Several hydrologists and agricultural engineers who now live outside Cuba but who were involved in irrigation projects in southwestern Cuba have related that the technical cadres were well aware of the dangers inherent in extracting underground waters at excessively high rates, but their advice was often ignored by managers of agricultural enterprises intent on meeting centrally dictated production guidelines.[13] Specific instances of such mismanagement include the poorly regulated use of very large pumps with deep wells and the inefficient allocation and use of irrigation waters.

The decision to continue to pump excessive amounts of water, in turn, may have been dictated in part by the inability to increase the volume of water in reservoirs in La Habana and Matanzas Provinces as potential sites where additional dams and water reservoirs could be constructed became scarce. The extensive construction of dams and water reservoirs, particularly in western Cuba, may also explain the increasing emphasis being given to the construction of microdams.

Low Economic Yields of Irrigation Projects, Water Misuse, and Environmental Damage

Atienza Ambou and her colleagues (1995, 14–15, 18) have documented some of the major problems confronted during the implementation of socialist Cuba's irrigation policies and why these efforts have only met with limited success (in terms of increasing agricultural production). These problems include inadequately maintaining irrigation systems; introducing irrigation in areas with limited agricultural potential while neglecting soil improvement preparations (e.g., not leveling irrigated soils); ignoring agricultural technical requirements and timetables; and poorly implementing irrigation schedules, including lax measurement of water flows. They note, in addition, that irrigation in socialist Cuba has been plagued by the construction of irrigation canals too large for requirements; the failure to provide secondary water distribution systems; the failure to assess irrigation requirements by type of crop in relation to humidity levels, type of soil, climatic characteristics, and other considerations; mismanagement of irrigation plans; the failure to develop an adequate irrigation-supporting infrastructure; noncompliance with planning directives; delays in soil preparation and underutilization of irrigation equipment; an insufficient and unstable workforce; and, in some cases, lack of irrigation water. Many of these problems occurred despite the fact that the Cuban government devoted considerable resources to evaluating soil characteristics in relation to crop types and irrigation needs, as documented in several sources (e.g., Ministerio de Agricultura 1984a,b,c).

Specific problems involved in implementing irrigation directives that have plagued the rural sector account in many instances for the low productivity of Cuban agriculture and for the environmental damage arising from poor water management (Atienza Ambou et al. 1995, 19). Of note among these problems are the following:

▶ The inability to use a comprehensive technical approach in agriculture to efficiently coordinate soil preparation tasks, weeding, fertilization, irrigation, and drainage.

▶ Inefficient and inadequate selection of irrigation technique according to soil type.

▶ The inability to introduce more efficient irrigation techniques, since these require the allocation of substantial financial resources and considerable time.

Soil Salinization and Drainage Problems

Poor irrigation practices and high evaporation rates from man-made reservoirs and irrigation ditches are contributing to the salinization of Cuba's soils. Socialist Cuba is not unique in this regard. The World Bank (1992, 57) reports that in many regions of the developing world "salinization and waterlogging . . . are eating away at the productivity of irrigation investment." It notes further that "the problem is substantially greater in tropical developing countries, where soils, rainfall, and agricultural practices are more conducive to erosion and where many reports have found rates of soil loss well above the natural rate of soil formation" (56).

Significant expanses of Cuba's agricultural lands have been degraded by mineral deposits. It is estimated that 1 million hectares, or about 14 percent of the country's agricultural surface (6.7 million hectares), have excessive salt deposits (Estudio 1991; see also Chapter 4).[14] Of these, about 600,000 hectares with light to modest salinization levels could be reclaimed with appropriate techniques and the use of salt-tolerant crops. Badly salinized soils could be improved if treated with organic and mineral compounds. The Cuban regions with the highest salt concentrations are in Guantánamo and the Cauto River valley. The salinization of these regions is a result of human intervention as well as natural features. Waterlogging associated with the flooding from rapid rainwater runoff, the relative impermeability of the soils, and inadequate drainage are important determinants of salinization in these regions (Exhorta el partido 1985). Saline soils are found in many other regions, including Ciego de Avila Province where, because of its karstic aquifers, the danger of seawater infiltration is regarded as considerable. Large tracts of agricultural land in western, central, and eastern Cuba also exhibit a high potential for secondary salinization.

Ambitious irrigation plans have not been accompanied by required complementary drainage facilities to prevent soils from accumulating excess waters. Without proper drainage, salinization of the soil occurs. In the absence of drainage water the minerals deposited by irrigation remain in the soil as the water is absorbed or evaporates. It had been reported that drainage facilities in flood-prone areas were sorely lacking and that irrigation canals frequently were activated before adequate drainage measures were in place (Peñalver 1979; see also, Atienza Ambou et al. 1995).

The 1981–1985 plan called for a national census of drainage areas; development of a taxonomy of flood-prone soils; basic research on the country's aquifers; evaluation of the potential of salt-resistant plants; technical assessments of the economic potential of soils that could be protected against flooding; and studies about the proper maintenance of drainage facilities. Even if it had been possible to acquire the basic drainage know-how anticipated under the 1981–1985 plan, it is highly unlikely that much could have been accomplished in the way of constructing a drainage infrastructure during the 1986–1990 quinquennium. With the end of the Soviet subsidies and the special period, it is safe to assume that most drainage facility work came to a halt, although during the early 1990s, efforts were renewed to expand the country's water reservoir system in support of a drive to attain food self-sufficiency (Atienza Ambou et al. 1995). The likely further neglect of the drainage infrastructure could well be aggravating the salinization problem. Overall, however, it is difficult to assess the consequences of these developments for the preservation of Cuba's soils, but it is likely that the extraordinary expansion of irrigation—even during the special period—in the absence of adequate drainage facilities may have contributed and continue to be contributing to the further salinization of the country's soils.

Salinization of Underground Waters

By 1976, Cuba was already reporting that the country's major environmental problem was the intrusion of saltwater into coastal areas (Report of Cuba 1976, 58). A 1980 report remarks that the scope of this problem was alarming. All along Cuba's southern coast the inland intrusion of saltwater reached between 2 and 15 kilometers (Shayakubov and Morales 1980, 10–11).[15] Official sources attribute this environmental damage to the indiscriminate use of underground water for irrigation prior to the revolution. This argument is not convincing, however, since prior to the revolution irrigation was used relatively sparingly in Cuba. Whereas in 1958, 160,000 hectares were being irrigated, by 1982 land under irrigation had increased to 900,000 hectares, or by a factor of six (Riego 1982, 63).

The diversion of rainwater into reservoirs and a high rate of extraction of underground freshwater is associated with the salinization of aquifers along the coastline. The best documented case concerns the aquifer bordering the low-lying coastal southern region of La Habana Province (Salazar 1991). This aquifer supplies water to some of Cuba's richest agricultural regions (e.g., Güira de Melena) and the city of La Habana, contributing 200 million cubic meters per annum of water for human and agricultural uses.

In 1985 construction began on an underground dike (Dique Sur) of approximately 100 kilometers in length, running from Majana in the west to Batabanó in the east, with a width of 7 meters and a depth of 2–4 meters (portions of the dike were still under construction in the mid-1990s). Covering some 184,000 hectares, about 70 percent of which are mangroves, this region is reputed to be one of the country's most heavily affected by salinization, particularly the underground waters in the Güira de Melena area (Lezcano 1994a). The objective of the dike is to arrest the inland infiltration of saltwater—and prevent freshwater losses to the sea—by impeding water flows in either direction, thus contributing to the replenishment of the aquifer with freshwater. The design of the dike includes drainage canals to facilitate the flow of excess surface water during the rainy season. In a 1989 address President Castro referred to the Dique Sur:

> The dike we are building in the South of La Habana province, near the coast, is aimed at protecting underground water, preventing them from salinization and from escaping to the sea. The dike being built East of Batabanó should protect some 300 million cubic meters of water. It is a novel project, and we are considering raising the height of the wall, as there is still some water loss. These are techniques to protect aquifers and make available larger quantities of water to meet agricultural and population needs. Afterwards we are going to proceed building a dike from Batabanó toward the East—the project I described above is West of Batabanó—toward the Zapata Swamp, to try to protect some 200 to 300 million cubic meters of additional fresh underground water from going to the sea. (Castro 1992, 74–75)

It is claimed in the official press that, by 1991, the Dique Sur had helped reduce the freshwater runoff by 90 percent, and the level of salinization in the aquifer is reported to have declined from 4,440 milligrams per liter in 1982 to 1,000 milligrams per liter in 1991, making the water suitable for human consumption (Salazar 1991). This claim is questionable, since contamination of underground water resources is very difficult to reverse and entails a long-term process.

There is a growing body of evidence indicating that the Dique Sur, far from stopping or reversing environmental damage, has wrecked the coastal ecosystem of southwestern Cuba. According to several accounts, the dike has contributed to the waterlogging of vast expanses of agricultural land; is an important factor behind recurring floods in the region; and has emaciated the coastal mangrove ecosystem, leading to erosion along the coastline.[16] José Carlos Lezcano (Lezcano 1994a), a physical geographer specializing in coastal climate and oceanography who until the early 1990s was employed by the Cuban Academy of Sciences, has described the environmental effects of the Dique Sur:

> The Dique Sur certainly diminished salt water intrusions along the southern coast but, in its place, gave rise to increases in chemical and organic contamination, causing the explosive appearance of several water-borne diseases, leading to their unprecedented propagation, causing cracks in the floors of houses, destroying tens of square kilometers of mangroves, and causing flooding in places where these floods were not remembered. . . . Despite these problems, the worst is yet to come. A tropical hurricane passing through the south of La Habana province, one of the most hurricane-prone areas in the Caribbean, would provoke floods of a magnitude 20% to 30% greater than in the past, with the potential to destroy places like Surgidero de Batabanó, Guanimar and Cajío, among others, since the storm's sea water surge would not drain normally, given that the dike interferes with the natural hydraulic compensation process, and since in the absence of the mangroves, the waves' effects would not be dissipated. Thus, sea flooding could reach as far as ten kilometers inland, aside from the effects of the rains. . . . Unfortunately, the Dique Sur is just one example of [hundreds of documented] cases of profound and indiscriminate transformations of the natural environment across the island. (1995, 403)

Not noted by Lezcano in his summary is that the destruction of mangroves also obliterates critical habitats and leads to the loss of biodiversity and genetic resources (Food and Agriculture Organization 1994, 203), a point eloquently made by Carlos Wotzkow (1998), a Cuban naturalist. In Cuba this should be of special concern given the high endemicity of many of its life forms.

The location of the dike and of the aquifer it is intended to protect suggests that the saltwater intrusions may have resulted from three main causes: (1) the diversion to reservoirs of groundwater that normally would help replenish coastal aquifers; (2) the continued excessive drawing of aquifer waters to serve the needs of the city of La Habana; and (3) the increasing use of aquifer water for irrigation. Between 1970 and 1988, sixteen reservoirs, with a total capacity of 458 million cubic meters of water, were built in La Habana Province (Comité Estatal 1989, 175). In the neighboring province of Pinar del Río, during roughly the same time period, twenty-one dams were built, with a total storage capacity of 832 million cubic meters. Some of these reservoirs were built to provide water to urban populations, but most were intended to irrigate crops with high water consumption, such as sugarcane, citrus, and rice.

For example, to increase the country's self-sufficiency in rice, one of Cuba's main food staples, large-scale cultivation of this very water-intensive crop with a fairly high tolerance for salty waters has been expanded in recent years (Colina and Peláez 1976). Although under the right circumstances rice cultivation may contribute to the desalinization of the soil, it could well aggravate the problem if proper agricultural practices are not followed or if poor quality waters are used for irrigation (Martín Alonso 1976; Sotolongo and Abreu 1992, 165–66). Similarly, citrus cultivation, also a water-intensive crop (Bianchi Ross 1985, 24), has also increased phenomenally. Many of the country's citrus plantations are located along the southern coast of central and western Cuba. The very large water needs of rice and citrus have required the construction of a network of reservoirs, which have reduced the amount of water flowing to the coastal aquifers of Pinar del Río and La Habana Provinces. Furthermore, since the potential is limited for increasing the capacity of reservoirs in these two provinces—particularly in La Habana Province, whose landscape is relatively flat—the needs of sugarcane, citrus, and rice have had to be served with water drawn from aquifers. Due to the salinization of the soil in these and other coastal regions, fields formerly planted with crops and heavily irrigated have been abandoned (Wotzkow 1998).

Last, but by no means least, excessive water pumping from the aquifers serving Ciudad La Habana Province and urban agglomerations in La Habana Province are certain to have contributed to the salinization of underground waters. Between 1954 and 1964, the water table surrounding the basins of the Ariguanabo and Almendares Rivers dropped 6–8 meters (Egorov and Luege 1967, 24). In 1967 the Ariguanabo watershed supplied water to the towns of Bauta, Corralillo, Bejucal, and San Antonio de los Baños. The water consumption of these towns at the time was 55 million cubic meters a year. In addition, a textile plant in the region used 5 million cubic meters, irrigation, 30 million cubic meters, and other agriculture, 1.5 million cubic meters, with a total demand reaching 99 million cubic meters a year. Yet the replenishment capacity of underground waters in the Ariguanabo watershed was only 77 million cubic meters a year (Egorov and Luege 1967, 51).

At the time, the Almendares watershed and the Cuenca Sur, the main water sources for the La Habana metropolitan area, were experiencing annual water deficits. In the former, 221 million cubic meters were extracted annually while the aquifer received 175 million cubic meters annually, for an annual deficit of 46 million cubic meters (Egorov and Luege 1967, 54). In the Cuenca Sur, an area located in southern La Habana Province along the sea that provided 50 percent of the water consumed in metropolitan La Habana, the annual water deficit reached 48 million cubic meters (Shayakubov and Morales 1980, 10–11). In southern sectors of the Cuenca Sur between 1954 and 1964, the water table had dropped 0.5–0.8 meters, and in some northern sectors by as much as 4 meters. In many locations the underground water level was below sea level (Egorov and Luege 1967, 58–59). The conclusion of the hydrologists studying the situation was that if extraction rates remained at those levels, the inevitable result would be a continuous drop in the water table and/or the eventual exhaustion of underground water stores. As we have pointed out, diverting surface waters to irrigation and other uses, population growth, expanding the availability of urban water, and the absence of measures to reverse deteriorating water resources could only have resulted in significantly aggravating the water supply situation in western Cuba. In this context the Dique Sur may be seen as a misguided effort to arrest the decline of the rich karstic aquifer running from Pinar del Río Province to Matanzas Province.

Contamination of Surface and Underground Waters

Surface waters have been contaminated by industrial wastes and by the runoff of chemicals associated with the use of increasing amounts of chemical pesticides and herbicides. There is ample evidence of dumping of industrial wastes in rivers and bays (see, for example, Dávalos 1984a,b,c,d; Emprende 1985) and of increases in the release of chemical agricultural inputs. Arturo González Báez and Sigilfredo Jiménez Hechevarría (1988b, 18–24) have pointed to sugar mills as the main source of pollutants, although untreated urban sewage, industrial pollutants, and agricultural chemical inputs are also to blame. Seventy percent of the gross weight of the sugarcane processed in the mills is water, and another 15 percent is bagasse (Varela Pérez 1976). According to independent Cuban environmentalists, by the 1980s, 13,300 cubic meters of urban sewage and 6,015 cubic meters of industrial wastes were dumped into the Almendares River and its tributaries every 24 hours (Environmental Summary 1997). Of even more concern is the contamination of underground water by these same sources of pollution. The World Bank's global review of environmental issues in development concluded that "it is often more important to prevent contamination of groundwater than of surface water. Aquifers do not have the self-cleansing capacity of rivers and, once polluted, are difficult and costly to clean" (World Bank 1992, 47).

The reduction in water volume carried by many rivers has given rise to adverse environmental consequences, particularly during the dry season. In Cuban rivers with little or no runoff during the dry season, "sewage volume can surpass several times the stream flow" (Report of Cuba 1976, 54). The diversion of surface waters for irrigation and other uses is implicated in the increased pollution of many of Cuba's rivers. Reduction in water flow aggravates the old problem of contamination of rivers, since the rivers lose part of their capacity to carry away or dilute contaminants. Surface pollution also eventually reaches underground water stores. For decades, sewage and industrial contaminants from urban areas and sugarcane waste products generated by sugar mills in rural areas were directly discharged into rivers and streams. This problem is reported to have affected numerous rivers, including the Almendares (Emprende ciudad 1985; Guma 1989), Luyanó (Gómez 1985), and Quibú (Hernández Pardo 1975), all in the metropolitan La Habana region. The effluents flowing into these rivers and eventually to the coast have been blamed for the contamination of beaches to the west and east of La Habana Bay (Lezcano 1994b). Another documented example of polluted rivers includes the three that flow into or in the immediate vicinity of the city of Guantánamo in eastern Cuba. The Guaso, Bano, and Jaibo Rivers are said to be heavily contaminated by raw sewage, industrial discharges from an alcohol plant, and other residues dumped into their waters (Contaminación afecta 1997).

Results of local studies reported in the Cuban technical literature indicate that the mismanagement of water resources and the inadequate treatment of industrial, agricultural, and domestic discharges have adversely affected the quality of water stored in reservoirs. In the municipality of Segundo Frente in Santiago de Cuba Province, for example, these problems were aggravated by an insufficient supply of water caused by poor workmanship in the construction of a microdam and its subsequent mismanagement, which have led to a higher-than-expected rate of sedimentation. To make matters worse, the water treatment capacity of industrial facilities in the area is insufficient given the volume of toxic discharges, and frequent interruptions in electric supply have caused untreated pollutants to be discharged into water bodies used for the municipality's water supply (Quintana Orovio et al. 1992).

Similar problems are reported in other parts of the country as the number and capacity of water reservoirs has increased, river flows have declined, and the discharge of organic material and other pollutants have increased. Reduced water flow also contributes to a higher sedimentation rate in rivers. In Cuba the danger of contaminating underground waters is high since the karstic aquifers found in much of the country facilitate the flow of large amounts of subterranean waters. In the Cauto River basin, for example, 652 sources of pollution—industrial, cattle farms, and urban sewage—produce large volumes of contaminants (Portela 1997).

Some reports also suggest the dumping of organic and industrial wastes in many of the thousands of caves found in Cuba's karstic soil formations. Many of these caves, often part of underground water reservoirs, reputedly are used by government-owned agricultural and industrial enterprises to dispose of some of the wastes they generate during their production processes. It is alleged that the Candela cave near Güines, in La Habana Province, for example, has been used as a dump for tons of industrial wastes, as well as dead livestock. Other reports by speleologists not associated with government institutions indicate that in other caves used as dumps in La Habana Province they have found strychnine, caustic soda, and potassium. These substances have been discarded by sugar mills and industrial plants. Independent speleologists have also expressed concern about the increasing use of caves by the military as storage facilities for armaments, explosives, and chemical products. Cave ecosystems are fragile and vulnerable to large-scale human interventions. Still other reports made by speleologists allege that the level of the water table in caves they have visited in Ciego de Avila Province has declined substantially, and that underground waters serving important urban centers such as the city of La Habana are heavily contaminated by chemical and biological agents (Ecología/Cuba 1996).

There have been instances as well of official inaction following reports of high levels of pollution in water reservoirs principally assigned for human uses. This is the case with the Guirabo Dam in Holguín Province. Despite local complaints and technical studies by competent authorities about major accumulation of human and industrial wastes in this reservoir, few or only partial corrective measures were taken to redress the situation. As a result, there have been outbreaks of disease and reports of contaminated fish being sold to the population (Contaminación 1990, 1).

Contamination and Deterioration of Coastal Regions

While the contamination of La Habana Bay (United Nations 1985) and others (Schlachter 1990; see also Chapters 7 and 9) has received a great deal of attention, less well studied are discharges of untreated pollutants and interference with natural water flows that have contributed to considerable environmental damage along Cuba's coasts. With lower volumes of water, and the continuous discharge of organic matter into rivers and other water bodies, oxygen levels in the water are being depleted. Water flows reaching the ocean with high densities of contaminants damage estuary breeding grounds and coral reefs (Sáenz 1990; World Bank 1992, 49–50). As in most of the Caribbean, riverine estuaries, coastal lagoons, and mangroves have been damaged by upriver human interventions (see the prior discussion regarding the environmental consequences of building the Dique Sur).

Emilio del Barrio Menéndez (1990) reports that as a result of a persistent drought during the late 1980s and the diversion of water to reservoirs, the amount of freshwater reaching Cuba's coastal lagoons—economically important since they serve as hatcheries for shrimp and other species—has declined by 2,243 million cubic meters annually. As a result, many of these lagoons have dried out or are in the process of doing so. In the thirty years from 1960 to 1990, 9,800 hectares of coastal lagoons were damaged. Revealingly, 61 percent of the damage occurred between 1971 and 1985, the period of accelerated reservoir construction. Barrio Menéndez also reports that draught and the diversion of surface waters has also had an adverse effect on mangroves and has reduced the flow of organic substances reaching coastal areas. Organic material reaching the coastal lagoons is an important food source for juvenile shrimp (Rehabilitan 1991).

Water diversion to reservoirs is also likely to be implicated in the virtual destruction of the oyster bed and a major decline in the fish catch in the Casilda coastal region of southern Santa Clara Province (Dávalos 1984a,b,c,d). The oyster bed was at the mouth of the Agabama-Manatí River, which in 1984 alone received 50,000 cubic meters of industrial residues from the Pulpa Cuba paper plant, in addition to water from the FNTA (old Trinidad) sugar mill. The Pulpa Cuba plant, in operation since 1959, was a source of contaminants of the Agabama-Manatí River for many years, but the environmental damage to the Casilda coast did not become alarming until the 1970s. Between 1960 and 1972, in Villa Clara Province, four dams with a combined capacity of 780 million cubic meters were completed. The loss of marine life may have resulted as a reduced flow of highly contaminated water reached the coast. By 1989, four other dams were under construction in this province (Comité Estatal 1989, 175, 177).

A report prepared by CITMA in 1995 contains statistics on the magnitude of the pollution from discharges into coastal areas. A study of the 117 largest sources of pollution in the nation revealed that 247,000 tons of biodegradable organic materials were being dumped annually into coastal areas. This level of pollution, according to the report, was equivalent to the expected discharges of a population nearly 50 percent larger than that of Cuba (Ministerio de Ciencia 1995, 36).

Another documented instance of coastal pollution caused by the release of organic materials was reported in 1988. According to this account, torula yeast discharges from a factory associated with a sugar mill and the resultant contamination were blamed for the destruction of an important fishery located along the northern coast of Ciego de Avila Province. As early as 1980, fishermen in this area began reporting a large drop in the fish catch in the Jaruco area, extending about 9 miles beyond Cayo Campo. The fishermen also reported that pollution was so severe on the sea floor that their nets were being damaged. The problem was traced to raw waste dumping by the torula yeast factory, a situation that could have been prevented by an oxidation lake or some other mechanism to decontaminate the industrial waste. The local provincial fishing enterprise is said to have reported the problem to the national environmental commission, which, despite debating the issue, apparently failed to take any corrective action (Contamination Affects 1988). This incident poses a potential threat to the tourist industry, since the polluted area lies in a region slated for tourism development. The extensive damage to the mangroves in southern La Habana Province associated with the construction of the Dique Sur was noted earlier.

Although the evidence is mostly fragmentary, there are sufficient grounds to conclude that the coastal regions of Cuba have been degraded to an unknown extent by the hydraulic development policies pursued under socialism in Cuba. Further damage has been attributed to the untreated discharge of human, agricultural, and industrial waste, as well as by the by-products of mining operations, mainly in eastern Cuba (see Chapter 7).

Summary and Implications

Water policies pursued by the revolutionary authorities have intensified environmental stress in Cuba. Whereas many problems associated with management of water resources were already in evidence by the 1950s (e.g., too rapid extraction of underground waters in the Habana region), the efforts of the last thirty years to expand the urban water supply and increase irrigation has jeopardized the sustainability of some of Cuba's main water sources. The most worrisome environmental consequences of the water policies of the socialist government appear to be associated with overpumping of underground waters and an extensive dam construction program and their implications for the salinization of aquifers and the degradation of coastal areas. Both the mineral load in many of Cuba's aquifers and soil salinization seem to be on the rise.

Major questions remain as well regarding the extent to which Cuba has realized economic benefits from the major hydraulic development program in which it embarked nearly four decades ago and that consumed hundreds of millions of pesos. Despite this sizable investment, in the 1990s total and per capita agricultural production in Cuba—as well as agricultural yields for many crops (e.g., sugarcane) harvested in the country—is lower than in the 1950s or early 1960s. In addition, the seasonal water scarcity problem that the extensive dam construction program was supposed to address does not seem to have improved much despite the manyfold expansion of the country's water reservoir capacity.

✑ 6 ✑

Forestry and Agroforestry

THE ECOLOGICAL, environmental, and social values of forests are increasingly recognized, as is the disregard with which they have been treated throughout much of human history. The UN Food and Agriculture Organization (FAO) classifies the actual or potential benefits of forests in terms of their direct, indirect, and option and existence economic values (FAO 1995). Although some of these values can readily be measured in monetary terms, others cannot. According to the FAO classification, forests are valued directly for their consumptive uses (e.g., lumber, pulpwood) and nonconsumptive uses (e.g., recreation). Their value can also be indirect and related to the functions they play in watershed and soil protection, fertility improvements, gas exchange and carbon storage, habitat and protection of biodiversity and species, and soil productivity on converted forest land. Further, forests are valued simply because they are there, or because of potential future uses. Historically, the value tradeoffs implicit in the economic uses of forest resources were seldom considered. In most countries forests were regarded as a natural resource to be relentlessly exploited or, just as often, as barriers to agricultural expansion. The immediate economic benefits accruing from the exploitation of forest resources overshadowed their environmental benefits. As scientific understanding regarding the role of forests in nature has been gained, and as societies began to assign true economic values to forest resources, these attitudes have begun to change. Whereas forests had often been viewed as

138

free goods to be exploited at will, they are seen today as a common patrimony whose societal value must be preserved. These attitudes represent a radical departure from the past, when, even if the common value of forests was recognized (e.g., watershed protection), private interests were allowed to have sway, either because of inadequate policy frameworks or the inability or unwillingness to enforce conservation measures.

Cuba is no exception. As will be shown here, the wanton destruction of the Cuban forests in the name of economic progress over several centuries occurred despite repeated warning calls. By the mid-1950s, Cuba's forest resources had dwindled to a small fraction of their former size, and some observers expressed fear that continued exploitation would result in their demise. Weak initiatives to preserve the forests met with little success, and concern about their preservation continued to mount.

In this chapter we analyze the fate of Cuban forests, explore the causes leading to their overexploitation, and review policy measures proposed and instituted by various governments—and their outcomes—to protect the forests and the country's domestic supply of timber and other tree products. We also examine related agroforestry issues as well as review selected topics associated with the planting of perennial fruit trees (e.g., citrus and coffee), whether as part of large-scale plantations or in smaller accumulations.[1] These topics are relevant to the forestry discussion since they shed light on how various development approaches have affected the country's stock of trees and the natural environment. We begin with an examination of the historical record regarding the pace and extent of deforestation in Cuba.

The Demise of the Forests

Reconstructing a robust historical series of the impact of human activities on the Cuban forests—in terms of land area forested, and deforestation and reforestation rates—from the time of first European contact to the present is a major challenge. But it is important to attempt to do so, since there are obvious discrepancies in the literature dealing with historical and contemporary estimates of forest cover. For purposes of this book, it is particularly important to determine how much of the original forest cover remained before the revolution and in the years immediately following 1959. Part of the problem is that many of the available estimates are based on rough figures derived by contemporary observers without the benefit of credible statistical information. Some historical estimates simply consist of guesses regarding the extent of forestation (e.g., the percentage of Cuba's territory remaining forested, or the converse, amount of land area no longer forested) or are based on historical statistical sources of uncertain validity.[2] For more recent periods, the estimates tend to be more credible: they rely on more dependable agricultural censuses (e.g., the 1945 agricultural census) or land use data that became available in the 1970s and 1980s presented, for example, in the various statistical yearbooks published by the Cuban government. Increasingly valid are recent estimates based on aerial and satellite photography.

Some of the discrepancies for the period between the 1920s and the 1950s can be attributed to definitional issues, including what is meant by *montes,* a recurring term in the Cuban literature.[3] Additional definitional issues arise from the nature of agricultural censuses and how land use data are tabulated. The 1945 agricultural census of Cuba (as summarized in World Bank 1951, 87), for example, provides a detailed land use data tabulation with reference to land in farms only, leaving unclassified a residue of 2.4 million hectares, or 21 percent of Cuba's overall territory. Other problems relate to how widespread were forests in Cuba at the time of first European contact and whether historical forest-cover estimates made allowance for the fact that forests were never ubiquitous in the country. Despite these limitations, the historical series we have reconstructed by compiling and whenever possible adjusting existing estimates provides a reasonable approximation of the fate of the Cuban forests that is consistent with the country's economic history.[4] The series is presented in Table 6.1.

Leví Marrero (1950, 107, 195) and Antonio Bucek (1986, 15) provide estimates indicating that in 1492, at the time of Columbus' arrival in the Americas, 60–75 percent of Cuba's land area was forested, the remaining land area being occupied by *sabanas* or *prados naturales* (natural meadows or fields).[5] Over the course of the next five centuries, close to 80 percent of the island's forest cover was lost. Cuba's trees were first felled to accommodate early colonial needs (e.g., human settlements and limited agricultural and livestock production) and to meet Spain's lumber requirements for shipbuilding and construction. Early European commentators marveled at the richness and quality of Cuba's fine woods. During the seventeenth and eighteenth centuries, the availability of lumber gave rise to a vibrant shipbuilding industry in La Habana, where many of Spain's seafaring vessels were built. The island's fine forest products were coveted in Europe and extensively used to build many of Europe's better-known architectural projects, such as El Escorial and Madrid's Royal Palace (Marrero 1950, 299; Moreno Fraginals 1964, 74).

Settlement of the island and exploitation of forests for production of lumber had only a relatively modest impact on the amount of forest cover lost by the end of the seventeenth century, with about 90 percent of the forests, or 5.1 mil-

**Table 6.1. Historical estimates of Cuba's forest cover and deforestation
and reforestation rates[a]**

Year	Forested area (in millions of hectares)	Percentage forested	Annual rate of de(re)forestation (in hectares)
1492	8.2	**72**	
1799			– 6,700
1800	7.4	**65**	
1819			– 13,400
1830			– 26,800
1844			– 53,600
1850	6.2	54	
1852	4.6	**40**	
1905–25			– 80,000 to – 100,000
1900	4.0	**35**	
1923		16	
1929		13	
1945	2.0 to 2.4	**18–21**	
1946–1948			– 13,400
early 1950s			– 40,470
1959	1.5	**14**	
1962	.9	8	
1959–1992			+14,000
1992	2.0	**18**	+103,500

[a]For explanations of discrepancies in secular trend, see text. Preferred estimates are in bold.
Sources: Various, see text.

lion hectares, remaining relatively intact (Westoby 1989, 131). Deforestation was a localized phenomenon, with virgin forests no longer being found on the plains region surrounding the city of La Habana, in parts of present-day Matanzas Province, or in areas adjacent to other early settlements. However, as Manuel Moreno Fraginals (1964, 76) notes, by the late eighteenth century, torching forests to open land for sugarcane and other agricultural uses was a daily occurrence in the Cuban countryside.

Following Haiti's successful slave revolt and the collapse of sugar production there, in the first half of the nineteenth century the loss of forest cover in Cuba accelerated dramatically as sugarcane acreage expanded rapidly. The collapse of the Haitian sugar industry marks the beginning of a period lasting slightly over a century (from about 1800 into the 1920s) when the wholesale destruction of Cuba's forests was carried out in the name of sugar production. Throughout this period, forests were cleared, usually by slashing and burning, to expand the acreage devoted to sugarcane cultivation. Forests were also harvested for firewood to fuel the sugar mill boilers. Moreno Fraginals (1964, 75) cites figures suggesting an eightfold increase in annual deforestation rates between the late 1700s and the mid-1800s: 6,700 hectares per annum at the close of the eighteenth century, 13,400 hectares in 1819, 26,800 hectares in 1830, and 53,600 hectares in 1844. By mid-century, and based on a contemporary estimate (José García de Arboleya as cited in Grupo Cubano 1963, 171), we calculate that Cuba had 4.3 million hectares of forests. This estimate, together with Bucek's (1986, 15) somewhat lower 1852 census estimate (see Table 6.1), confirm that at mid-nineteenth century, forests remained widespread, occupying 45–50 percent of Cuba's land area.

The second half of the nineteenth century saw further deforestation as sugar production continued to rise, though subject to marked yearly fluctuations. This period coincided with nearly thirty years of on and off warfare (1868–1878; 1879, 1895–1898) between Cuban insurgents seeking independence and Spain's colonial authorities. By the end of the nineteenth century, probably about 50 percent of the original forest cover present in 1492, or about 4.0 million hectares, remained (War Department 1900, 23), a figure consistent with García de Arboleya's and Bucek's estimates half a century earlier and with the limited deforestation that apparently occurred in the intervening period. This means that in 1900 about 35 percent of the island's land area was forested.

Explosive growth in sugar production resumed after Cuba gained its independence from Spain as foreign investment, particularly from the United States, flowed into the sector. The impact on the Cuban forests was devastating. As sugar production expanded eastward, mainly into the provinces of Camagüey and Oriente, it was accompanied by the indiscriminate burning of millions of hectares of virgin forests. *Marabú* patches were also cleared to plant sugarcane (Grupo Cubano 1963, 1067). Between 1902 and 1925, sugar output rose from 850,000 tons (close to the average of 933,400 tons for the 1891–1895 period, just before the 1895–1898 War of Independence) to 5.2 million tons, or by a factor of five (Thomas 1971, 1562–63). With technological developments that permitted the conversion of bagasse from a nuisance, abundant, and difficult-to-manage harvest waste product into an economical and readily accessible fuel, firewood was gradually displaced as the primary fuel in the sugar mills. Even then, forests continued to be burnt to clear land for planting sugarcane.

The 1931 census provides figures indicating that in 1923, 16 percent of Cuba's land area was classified as *montes*. This term, as used in the census, includes forests (*montes altos*, or forests in mountain areas, and *montes bajos*, or forests in the plains), as well as swampland (e.g., *ciénagas* and *manglares*, or mangroves) and other areas covered by vegetation, whether virgin or not (Memorias

1978, 28). Other estimates provided in the census suggest the uninterrupted felling of the country's forests; by 1929, according to this source, only 13 percent of Cuba's territory was covered by *montes* (Memorias 1978, 29). However, more robust estimates that became available years later indicate that the 1931 census report grossly overestimated the extent of deforestation.

Jack Westoby (1989, 131) reports that the 1945 agricultural census found that 1.3 million hectares, or 11 percent of the total land area, was still forested. A further 268,000 hectares, or 2.4 percent of the country's land, were covered by *marabú*. The estimate cited by Westoby, however, as noted earlier, does not take into account the 2.4 million hectares of Cuba's territory not included in farms. Assuming that between one-third and one-half (or between 790,000 and 1,188,000 hectares) of these hectares were forested, it can be estimated that between 1899 and 1945, forest cover had declined by between 1.6 and 2 million hectares, although most of the deforestation took place within the first twenty-five years of the century to make room for sugar plantations.[6] On the basis of these assumptions, we estimate that in 1945, 18–21 percent of Cuba's total land area remained forested. By assuming that most forest clearings occurred in the twenty-year period between 1905 and 1925, and allowing for uncertainties regarding the actual land surface still forested by mid-century (see Table 6.1), we estimate that in this interval, forests were being cleared at annual rates ranging between 80,000 and 100,000 hectares, or at a pace twice as high as in any other historical period. Some of the forest-cover loss can also be attributed to logging for domestic consumption and export: between 1899 and 1927, Cuban wood exports averaged between 1 and 2 million dollars annually (Grupo Cubano 1963, 490).

During the 1930s and 1940s, commercial lumbering was a further factor leading to deforestation. Also contributing to deforestation were large-scale open-pit mining operations, particularly those in northeastern Cuba (e.g., the Moa and Nicaro regions), where several large-scale mines were established in the 1940s (Reed 1992, 37). Yet another development leading to deforestation was slash-and-burn agriculture by landless subsistence farmers, mostly in the mountain regions of Oriente Province. But the major contributor to forest loss during this period was the use of wood as fuel. During the 1940s and 1950s, close to 70 percent of all domestic lumber was used directly as fuel or processed into charcoal (Marrero 1950, 301).

Excessively high timber harvesting rates also contributed to forest loss between 1945 and the early 1960s. In some years the forest cover may have been reduced at rates similar to those observed early in the nineteenth century, despite frequently voiced concern about the sustainability of the country's forest resources. In 1946–1948 forests were cut for commercial purposes (lumber and production of charcoal) at an annual rate of 13,400 hectares, when harvesting rates increased in response to high international prices for lumber (Marrero 1950, 303).

By 1962, the Comisión de Conservación de los Recursos Naturales (CCRN, Commission for the Conservation of Natural Resources) claimed that forests accounted for only 8 percent of the country's surface, or 880,000 hectares (Núñez Jiménez 1972, 342). This estimate was erroneously embraced by CCRN and Núñez Jiménez by misreading the results of the 1945 agricultural census, and perhaps the forestry figures shown in the World Bank study of the Cuban economy (World Bank 1951, 903). The error was further compounded by relating the forested land estimate to the total land area of Cuba rather than to the land area exclusively in farms. Later estimates indicate that the extent of deforestation was considerably less, but, as noted, it is difficult to estimate with any certainty how widespread the forests were in 1945 (our estimate suggests that forests accounted for 18–21 percent of the national territory at the time). Cuba provided an official estimate to the 1992 UN Conference on Environment and Development held in Río de Janeiro, indicating that forests accounted for 14 percent of the national territory in 1959 (COMARNA 1992, 29), or 1.54 million hectares. The forestry estimates for 1945 and 1959 suggest that during this interval Cuba lost between 460,000 and 800,000 hectares of forest; these estimates appear to be consistent with the annual deforestation rates recorded for this period and shown in Table 6.1.

The CCNR also noted that the remaining Cuban forests in 1962 had largely lost their ability to produce some of their finest woods due to selective harvesting, a practice underway for many years (Grupo Cubano 1963, 697–98; Marrero 1950, 298–309). The 1951 World Bank report, citing data from Cuba's Bureau of Forestry, observed that of the 200 tree varieties found in Cuban forests, only about a dozen had commercial value (World Bank 1951, 903). Other alarming conclusions of the CCNR report merely reinforced earlier assessments. One of the most serious was that deforestation was having serious adverse ecological consequences, since it was associated with extensive soil erosion and substantial watershed damage. Equally critical, the report concluded that at prevailing harvesting rates, the country's forest resources would be exhausted in five years.

Reversing the Deforestation Trend

Almost immediately after assuming power in 1959, the Castro government began an aggressive national reforestation program. In large measure this program responded to a grave national concern but also dovetailed with populist policy objectives to improve rural socioeconomic and employment prospects, particularly in mountainous regions, as well as with national development priorities outlined by Fidel Castro as early as 1953 (Castro 1972, 190). In the short term reforestation programs offered a relatively low-cost option to generate employment among those segments of the rural population that had harbored the revolutionaries during the armed struggle (e.g., the peasantry in the mountainous areas of eastern Cuba).

The reforestation strategy that the Castro government began to implement, however, had been bandied about in Cuba for years, with selected elements of the strategy beginning to be implemented in the early- to mid-1950s. A U.S. technical mission sent to Cuba in the early 1930s recommended, for instance, implementing "a forestry program [that] could undoubtedly develop the export of valuable timber, which would yield a greater profit than land now otherwise employed" and that, in addition, would produce "commercial returns and give added employment to workers . . . [as well as] assist in preventing soil erosion and in stabilizing the effects of rainfall" (Foreign Policy Association 1935, 463).

Similar themes were echoed by the 1950 World Bank Mission Report when it advised the reforestation of 3–4 million acres (1.2–1.6 million hectares) of land suitable for timber production to double the country's wooded land area (World Bank 1951, 907). This proposal was to be given priority in light of severe soil erosion and persistence of floods. The World Bank noted as well that "the scarcity of forests is so critical in many parts of the island that farmers lack sufficient wood for their small domestic fuel needs and similar farm uses" (World Bank 1951, 903). An FAO mission conducted in 1957 came to similar conclusions (Reed 1992, 37).

By 1992, according to official estimates, 18.2 percent of Cuba was covered by forests, natural forests accounting for 84 percent of the total, or 2 million hectares. Two-thirds (67.6 percent) of national forests were set aside as protected areas, and one-third (32.4 percent) was used for timber production (COMARNA 1992, 30).[7] The expansion of the forests was attributed to the joint effects of managing annual cutting rates (after first reducing harvesting rates from 150,000 cubic meters to 45,000 cubic meters) and the reforestation program (Westoby 1989, 132). Between 1959 and 1992, the net annual addition in forested land area approached 14,000 hectares. These gains, particularly considering the experience of other Caribbean and Central American countries, are commendable, Cuba being the only country in the region that reversed secular deforestation trends. In at least one South American country, however, a successful reforestation program is in place. In Chile, where forestry is a vibrant export industry, more than

110,000 hectares are reforested each year, whereas about 40,000 hectares are felled annually (Prado 1996, 8).[8] In Chile, however, natural, previously undisturbed forests continue to be harvested at a high rate, whereas in Cuba most harvesting is done in reforested sites, the remaining natural forests (discussed later) being few. In eight Caribbean and Central American countries studied by the World Bank, the forest cover declined 10–24 percent in the period 1981–1990 (Current et al. 1995a, 152). However, in only two of these countries—El Salvador (6.2 percent) and Haiti (1.3 percent)—were forests less prevalent than in Cuba. By comparison, Costa Rica (28.8 percent), the Dominican Republic (22.5 percent), Guatemala (39.3 percent), Honduras (41.2 percent), Nicaragua (50.8 percent), and Panama (41.1 percent) had much higher forest covers.

Between 1960 and 1990, as a result of the socialist government's reforestation program, 2.5 billion trees were planted in Cuba. By the late 1980s, COMARNA (1992, 30) claimed that for every hectare of forest harvested, 16.9 hectares of trees were being planted (or 110,000 hectares replanted for every 6,500 hectares harvested annually) (Reed 1992, 38). If the Cuban estimates are accurate, they suggest that in the 1990s, annual reforestation rates approximated the annual deforestation rates of the 1910s and 1920s (see Table 6.1). Early 1990s estimates of land area forested provided in a report on Cuba conducted by the Comisión Económica para la América Latina y el Caribe (CEPAL, Economic Commission for Latin America and the Caribbean) suggest that the forest cover has continued to increase, reaching 24.5 percent (or 2.7 million hectares) by 1995 (CEPAL 1997, tab. A.41). This estimate is not credible, however. It appears to depend on the reclassification of marginal land formerly being cultivated but abandoned during the special period because of the absence of imported agricultural inputs. Much of the reputedly additional land "forested" is likely to be covered by *marabú* or to be the site of incipient and only marginally viable reforestation projects.

In summary, the estimates in Table 6.1 suggest that Cuba lost its forests incrementally between the late eighteenth and the first quarter of the twentieth centuries. In the 1940s annual deforestation rates reverted to levels observed in the first quarter of the nineteenth century, as the Cuban forests continued to dwindle. Despite repeated claims that the forest cover had declined to less than 10 percent by 1959, later officials figures indicate that tree loss was less extensive than had been adduced: in 1992 COMARNA stated that in 1959, 14 percent of Cuba was still forested. Over the next thirty-three years, as a result of the reforestation policies implemented by the Castro government, the land area forested increased by 30 percent, or to 18.2 percent of the national territory, a level comparable to that estimated for 1945. A further 5.5 percent of the national territory not forested was covered by trees (small tree stands in agricultural cooperatives, small private farms, etc.) (Comité Estatal 1989, 185). The corresponding percentage for tree-covered land not in forest for 1945 is unknown.

In the remainder of this chapter we review the national debate regarding the fate and uses of forests and discuss some of the problems encountered in the implementation of land use and forestry policies. We also examine several issues surrounding the revolutionary government's reforestation policies, including the challenges imposed by the special period on Cuba's forest resources.

Land Use and Forestry Policies

Debate about what to do about the forests in Cuba began in colonial times. Despite concerns about the fate of the forests, preservation laws and initiatives to promote better forest management practices were largely ignored. Moreno Fraginals relates how feudal practices and the timber needs of the Spanish navy led to the inclusion of statutes in the Royal Decrees (Cortes de El Rey) governing the Spanish American colonies to protect Cuba's forest resources (Moreno Fraginals 1964, 75; see also Grupo Cubano 1963, 169–71). These laws prohibited land owners from freely disposing of trees in order to preserve the supply of fine wood for shipbuilding. This concern was pitted against the growing influence of the planter class who felt the Royal Decrees interfered with their plans to expand sugarcane cultivation into virgin lands.

After much debate, the sugarcane planters carried the day. In 1805 they were granted unencumbered access to their land, subject to some regulations that, as Moreno Fraginals notes, were never enacted. Concern about rapid deforestation led the Colonial Development Board (Junta de Fomento) in 1844 to offer a monetary incentive to the landowner who could present in 1848 a 4-*caballería* (53.6 hectares) plot planted with a three-year old stand of cedar, *majagua (Hibiscus tiliaceous-elatus)*, pine, mahogany, and other trees (Grupo Cubano 1963, 170). Marrero (1950, 299) notes that at mid-century, groups such as La Sociedad Económica de Amigos del País bitterly complained about the adverse consequences of deforestation. The Spanish colonial authorities took several measures beginning in the 1860s to protect Cuba's forests, waters, and ports, including the establishment of forest reserves (Wotzkow 1998, 150–51).

Alarm regarding relentless deforestation led to the enactment of several decrees in 1923 (Decree-laws 179, 318, 753, 772, 970, 979, and 179). Decree-law No. 753 (24 May) prohibited the felling of trees without prior authorization, mandated that 15 percent of land in private farms be maintained as forest preserves, and required that trees be planted along the banks of rivers. Decree-law

No. 753 also made it illegal to disturb *montes* within 20 meters of a river, spring, or seacoast, and set restrictions regarding cutting trees on hillsides (Memorias 1978, 29).[9] Decree-law No. 772 (24 May) prohibited the felling of royal palms and fruit trees (Grupo Cubano 1963, 698). Decree-law 970 (4 July) mandated the reforestation of cleared land (Grupo Cubano 1963, 490–91), prohibited harvesting certain wood trees, and required that, for other types of trees, three be planted for each one cut (Memorias 1978, 29). Decree-law No. 495 of 13 April 1926 set additional prohibitions regarding the felling of trees in private parcels in mountainous regions and regulated how trees were to be harvested. In May 1926 all forestry decrees were codified into a single law. The 1931 census report optimistically noted that, with this legislation, "Cuba is currently placed in an advantageous position due to its forestry legislation that should, in a short time, result in the recovery of 100,000 *caballerías* (or 1.34 million hectares) of land unsuited for agricultural pursuits that could become the foundation for the country's great future forestry wealth" (Memorias 1978, 29–30).

Additional forest protection laws were adopted between 1930 and 1941, including Decree-laws 487 (1930), 681 (1936), and 509 (1937) (Wotzkow 1998, 150–51). A Forestry School and several reforestation stations were established in 1933 (Marrero 1950, 309). These initiatives were accompanied by feeble and largely ineffective efforts to distribute free seedlings and to encourage plantings of rapidly growing species such as teaks and eucalyptus. Forestry preservation laws enacted later (e.g., Decree-law No. 681 of 31 March 1936) also required the establishment of forest preserves of no less than 100 *caballerías* (1,340 hectares) in each province, expanding the National Experimental Tree Nursery, and improving the Forestry School. The legislation also required government authorization to fell any trees and the payment of timber taxes; the proceeds of the taxes were intended to finance the reforestation program. In 1947 and 1948 the government collected 290,000 and 258,000 pesos, respectively, from forest licenses, taxes, and fines (World Bank 1951, 905–906).

In the 1940s further restrictions were enacted, this time to limit log exports. A governmental directive was approved in 1946 prohibiting timber exports for five years. By the early 1950s, and with the support of the Banco de Fomento Agrícola y Industrial de Cuba (BANFAIC, Cuban Agricultural and Industrial Development Bank) several privately sponsored reforestation efforts got underway. Private interest in the reforestation program was to be expected given the dominance of privately owned parcels in the logging industry.

In 1945, 91.2 percent (by value) of the wood harvested in Cuba came from privately owned forests; 1.4 percent from government-owned forests; and the remainder, 7.4 percent, from unauthorized logging on public lands. At the time, the logging industry employed some 10,000 workers in 200 mostly small saw mills. The reforestation efforts were partly induced by alarm over growing domestic lumber shortages. Average annual production of sawn wood had declined from 43 million square feet in 1931–1946 to 22 million square feet in 1950–1955 (Marrero 1950, 303), despite major production increases in 1946 (86.5 million square feet), 1947 (63.4), and 1948 (65.9). The production decline was most evident regarding Cuba's hard woods, the share of soft pine increasing as a percentage of the total output of wood products.

Reforestation efforts were mainly conceived, however, to assure the country's future ability to export wood products (Grupo Cubano 1963, 1068; Marrero 1950, 309), although Cuba was, paradoxically, at the same time a major importer of these products. In value terms, in the 1930s and 1940s, Cuba imported twice as much wood products as it exported (World Bank 1951, 904; Núñez Jiménez 1972, 351). This was not a new development. Even when most of Cuba's forests were undisturbed, considerable amounts of lumber were imported. During the mid-nineteenth century, for example, Cuba was a primary trading partner of the U.S. state of Maine, which supplied the bulk of the wood used to make crates for exporting sugar (Demeritt 1991). Although forests were abundant in Cuba, importing lumber made economic sense because of the costs of timber and transportation. Because of excessive localized deforestation, wood sources were far from the sugar mills; other factors, such as the lighter weight of sugar crates made with Maine's abundant white pine and the supposed ability of this wood not to affect the taste of sugar, made these lumber imports particularly attractive. Wood for tobacco boxes continued to be imported well into the twentieth century (Foreign Policy Association 1935, 463).

In retrospect most legislative initiatives to cope with the deforestation problem were blatantly ignored. The promising efforts initiated with BANFAIC support during the early 1950s did not have time to blossom before the onset of the profound socioeconomic and property changes that accompanied the 1959 Revolution. What is apparent, however, is that in the 1950s, Cuba, like many other countries, was beginning to implement policies consistent with the post–World War II global impetus for economic development then just getting underway.

Pre-revolutionary Cuba's experience regarding the inadequacy of regulations to protect forests—more than forty legal instruments could not prevent the destruction of forests (Atienza Ambou et al. 1992, 5)—was not unique in the region. Many countries have paid, and continue to pay, lip service to the protection of the forests. In Haiti, for instance, according to a World Bank study, although "over one hundred well-versed and well-intended laws and policies aimed at protecting natural forests and the environment have been enacted, few, if any, have been effectively implemented" (Current et al. 1995b, 176).

Forestry Policies of Cuba's Socialist Government: An Appraisal

To our knowledge, no systematic attempt has been made to evaluate the impact of the reforestation policies in effect in Cuba since 1959. We rely, for our purposes, on a broad definition of forestry that includes initiatives to preserve natural forests and restore formerly forested areas, inclusive of national parks, nature preserves, and tree plantations (for commercial purposes). Our analytical paradigm also includes agroforestry practices and selected issues related to the planting of perennial fruit trees.

Impetus for the new government's forestry program was provided by the Reforestation Law (Ley de Repoblación Forestal) of 10 April 1959. Article 10 created nine new national parks (in addition to the Sierra Cristal National Park), namely Cuchillas de Toa, Gran Piedra, Sierra Maestra, Escambray, Laguna del Tesoro, Los Organos, Guanacabibes, Ciénaga de Lanier, and Sierra de Cubitas (Núñez Jiménez 1972, 356–57; see also Ministerio de Ciencia 1995, 36–40). The law specified that the national parks individually should not have an extension of less than 500 *caballerías* (6,700 hectares). With domestic tourism in mind, the parks were to be made accessible to the public and be provided with utilities and hotel accommodations.

The location of these national parks was in part dictated by the distribution of remaining forests, which in itself was a function of state land-holding patterns prior to 1959. In pre-revolutionary days the Cuban state claimed ownership of 37,000 *caballerías* (or 495,800 hectares) of forests, 42 percent of which were located in the easternmost province of Oriente. In 1930, in Sierra Cristal National Park, in the Baracoa region, a 2,000 *caballería* (or 26,800 hectares) forest preserve where hunting and logging was prohibited had been established to protect the local flora and fauna (Marrero 1950, 309).

Municipal forestry parks were also called for by the 1959 Reforestation Law. These were to be created with support from the national forestry authorities. Antonio Núñez Jiménez, an influential voice in the conceptualization of the revolutionary government's forestry and ecological initiatives, revived the notion of replanting trees along the banks of the country's major rivers (to a depth of 100 meters), the focus of legislation enacted in 1923. Núñez Jiménez went further and proposed the development of a Gran Barrera Forestal (Great Forestry Barrier)—presumably running all along the spine of the island—to retain humidity, preserve water resources, address the problem of erosion, and moderate the country's climate. Grandiose as it was, his proposal was sensitive to geographic and economic regional variations and was premised on the use of approaches consistent with local circumstances (Núñez Jiménez 1972, 358–59). The barrier was to consist of rapidly growing tree species (e.g., eucalyptus, teaks), combined with other species, including fruit trees. By 1991, according to Maydel Santana (1991, 13):

> The Cuban National System of Protected Areas ... [had] over 200 protected areas that cover 12% of the country. However, only 1–2% of the country is strictly protected and some reserves appear to be too small to effectively preserve the biota they contain. It is estimated that Cuba has about two million hectares of forests, of which 1.7 million hectares (85%) consist of natural forests and the remainder of forestry plantations. Of these forested lands, national parks cover 5.1%, wildlife conservation areas 24.3%, water-shed protection areas 17.0%, coastal protection areas 19.2%, production forests 32.7%, and other categories 2.7%.

Reforestation

The post-1959 reforestation efforts were initiated in 1960 under the direction of the Instituto Nacional de Desarrollo y Aprovechamiento Forestal (INDAF, National Institute for Forestry Development and Use). Between 1960 and 1966, as part of the Plan de Repoblación Forestal (Reforestation Plan), 299 tree nurseries were established and 348 million trees planted (Núñez Jiménez 1972, 355–57). About one-third of the trees planted were eucalyptus (122 million), followed by pines (68 million), casuarinas (48 million), and other species (109 million). Seedlings produced in fruit tree nurseries established in each of the country's (then) six provinces were used to plant 1.6 million fruit trees between 1964 and 1965. By the late 1970s, the number of trees planted each year was roughly twice the number planted in 1960. Annual tree planting totals increased rapidly during the 1980s, exceeding over 100 million annually after 1982. Three times as many trees were planted in 1989 as in 1979.

The species distribution of trees planted changed appreciably over the years. Whereas in 1960, 92 percent of all trees planted were eucalyptus (35 percent between 1960 and 1966), by 1989 eucalyptus accounted for only 3 percent of all seedlings. As Cuba's cadre of professional foresters was trained and experience gained about Cuba's natural conditions, more informed decisions appear to have been made about which species of trees to plant and where. By the early 1990s,

Cuba had more than 1,000 forestry engineers and biologists, as well as close to 2,000 forestry technicians (Atienza Ambou et al. 1992, 6). The widespread introduction of foreign tree species, however, has been criticized by Cuba's naturalists on ecological grounds, indicating that a "lack of affection towards indigenous animals has underlaid the many decisions for introducing exotic species. Disseminated throughout the archipelago are whole forests of Australian pine and eucalyptus" (Silva Lee 1996, 140).

The early emphasis on planting eucalyptus trees was probably guided by ignorance and the tendency of the socialist government to make rushed and wasteful economic decisions, a characteristic found not only in forestry. In support of this interpretation Westoby notes that "much of the early [reforestation] effort was wasted as the result of elementary errors: seed of poor provenance, species ill-adapted to sites, over-emphasis on the number of trees planted, and neglect of subsequent tending" (Westoby 1989, 132). Comparable problems have continued to plague reforestation programs into the 1990s, as reported by an environmental overview published in La Habana in 1997 (Borges Hernández and Díaz Morejón 1997, 15–16).

The focus on eucalyptus at the beginning of the reforestation program may also have been dictated by the belief that reforestation with this rapidly growing species would help arrest soil erosion, a widespread problem that was assigned the highest priority (Núñez Jiménez 1968). As an FAO report has noted (Poore and Fries 1985, 21), "most eucalyptus are not good trees for erosion control. When young, they are very susceptible to grass competition, and to obtain good growth, clean weeding is necessary during the establishment period, which is undesirable on steep or eroding terrain. Even mature stands may be ineffective in halting surface run-off." Moreover, the heavy water consumption of eucalyptus has been found to contribute to wetlands deterioration in several countries that are now trying to eradicate them. The problem with the selection of eucalyptus as the main reforestation vehicle was not an isolated phenomenon. There is some evidence of poor species selection in other reforestation, reclamation, and agricultural projects. Plantings of casuarina trees, not naturally found on beaches, in the 1960s for shade and aesthetic reasons in Varadero, Santa María del Mar, and Guanabo (some of Cuba's most renowned beaches) were responsible for extensive erosion because their shallow and dense roots interfered with normal sand shifts. In the 1970s the government decided to uproot the casuarina trees planted a decade earlier on these beaches (Pagés 1981, 4). Rene Dumont (1970a, 135, 137) reported poor planting practices and high tree mortality associated with the large-scale development of citrus plantations, as well as with the planting of hardwood trees in coffee areas. Another reported instance of poor reforestation practices is provided by Richard Levins. In the mid-1970s, he notes, "the Institute of Botany refused to work with the Forestry Institute on its plan for terracing mountainsides in Pinar del Río, planting monocultures of teak or hibiscus and clear-cutting of trees. They saw the plan as too vulnerable to pest problems and provoking massive erosion" (Levins 1993b, 57).

The most recent emphasis on the planting of conifers—the leading refor-estation species in the 1980s—may also pose more environmental concerns. In southern Florida, for example, an environment comparable to many parts of Cuba, it has been decided to remove Australian pines because of their inability to withstand high winds and the hazard they pose during hurricanes by blocking waterways and increasing flooding potential. Concern continues to be expressed to this day about Cuba's marked tendency to rely on a single tree species, not always well-adapted to local conditions, for many of its reforestation efforts (Wotzkow 1998, 35).

Effects of the Reforestation Program: A Quantitative Assessment

A gross estimate of the effects of Cuba's reforestation program can be made by relating the total number of trees planted between 1959 and 1989 to the increase in number of forested hectares, net of forested area harvested. Our esti-mating procedure is summarized by the following equations:

$$FA(t + i) = FA(t) + RA(t + i)$$
$$RA(t + i) = TP/D - HA,$$

where

$FA(t + i)$ = forested area in 1989 (in hectares)
$FA(t)$ = forested area in 1959 (in hectares)
$RA(t + i)$ = area reforested between 1959 and 1989 (in hectares)
TP = total number of trees planted between 1959 and 1989
D = density of tree plantings (in trees per hectare)
HA = number of hectares harvested between 1979 and 1989.

The net gain in forested area is the difference between the land area forested in 1959 and the area in state forest enterprises in 1989 (inclusive of virgin and reclaimed forests), minus an estimate of the forested area harvested between 1979 and 1989.

In 1959, according to COMARNA, 1,540,000 hectares of Cuba's land were forested. There are two estimates of forested land area in 1989. The *Anuario Estadístico* (Comité Estatal 1989, 216) reports that forests in state forestry enter-prises accounted for 2,273,300 hectares in 1989. COMARNA (1992), however, provides data suggesting that the increase in forests was considerably smaller, slightly over 2 million hectares. The differences between the 1959 figure and the two 1989 estimates entail a gain of 730,000 and 500,000 hectares of forest cover, respectively, over the intervening period. For estimating purposes, we assume that 65,000 hectares of forests were harvested between 1979 and 1989.[10]

By assuming two different forestry plantation densities (1,500 and 2,000 trees per hectare),[11] relating these densities to the 2.5 billion trees COMARNA claims have been planted in Cuba, and accounting for the forest area harvested between 1979 and 1989, the success of the reforestation program can be calculated as ranging between 27 and 53 percent. In other words, 27–53 percent of trees planted survived to maturity, tree mortality rates being in the order of 47–73 percent. Both estimates are considerably higher that the only official tree mortality estimate of 40 percent we have been able to identify in the Cuban literature (Gómez 1979, 5). Mortality rates of this magnitude are not out of line with international experience, especially in situations where seedling care and early weeding practices are not given the required attention.[12] For a reforestation project to succeed, effective weeding is essential. During the first two to three years of a project, as many as three to four weedings a year are necessary (Committee on Selected 1982, 143). That proper seedling care procedures were not followed in Cuba, at least during the initial years of the reforestation program, is consistent with Westoby's observations (cited earlier) and with what is generally known about the poor follow-up associated with many of Cuba's agricultural practices. These estimates suggest that Cuba's forestry accomplishments have been achieved at a very high economic cost.

Socialist Agrarian Development Policies Leading to Tree Losses

A balanced assessment of the accomplishments of the reforestation program must evaluate the adverse impact that the socialist rural development model implemented since the early 1960s had on the 1959 tree stock. This model was largely premised on the establishment of large-scale state farms and agricultural cooperatives. There are ample reasons to believe (see Dumont 1970a,b) that large-scale Soviet-style farming led to the destruction of countless trees. Rene Dumont, a well-known French rural development expert, described what happened as brutal, "because not even a tree that could provide shade or serve other purposes was left" (as cited by Nelson 1972, 94).

Land clearing on a vast scale was underway in the 1960s as the government embraced an agricultural development approach based on several pillars, including widespread mechanization, collectivization, and extensive land use. To prepare vast tracts of land for the use of mechanized equipment, small stands of mature and productive trees were obliterated. Trees were uprooted to permit heavy tractors and combines to operate unobstructed.[13] Also contributing to the destruction of localized tree stands was the consolidation of small farms into agricultural cooperatives. In pre-revolutionary days Cuban peasants relied on traditional agroforestry practices (Current, Lutz, and Scherr 1995a, 153–54), planting trees for fruit, shade, fencing, wood, cooking fuel, and other purposes for domestic consumption and as cash crops. Fewer and fewer tree groves were left standing as collectivization policies led to the consolidation of small farms into state farms and cooperatives, peasants were relocated to planned urbanized communities, and large-scale mechanization was introduced. Emilio Farrés Armenteros, chief of the Fruits Division of the Ministry of Agriculture, stated in June 1997 that "during the 1960s, excessive mechanization and the construction of hydraulic complexes resulted in the elimination of 'obstacles,' that is, fruit groves that hindered the use of machinery" (Rivero 1997).

The land clearing tasks were assigned to mechanized military brigades, the notorious "Che Guevara columns." The modus operandi of these brigades was to drag a chain between two tractors or army tanks, pulling along the way any vegetation they encountered, be they brushes, *marabú*, or small and large trees. Thirty-six such units were operating throughout the country in December 1969, each equipped with twenty pieces of heavy equipment and manned by 117 men (Nelson 1972, 94–95). One astonishing result of this policy was to nearly obliterate royal palm trees, Cuba's national symbol, from much of the countryside, leading some observers to note that in the 1990s "most state farms are devoid of palm trees" (Deere et al. 1994, 225). These same observers note that lack of shade trees is interfering with the implementation of the Voisin cattle grazing system (215). This system is based on feeding a herd primarily in pastures that are fertilized with the cattle's own manure. Pasture productivity is maintained by periodically rotating the herd from one grazing area to the next through the use of shifting movable electrified fences. Supplementary feeding, transportable water sources, and a relative abundance of strategically distributed shade trees are essential components of the Voisin grazing system. Many shade trees that formerly grew in grazing areas fell victim to the Che Guevara columns in the 1960s.

Compounding the detrimental effects of the agricultural strategy was the decision to bring under cultivation marginal agricultural land. To bring these lands into production, they had to be cleared of trees and shrubs (Dumont 1970b, 39). Particularly damaging to trees and forest cover were agricultural policies of the mid- to late 1960s, when most agricultural land resources were assigned to the failed goal of producing 10 million tons of sugar in the 1970 harvest. Expanding the area under cultivation, however, was a constant of national agricultural policy during nearly three decades. In the absence of foreign subsidies (after 1990), when critical agricultural inputs were no longer available, many of the reclaimed lands were abandoned. This process has been documented with regard to citrus plantations (Spreen et al. 1996, 20) where the number of hectares planted declined by 17,000 hectares between 1990 and 1993, or by 12 percent, primarily through a contraction in the number of hectares replanted with citrus and an increase in the number of hectares from which citrus trees were removed.

To partially compensate for the destruction of traditional fruit tree groves because of the expansion of sugar production, and as part of the country's development strategy, the government embarked on a program to develop fruit tree, coffee, and cacao plantations. Resources were assigned to develop mango and guava tree plantations, as well as to expand coffee and cacao plantations in mountain areas, often as part of agroforestry projects. In some cases ill-advised attempts were made to develop coffee plantations in regions poorly suited for this permanent tree crop. Best known was the directive during the 1970s to develop a "green belt" of coffee plantations and other crops in the lowlands surrounding the city of La Habana to make the city self-sufficient in food. This project was eventually abandoned, the coffee trees never having borne fruit. The waste of economic resources on this project alone was very large.

The success of planting efforts of fruit-bearing trees can be partially judged by analyzing statistics on the number of hectares planted with four fruit trees (mango, guava, coffee, and cacao) for 1970 and 1975, and for 1978 to 1989. Increases in the number of hectares planted with mangoes and guavas during the late 1970s and 1980s can be attributed to efforts to reverse the damage done to fruit tree stands during the 1960s. An important characteristic, likely to differ from pre-revolutionary patterns, is that most mango and guava trees were planted in large stands, as opposed to small-scale groves, as suggested by differential growth patterns between the state and private sectors. This pattern is consistent with the socialist model of organizing agricultural production on a large scale, in part to facilitate the use of chemical inputs and mechanization. Particularly telling are the major fluctuations observed from year to year in the number of hectares planted in the state sector. Fluctuations of this magnitude suggest poor planting practices or shifting agricultural priorities. In any event, they are indicative of a vast waste of resources and may also explain (together with exports) the persistent fruit shortages reported in Cuba. The environmental underpinnings of these trends, if any, as well as their consequences, remain to be analyzed.

The trend for coffee is consistent with the gradual depopulation of Cuba's mountain regions (up to the early 1990s) and with the failed attempts to expand coffee production into poorly suited areas close to major urban centers. The area planted with coffee peaked in 1979, gradually diminishing through 1989; the decline amounted to 25 percent over this period. The decline is especially noticeable in the private farming sector, although in 1989 over half of coffee production remained in private hands. Rural flight to urbanized localities is likely to be implicated in the decline of coffee plantations. Former coffee areas were the target of reforestation, given that coffee has traditionally been planted in association with other trees that provide shade and protection to coffee plants and also contribute to soil improvements. The declining trend is certain to have been reversed in the 1990s during the special period with the push to increase agricultural production in the mountains as part of the food self-sufficiency program. Cacao plantings remained relatively constant over the 1970–1989 period, except that the share of these plantations in the state sector increased. Cacao plants often are also found in close association with larger shade trees.

Also harshly criticized by Dumont were projects begun during the 1960s to restructure pastures in livestock farms according to inflexible geometrical designs. To achieve these designs, it was necessary to remove extensive living tree fences (live fenceposts) and, in many instances, trees planted along river banks (Dumont 1970a, 123–27). Some of these trees could well have been planted decades earlier as part of erosion control projects. The number of trees that were destroyed is unknown, but it was probably in the millions.

Special concern has been voiced about the difficulties of reclaiming the open-pit nickel mining areas found in eastern Cuba. Open-pit mining completely removes the topsoil, thus leaving behind a "lunar landscape" denuded of vegetation (Reed 1993, 32). By the late 1980s, after four decades of mining operations, 200 square kilometers of land (or 20,000 hectares) in the Moa-Nicaro area were totally degraded (Bucek 1986, 15). Gail Reed (1993, 38), using a much lower estimate of amount of land degraded by open-pit mining, claims that by the early 1990s some 3,000 of 11,000 degraded hectares had been reclaimed. Similar concerns have been expressed regarding the mining of peat deposits in the Lanier Swamp area in the Isla de la Juventud. To lay the peat to dry under the sun, whole forest patches have been denuded by bulldozers. Due to the prevailing soil characteristics in this area, reclamation efforts will be virtually impossible (Gort et al. 1994, 52).

Although the evidence is only partial, there is reason to conclude that development projects carried out in several of Cuba's mangrove regions also had a detrimental impact on the natural vegetation, trees included. Perhaps the more damaging were projects to increase the agricultural and tourism potential of the Zapata Swamp in south-central Cuba (Wotzkow 1998). The FAO (1994, 84) has called attention to the adverse ecological consequences of road construction and other human activities in mangrove areas of the country, an observation also made by A. Borhidi (1991, 453) in connection with the Zapata Swamp in his exhaustive study of Cuba's vegetation ecology. Included here are environmentally damaging agricultural activities—such as inappropriate drainage, pollution, grazing, and use of marginal areas—as well as activities related to charcoal production and peat extraction. The unfavorable effects of these practices are corroborated by data reviewed later from a study on the conservation status of the major ecological regions of Latin America and the Caribbean, Cuba included (Dinerstein et al. 1995).

Ecological Conservation Status: Findings of a Regional Assessment

A 1995 World Bank/World Wildlife Fund conservation assessment of the major ecoregions of Latin America and the Caribbean provides an overview of the conservation status of some of Cuba's natural ecoregions, identifying their degree of vulnerability and protection needs (Dinerstein et al. 1995). The study is particularly revealing since it provides an independent evaluation of the status of Cuba's natural regions within a regional comparative framework. An ecoregion is defined as a "geographically distinct assemblage of natural communities that (a) share a large majority of their species and ecological dynamics; (b) share similar environmental conditions; and (c) interact ecologically in ways that are critical for their long-term persistence" (124).

According to these criteria, of the five ecoregions into which Cuba is divided (see Table 6.2), two—the tropical broadleaf dry forests (formerly extending across much of Cuba) and the flooded wetlands (the Zapata Swamp)—are considered to be endangered, and three—the moist broadleaf forests found in the country's highest altitudes, the coniferous pine forests (Pinar del Río and Isla de la Juventud in the west), and the cactus scrub (in the southeast)—are considered to be vulnerable. The endangered and vulnerable classifications are intermediate between the classifications of extinct (or completely converted from natural habitat) and critical, on the one hand, and relatively stable or relatively intact, on the other (Dinerstein et al. 1985, xvi).

The study's taxonomy is based on five indicators of landscape integrity related to the maintenance of ecological processes and biological diversity (total loss of original habitat, number and size of blocks of original habitat, rate of habitat conversion, degree of fragmentation or degradation, and degree of protection). Most forests protected at the present time are the broadleaf moist forests (probably because they are the ecoregions of more difficult access) and least are the broadleaf dry forests, pine forests, and cactus scrub. But even in the most protected areas, human activities, including mining and logging opera-

Table 6.2. Conservation assessments of Cuba's ecoregions

Tropical moist broadleaf forests

Moist forests: Vulnerable: 20,069 square kilometers.

The moist forests of Cuba, and those of the Greater Antilles generally, maintain exceptionally diverse insular biotas with many regional and island endemic species in a wide range of taxa. Cuba, in particular, has a rich moist forest flora. The Greater Antilles are notable for numerous unusual relict species and higher taxa. Expansion of cacao, coffee, and tobacco production are serious threats in some areas.

Tropical dry broadleaf forests

Dry forests: Endangered: 61,466 square kilometers.

Clearcutting and selective logging, charcoal production, frequent burning, and slash-and-burn agriculture pose threats to the ecoregion.

Tropical and subtropical coniferous forests

Pine forests: Vulnerable: 6,017 square kilometers.

The pine forests of Cuba and Hispaniola support a number of endemic plant and animal species. Mining, citrus plantations, grazing, and logging severely threaten the ecoregion. Exploitation of threatened parrot population occurs in western portions of the ecoregion .

Flooded grasslands

Wetlands: Endangered: 5,345 square kilometers.

The Zapata Swamp on the southern coast of Cuba is noted for its large size and endemic species. Draining and agricultural expansion, agricultural pollution, charcoal production, grazing, peat extraction, and exotic invasions all pose severe threats to the ecoregion.

Deserts and xeric shrublands

Cactus scrub: Vulnerable: 3,044 square kilometers.

Grazing, woodcutting, and the conversion and resource exploitation associated with increased urbanization pose threats to the ecoregion for the foreseeable future.

Source: Summarized with minor modifications from Dinerstein et al. 1995, 86, 93, 96, 100, 103.

tions, as well as the presence of permanent plantations (e.g., coffee, citrus), are having adverse ecological effects. In relative conservation terms Cuba is not doing significantly better or worse than neighboring nations. Nevertheless, the Cuban forests continue to be threatened by human activities—in particular, excessive logging, expansion of agroforestry activities, and, surprisingly, slash-and-burn agriculture.

Not mentioned in the study is that these threats have intensified in recent years as a result of the food self-sufficiency policies being pursued by the Cuban government during the special period, most of all by the drive to increase food production in mountainous areas and more intensively harvest forests to produce domestic fuel and charcoal. Particularly notable examples are the widely publicized Turquino and Manatí plans, which seek to relocate farmers to mountainous regions by giving them private plots of land to produce several commodities (e.g., coffee and honey) and, in some cases, engage in forestry activities (e.g., reforestation of denuded areas and reclamation of areas affected by mining activities) (Gersper et al. 1993, 22). These plans were conceived in part to address low agricultural production levels in these areas, their success being predicated on expanding the labor supply, introducing seldom-used technologies, developing new types of farming arrangements, and granting farmers access to individual plots of land (Cuban Commission 1996, 11–12).

Increased human interventions in these areas, some with only limited agricultural potential, may be undermining preservation and reforestation efforts, despite the avowed preservation intent of the mountain development initiatives (Ministerio de Ciencia 1995, 38–41). Preservation and development objectives are clearly contradictory. In the Escambray Mountains of central Cuba, for example, targets have not been met due to poor road conditions, water quality, and other factors (Cuban Official 1996, 4–5). The nature of the threats varies depending on the ecoregions in question. They are less severe in the more inaccessible mountain regions, but even here they are so in selected areas (see Table 6.2). Moreover, many of the reforestation efforts carried out on the island have been conceived solely for their economic potential while ignoring the ecological role of forests in helping preserve the natural plant and animal life (Gort et al. 1994, 52).

The findings of the World Bank/World Wildlife Fund study clash with the nearly idyllic description of conservation and reclamation policies portrayed in official Cuban government documents. Official interpretations are also at variance with the perspective provided by an FAO document with respect to the National Forestry Action Plan elaborated by Cuba in the early 1990s. The findings of this document are reviewed in the next section.

Current Forestry Priorities

Cuba has requested international assistance to broaden the national forestry effort (Reed 1993, 38) and, specifically, to implement the National Forestry Action Plan (NFAP). The NFAP was developed on the heels of the 1992 International Conference on the Environment. The objectives of the NFAP (FAO 1993, 197) are:

▶ restitution of the forest cover and reconstruction of degraded natural forests with a view toward protection and production;

▶ sustainable management of forest resources for the production of wood products and the protection of fragile watersheds and ecosystems;

▶ increasing production and diversifying it, including developing integrated forest industries;

▶ intensive use of forest biomass to produce charcoal and fuelwood;

▶ recuperation of degraded ecosystems;

▶ application of management techniques to protected and special areas for the benefit of local populations and to protect biodiversity; and

▶ strengthening of research and training institutions.

Driving the NFAP is what is judged to be the limited potential of the national forestry industry because of a dearth of raw materials, obsolete technology, and inadequate infrastructure, all of which contribute to Cuba being forced to import 55 percent of its wood products needs (FAO 1993, 196). Implementing the NFAP, which consists of eighteen programs, would require a domestic investment of 61.9 million pesos (US$81.4 million at the then-official exchange rate), and external assistance to the tune of US$34.7 million. The FAO concluded, however, that "lack of response from potential donors has not made possible to convene an international Round Table . . . to discuss the implementation of the NFAP" (FAO 1993, 197).

The NFAP may in part be seen as a response to the economic crisis engulfing Cuba since 1990. Cuba used to import much of its lumber from the Soviet Union. During the 1980s, the Soviet Union had a shortage of lumberjacks and therefore lumber shipments to Cuba were irregular. To address this problem, in December 1986 Cuba and the Soviet Union created a joint enterprise to exploit Soviet Far Eastern timber resources for export to different countries, including Cuba. As many as 400 Cuban lumberjacks were assigned to these activities (Pérez-López and Díaz-Briquets 1990, 287–88). The outcome of this venture has not been documented, but we assume that it concluded with the collapse of the Soviet Union, if not before. Greater reliance on, and better management of, national forests resources could mitigate supply shortfalls and abate the overexploitation of forest resources during the special period.

There is mounting evidence that forestry conservation efforts have experienced a setback as lumbering rates in the 1990s increased in response to the end of wood and fuel shipments from the former Soviet Union. Shortages of home cooking fuels have been partly compensated by increasing supplies of domestically produced charcoal and fuelwood. Government officials have "sharply criticized the indiscriminate chopping down of trees and theft and misappropriation of the wood," as the unavailability of fuels and other supplies has led to unregulated use of forest products (Official Urges Measures 1994, 3).

Attempts to increasingly rely on forest products to attain self-sufficiency extends to the planting of "power forests," or stands of rapidly growing trees to produce charcoal. A 1994 report indicates that this program has run into problems: the locations in which planting goals have been met are too far from where charcoal supplies are needed (Roundup of Economic Developments 1994, 9). There is also concern about intensified logging pressures on regions that have traditionally served as charcoal sources (National Public Radio 1995). Further, tourism construction projects are likely to have sustained (or even increased) overall demand for domestic lumber, despite a cutback in housing construction, given the end of wood imports from the former Soviet Union.

As a result of these developments and the apparent lack of reforestation inputs, efforts to further increase forested areas are faltering (Effects of U.S. Blockade 1995, 8), not only due to higher harvesting rates, but also because reforestation rates have declined. In 1992, for example, only 500,000 seedlings were planted, although the target was to plant 4 million trees (Roundup of Economic Activity 1992, 3). Fires have also decimated some of Cuba's forests in recent years. In February 1997, 130 hectares of a pine plantation in the Sierra Cristal Mountains in eastern Cuba suffered extensive fire damage (Alfonso 1997a).

Summary and Implications

The forestry sector is one of the few in which the policies instituted under socialism in Cuba can be judged to have met with considerable environmental success, as the country's total land area forested increased significantly. These achievements did not come about cheaply, however, since much experimentation and poorly implemented reforestation programs wasted numerous resources. Efficiency in the reforestation effort does not appear to have been a major priority given Cuba's highly subsidized economy. There are reasons to believe, however, that while forest cover increased, the country also lost many trees as a result of some of the major initiatives undertaken under the socialist government agricultural and water development policies (described in Chapters 4 and 5), and more recently, under the drive to promote the tourism industry (see Chapter 10).

During Cuba's current economic crisis, pressures on Cuba's natural resource base, including its forests, have increased as a consequence of special period policies. As we write, it is apparent that some of the gains of the last four decades are being reversed under pressure from economic events. Some of these policies emanate from the national drive to attain food self-sufficiency and others from the economic difficulties associated with the curtailment of the ability to import. The policy to encourage resettlement in coffee-growing mountainous regions, for example, adds environmental pressures to heretofore relatively undisturbed areas. Even primitive subsistence agricultural activities will continue to damage the forests as peasant families push further into the country's forested areas. The same is true regarding charcoal production, a pursuit that has received renewed emphasis given the national shortage of commercial home cooking fuels. The housing shortage and the boom in the tourism industry are also of major concern: in the absence of imports, domestic lumber must be used for construction projects. The dearth of inputs, finally, must also be having a deleterious impact on reforestation efforts as even the import of the most basic reforestation inputs has been curtailed.

✠ 7 ✠

Industrial Pollution

CONVENTIONAL WISDOM holds that environmental deterioration is an inevitable by-product of economic development. However, the experience of the industrial world in recent years suggests that "economic growth can be reconciled with environmental management" (Pearce and Wardford 1993, 15).

Industrialization and urbanization tend to put additional pressure on natural resources and generate environmental disruptions that every economic system must be able to address. Environmental problems in the former Soviet Union were somewhat different in character from those of market economies. In many respects the rapid industrialization embodied in the Stalinist development model was responsible for the severe environmental situation in that country; as one analyst observes, "in the strive for rapid industrialization expansion, the leadership pursued an extractionist policy in the use of natural resources. There was little room for environmental concerns as over-ambitious plans and the focus of the annual plan implied a short-sighted policy. As this policy resulted in the creation of massive industrial conglomerates, based on the exploitation of local natural resources, several types of environmentally harmful activities were concentrated in some regions" (Satre Ahlander 1994, 151).

The imposition of the Stalinist development model resulted in even greater stress on the environment of the socialist countries of Eastern Europe. These countries were less richly endowed with mineral resources, empty space, water, and other natural resources than the Soviet Union. The "big is beautiful" men-

164

tality of the Soviet Union during the industrialization in the 1930s, which focused on heavy industry, brought about the wasteful use of natural resources, an emphasis on quantity rather than quality, and disregard for the "external" effects of production decisions. Urbanization was rapid and poorly planned. A cumbersome and centralized economic planning system encouraged waste of natural resources and highly inefficient use of energy in the productive processes, resulting in increased emission of pollutants (DeBardeleben 1991, 3).

A recent report on environment and health in Central and Eastern Europe prepared for the Environmental Action Programme for Central and Eastern Europe (Hertzman 1995, 16–17) identified the following pollutants—largely associated with the industrial sector—as having the greatest impact on human health in the region: (1) lead in air and soil from emissions from lead and zinc smelters and, in certain cities, emissions from transport due to the use of leaded fuels; (2) airborne dust from coal burning in household furnaces, small-scale enterprises, power and heat plants, and metallurgical plants; and (3) sulfur dioxide and other gases from power and industrial plants, and households burning high-sulfur fuel oil. Secondary pollutants included nitrates in drinking water from inadequately maintained/designed or improperly located rural septic tanks, feedlots and agricultural enterprises, and incorrect fertilizer application; contaminants in food from the inappropriate handling or disposal of lead dust, heavy metals, pesticides, and toxic chemicals; and other contaminants in water from the inappropriate handling or disposal of water contaminated with arsenic, viruses and bacteria, pesticides, and so on.

Cuba's adoption of the Stalinist economic development model resulted in environmental stresses on the island similar in many ways to those experienced by Eastern Europe. The industrialization strategy was based on an unbalanced growth model, which gave priority to investments in heavy industries (e.g., metalworking, chemicals, construction materials, energy) that would provide linkages to the rest of the economy and support the expansion and modernization of the sugar industry and other export industries (Rodríguez Mesa 1980, 174–75).

Cuban planners, like their counterparts in other centrally planned economies, had a strong preference for large industrial complexes; location decisions tended to respond to regional development priorities rather than to economic or environmental soundness criteria. Despite these misguided policies, environmental disruption has not been as severe in Cuba as in Eastern Europe and the former Soviet Union because of several mitigating factors. First, as is discussed in Chapter 2, internal migration to the capital city was controlled, and an explicit policy of distributing industries throughout the island encouraged the growth of secondary cities. Second, Cuba's geography and climate flushed pollution away from the island. Not contiguous to other countries, Cuba did not receive pollution from neighbors. Short rivers flushed pollutants into the ocean. Relatively flat terrain and air currents took airborne pollutants away from the national territory and prevented pockets of stationary polluted air common in Eastern Europe. Third, despite industrialization efforts, Cuba remained primarily an agricultural country. Since the 1980s, tourism has also become an important industry. Thus, Cuba has been spared from the worst environmental problems associated with heavy industrialization faced by Eastern Europe and the former Soviet Union—such as air pollution from large industrial complexes—except for some specific cases described in this chapter. As discussed elsewhere in this book, agricultural and tourism development have not been benign from an environmental point of view, but the nature of the stresses they placed on the environment and the natural resource base are different from those associated with heavy industrialization.

COMARNA Vice President Ferrer has described Cuba's environmental problems as follows: "We're privileged in a certain sense—because we don't have large petrochemical industries, so we don't have the atmospheric and water pollution problems of other countries. And the fact that we are a long, thin island means that whatever pollution we have only lasts two or three hours, not two or three days. At the same time we are conscious of our responsibility to decrease the pollution we do have—for example, some of our thermoelectric plants, whose technology is very old. These plants produce sulfur, nitrogen, and carbon oxides which seriously concern us, and which we are attempting to bring under control" (Reed 1993, 31).

In this chapter we review the impact of Cuba's industrial sector on the environment and the conservation of natural resources. The review focuses on industries that are deemed to be the most serious polluters because of their size or the intensity of their emissions. Industries that fall in this category include sugar and derivatives; nickel mining and processing; cement; oil production, transportation, and refining; and other industries. The information underlying this review originates from many different sources with varying degrees of reliability. It may also refer to specific events and problems that existed in the past and may have been addressed by some form of government action. For these reasons, caution should be exercised in drawing conclusions about the current environmental situation from the review.

The Cuban Industrial Sector

In 1988, the latest year for which data are available, Cuba's industrial sector consisted of 836 state enterprises, 700 of which were controlled directly by national authorities (state committees, ministries, national institutes) and the rest by municipal or provincial authorities. The latter enterprises tended to be primarily engaged in foodstuffs production, printing, and woodworking (Figueras 1994, 103–104). Consistent with the Cuban government's policy of minimizing rural–urban migration, new industrial enterprises were located throughout the national territory and particularly in the relatively less developed and more populated eastern half of the island. In 1988 the eastern provinces accounted for 30 percent of the output of the nonsugar industrial sector, compared to 16 percent in 1962, whereas the corresponding shares for provinces surrounding Havana were 46 percent and 70 percent, respectively (Figueras 1994, 105).

According to data from the official statistical yearbook, in 1986 Cuba had 827 enterprises in 22 industrial sectors. The bulk of the enterprises corresponded to the production of nonelectrical machinery (150), sugar (148), and foodstuffs (145). Other sectors well represented in terms of the number of enterprises were production of construction materials (46), beverages and tobacco (43), chemicals (42), and metal products (30).

Using employment level as an indicator of size, the Cuban industrial sector in 1986 had 499 enterprises employing up to 1,000 workers, 190 employing between 1,001 and 2,000 workers, 93 employing between 2,001 and 3,000 workers, 19 employing between 3,001 and 4,000 workers, and 26 employing over 4,000 workers. The 138 largest industrial enterprises—those employing over 2,001 workers—corresponded to electricity generation (3), fuels (1), ferrous mining and metallurgy (1), nonferrous mining and metallurgy (6), nonelectrical machinery (6), electrical machinery and electronics (1), chemicals (2), construction materials (6), textiles (6), apparel (9), leather (2), sugar (75), foodstuffs (8), fisheries (3), beverages and tobacco (7), and other industrial activities (5).

As will be discussed, enterprises in the sugar, nonferrous metallurgy (particularly nickel mining and processing), cement production, and oil production and refining industries are significant sources of air, water, and land pollution. Specific enterprises in other industries are also significant polluters. According to UN statistics (1992a, 748), emissions of carbon dioxide (CO_2) from fossil fuel combustion and cement manufacturing in Cuba were roughly 3.8 million tons in 1960, 5.1 million tons in 1970, 8.4 million tons in 1980, and 9.4 million tons in 1991.

Sugar and Derivatives

Sugar production remains Cuba's most important industrial economic activity. The sugar industry produces raw and refined sugar as well as numerous other derivatives, such as alcohol, paper, bagasse boards, and yeast. In the mid-1990s the sugar and derivatives industry consisted of the following enterprises (CONAS 1994, 81): 156 sugar mills, seventeen sugar refineries, thirteen alcohol distilleries, four paper factories, five bagasse board factories, eleven torula yeast factories, and one dextran factory. Sugar mills and sugar derivatives factories are located throughout the island, although they tend to be concentrated in the eastern half of the nation, where sugarcane agriculture predominates.

The sugar industry—with the bulk of the industrial stock and comprising the largest industrial enterprises in the island (54 percent of the larger industrial enterprises in 1986)—is an important source of pollution. Economist Antonio Morales Pita (1995, 132) deems the Ministry of the Sugar Industry as the principal source of pollution in the country given that its facilities are located throughout the island and the "aggressiveness" (toxicity) of its effluents and emissions. According to scientist José Oro (1992, 55):

> The sugar cane industry is a significant contributor to the high level of pollutants in Cuba, launching large amounts of smoke, soot, and steam into the atmosphere. Since it uses the waste pulp "bagazo" as fuel, it also emits a great variety and amount of toxic industrial wastes and molasses into the surface and groundwater systems. However, the main ecological impact of the sugar cane industry is related to the production process of derivatives such as torula, bagasse boards, paper, furfural and furfurilic acid, high molecular weight alcohol, and recently lysine (a protein for feed production), which lack the necessary pollution controls.

Raw Sugar Production

The development strategy crafted during 1960–1961 by Cuban officials, with the technical assistance of experts from abroad, stressed agricultural diversification away from sugarcane and rapid industrialization.[1] Accordingly, large sugarcane estates were cleared and replanted with other crops—rice, fruits, vegetables—and several sugar mills were dismantled. Sugar output declined sharply, falling to 3.8 million tons in 1963 compared to 6.8 million tons two years earlier. Sugar exports declined proportionately, contributing to a severe balance-of-payments crisis.

To overcome the economic crisis, Cuba shifted gears and redefined its development strategy to give agriculture—and sugar production in particular—a central role. Resources were redirected to the sugar industry, in support of a plan to produce 10 million metric tons of sugar in the 1970 *zafra* (sugar harvest). The much-publicized 10-million metric ton target for 1970 was not reached (production was about 8.5 million metric tons); in the aftermath of the 1970 *zafra* and the severe economic dislocations that occurred as other sectors of the economy were neglected in pursuit of the sugar objective, policies emphasized more stable and efficient sugar production, with gradual increases in output. Efforts were made to increase industrial yields, reduce fuel consumption by sugar mills, broaden the scope of mechanization in cutting and loading sugarcane, and improve on timeliness of export deliveries. In the 1970s and 1980s Cuba overhauled most of its sugar mills and built about ten new ones, the first mills built on the island since 1927. Sugar output expanded in the 1970s and 1980s, averaging about 7–8 million metric tons per annum.

Agriculture

A large sugar industry such as Cuba's requires prodigious volumes of sugarcane as input into the industrial process. As discussed in Chapter 4, in 1989 nearly 2 million hectares or 31.5 percent of Cuba's agricultural land was devoted to sugarcane, compared to 1.1 million hectares or 18.8 percent of agricultural land in 1945. An industrial yield of 11 percent—roughly the average yield obtained in Cuba in the period 1960–1989—means that the agricultural system must deliver 100 tons of sugarcane to sugar mills in order to produce 11 tons of raw sugar. Since raw sugar production averaged about 7.4 million tons per annum over the period 1980–1989, sugar mills ground 65–70 million metric tons per annum of sugarcane during the 1980s.

One of the significant achievements of Cuba's socialist government regarding sugar production is the high degree of mechanization of agricultural activities: cultivation, harvesting (cutting and loading), and transportation of sugarcane. Mechanization of sugarcane agriculture meant not only developing or importing machinery to cultivate, cut, load, and transport cane, but also laying out, preparing, and planting fields with sugarcane varieties that would enhance the operation of the equipment. Because sugarcane was grown in pre-revolutionary Cuba largely by small operators (*colonos*) who sold their output to neighboring mills, fields were not suitable for large-scale harvesting. Mechanization required consolidating individual fields into larger areas where harvesting equipment could be used more efficiently. This led, in many instances, to leveling sloped areas and eliminating roads, forest cover, drainage ditches, creeks, wetlands, and other obstacles that might impede the continuous operation of machinery. In addition, varieties of sugarcane that tend to have straighter stalks, and therefore are more suitable for mechanical harvesting, have been planted. To increase the efficiency of mechanical harvesting equipment, it became commonplace to set cane fields on fire prior to harvesting in order to eliminate excess sugarcane leaves and straw.

Environmental disruptions associated with the mechanization of the sugar harvest—e.g., elimination of creeks and wetlands, leveling of tree fences, soil compaction from heavy equipment, waterlogging caused by improper irrigation drainage, generation of smoke and soot from burning sugarcane fields—are well documented in the literature, although their magnitude is difficult to assess.

Although not related to industrial pollution, environmental disruption associated with the *roya* (rust) epidemic that hit Cuban sugarcane fields in 1979–1980 illustrates the carelessness with which socialist Cuba made key decisions that affected the environment. Cuban sugarcane production was heavily affected by an epidemic of *roya*, a fungal disease that tends to dry up sugarcane stalks and reduce juice content. The disease was first observed in Cuba in 1978 in plantings of B-4362, one of the more popular varieties of sugarcane cultivated on the island at that time. B-4362 was one of the preferred varieties because of its early maturation (which extended the time during which mechanical harvesters could be used) and high sugar content. According to a former official of the Cuban Ministry of the Sugar Industry, in 1972 a Cuban technician conducted a study of B-4362 and concluded that because of its high susceptibility to *roya,* it should be planted in small areas, far apart from one another, so that a possible outbreak of the disease could be controlled (Llovio-Menéndez 1988, 311). The warning was ignored and large areas were planted with B-4362 to maximize sugar production. *Roya* spread through cane fields in 1979–1980, with agricultural losses from B-4362 plantings ranging from 10 to 80 percent and averaging 30 percent nationwide. According to President Castro, the *roya* epidemic affected one-third of Cuba's sugarcane lands, significantly reducing output and translating into a loss of $400–600 million in convertible currency in potential sugar exports.

Industrial production of sugar

Sugar production is an energy-intensive process, particularly those steps that involve condensing sugarcane juice (*guarapo*) by heating. Sugar is produced in Cuba from sugarcane through an industrial process whereby: (1) crushing machines burst sugarcane stalks to extract cane juice, (2) the cane juice is heated and filtered, (3) the clear juice is run through evaporators and vacuum pans to remove excess water until a thick syrup is formed, and (4) sugar crystals are separated from the syrup by centrifuging. Harvesting and transporting sugarcane to mills require a great deal of fuel, as does transporting sugar to export terminals.

Cuba's sugar industry is largely energy self-sufficient because of its use of bagasse (*bagazo*), the moist mass of stalks and leaves left behind when cane is ground to extract its juice. Because its bulkiness and low caloric value make it uneconomical to transport, bagasse is used as a fuel exclusively by sugar mills, although some small portion (under 5 percent) is used for industrial, nonfuel purposes such as production of paper and paperboard. In addition to consuming virtually all of the bagasse produced, sugar mills also use fuelwood and oil products when available bagasse is not sufficient to cover their needs.

In Cuban sugar mills bagasse is generally burned directly after the sugarcane is crushed. This method avoids a number of steps associated with removing, storing, and again feeding the bagasse to the boilers but results in bagasse with a much higher moisture content (about 50 percent) than if it were sun or steam dried. The relatively low caloric value of high-moisture Cuban bagasse (the technical literature equates the caloric value of 5.6 metric tons of bagasse with that of 1 metric ton of medium-quality oil) means that boilers must process huge quantities of it, launching into the air large quantities of smoke, soot, and steam, which affect nearby areas. Sugar mills store bagasse in open-air piles, creating the risk of runoff of fermented juice to nearby streams and of autocombustion.

Cuban sugar mills are notorious for polluting wetlands, streams, reservoirs, and ocean fronts with liquid waste products such as molasses, filter mud (*cachaza*), bagasse composts, and fermented liquid wastes. Industrial wastes from sugar mills affect water quality in general and marine resources in particular. During a review of the fishing industry conducted by the National Assembly of People's Power in December 1985, President Castro expressed surprise when informed that shrimp no longer lived in Nipe Bay because of the high level of pollution in that body of water. Castro stated that he remembered fishing for shrimp in that area during his youth and inquired about the reasons for the pollution of the bay. He was informed that the chief sources of pollution were a torula yeast factory and three sugar mills, the "Fernando de Diós," "López Peña," and "Guatemala" sugar mills (Aprobados los informes 1985, 3). At the same session of the National Assembly, the minister of the Fishing Industry reported that the "Loynaz Echavarría" sugar mill had been dumping wastes into the Nipe and Sabanilla reservoirs, affecting fishing in those locations (Aprobados los informes 1985, 3).

Talking to a journalist in 1991, the vice president of COMARNA recounted how in May of that year, the "Grito de Yara" sugar mill dumped waste products on three separate occasions into the Cauto River, destroying clam beds at the mouth of the Cauto and affecting shrimp farms downriver (Gutiérrez 1991, 28).

The COMARNA official stated that the Ministry of the Sugar Industry was already making the necessary investments to eliminate waste discharges from the 10–12 sugar mills that followed this environmentally harmful practice. Nevertheless, in January 1997 the Cuban press reported that pursuant to an emergency plan being developed to reverse pollution in the Cauto River, the thirteen sugar mills located in that river's basin pledged that during the 1996–1997 *zafra* they would not dump waste materials "that damaged the Cauto." The journalist indicated that it was not clear what specific measures several of the sugar mills ("Los Reynaldo" and "Julio Antonio Mella" in Santiago de Cuba Province; "Antonio Maceo," "Cristino Naranjo," and "Urbano Noris" in Holguín Province; and "Grito de Yara" and "José Nemesio Figueredo" in Granma Province) would take to stop the practice of dumping waste products into the Cauto and that one of the mills—"Majibacoa" in Las Tunas Province, one of the new mills built during the 1980s—would have its effluent discharges measured and considered as one of its indicators of efficiency (Batista Valdés 1997, 10). Economist Antonio Morales Pita (1995, 132) reported on collaboration between the Ministry of the Sugar Industry and of the Fishing Industry to treat liquid discharges from sugar mills and related plants so that they do not harm shrimp farms in Cuba's southeastern coast. A COMARNA official, writing about the application of "clean technologies" in Cuba, has referred to the use of liquid effluents from sugar mills to irrigate fields, in a method called "ferti-irrigation." In so doing, "one of the principal causes of water pollution is removed (because the burden on lagoons, traditionally used for waste treatment is reduced) at the same time that agricultural yields rise" (Acosta Moreno 1989, 43).

Table 7.1 lists fourteen sugar mill complexes identified by Oro (1992) as being among the most severe polluters of Cuba's environment because of their practice of discharging molasses and wastes into open ditches, some of which empty into streams and rivers and others directly into the ocean.

Sugar Derivatives

Enterprises producing sugar derivatives tend to be located near sugar mills, sometimes sharing facilities with the mill. This is the case with all but one of the seventeen existing sugar refineries and the majority of the other derivatives plants. As is illustrated in Table 7.1, waste products from sugar derivatives plants add to the environmental harm caused by pollution from sugar mills.

The most economically significant sugarcane derivatives produced in Cuba are those obtained from bagasse, particularly paper and particleboard. Experimentation and small-scale production in Cuba goes back to 1915, when kraft paper was produced at a plant attached to the "Preston" (currently "Guatemala") sugar mill in Holguín Province; particleboard was first produced in Cuba in 1928.

In 1959 Cuba produced a broad range of bagasse-derived products: newsprint, particleboard, acoustical tiles, bagasse-pulp fine papers, and office furniture. In particular, three bagasse pulp and paper plants were in operation with an annual capacity of 53,500 tons of paper and paperboard. In the 1970s Cuba built an experimental plant to produce pulp and newsprint from bagasse; the plant, constructed with technical and financial assistance from the UN Development Program, was adjacent to the "Pablo Noriega" sugar mill in Quivicán, La Habana Province. A large plant capable of producing bagasse pulp and writing and printing paper—named "Panchito Gómez Toro"—was completed in 1986 adjacent to the "Uruguay" sugar mill, in Sancti Spíritus Province, as were also several particleboard plants.

A study of effluents from Cuban pulp and paper plants conducted by a group of scientists in the late 1970s (Barquinero et al. 1981) revealed that water used in the production process when returned to rivers or oceans contained a variety of organic and inorganic substances, some associated with pulp and paper production and others with chemicals used for whitening the paper; effluents also contain residues of fibers, ash, clay, limestone, acidic materials, and alcohols. Specifically, the study found the following:

▶ Storage and handling of the raw material—bagasse—is a significant source of pollutants. As the pulp and paper plants work throughout the year while sugar mills only do so for a few months, bagasse produced during the harvest must be stored and introduced into the pulp and paper plants at the appropriate time. Two methods are used for storing bagasse: (1) forming it into bales, which are stacked and covered with sheets of plastic; and (2) making large piles of the raw material.

▶ Baled bagasse tends to become very dry and give up large amounts of dust when the bales are broken up to begin the production process. It is common to wet the bales to reduce the dust; the water than runs off from this process contains a heavy concentration of organic matter that raises the oxygen level of streams and reservoirs. Piled bagasse is treated continuously with fresh water to remove airborne dirt and dust; the water that drains from the piled bagasse contains a heavy concentration of organic matter.

▶ Effluents from the pulp and paper production processes contain high concentrations not only of organic matter but also of chemicals such as caustic soda, chlorine, sulfates, sodium hydroxide, and sodium carbonate, which have an adverse effect on marine life and vegetation.

Table 7.1. Main sources of pollution—sugar mills and sugar derivatives plants

Facility	Range of ecological impact of pollution	Type and magnitude of pollution
"César A. Sandino" sugar mill Mariel, Havana Province	Mariel and the village of Cabañas, including Cabañas Bay	Smoke, soot, noise, and low-frequency vibration. Discharge of all types of waste into an open ditch running to Cabañas Bay
"José A. Echevarría" sugar mill "Técnica Cubana" paper factory Cárdenas, Matanzas Province	City of Cárdenas and Cárdenas Bay	Smoke, soot, vibration. Discharge of molasses and technological waste from both the sugar mill and the paper factory in an open ditch running to Cárdenas Bay. Danger of autocombustion of the dry waste pulp
"Sergio González" sugar mill Waste pulp board factory Colón, Matanzas Province	Colón area, but significant	Smoke, soot, noise, and vibration. Discharge of molasses and technological waste in an open ditch. Fires from autocombustion of dry waste pulp. Unpleasant mephitic exhalations. Discharge of waste from the board factory
"Camilo Cienfuegos" sugar mill Waste pulp board factory Santa Cruz del Norte, Havana Province	Santa Cruz del Norte area, but very extensive	Smoke, soot, noise, and vibration. Noise and vibration level is very high. Discharge of molasses and technological waste in an open ditch running to the Santa Cruz River. Danger of fires from autocombustion of dry waste pulp. Very unpleasant mephitic exhalations. Discharge of waste from the board factory and the steel workshop
"Granma" sugar mill Coliseo, Matanzas Province	Coliseo area, but intensive	Soot, steam, and smoke. Discharge of molasses, alkaline, catalysts, and other technological waste into an open ditch running to the surface and subsurface hydric systems. High incidence of mephitic smells and insects. Danger of fires from autocombustion of dry waste pulp
"Manuel Martínez Prieto" sugar mill Marianao, Ciudad de la Habana Province	Marianao area, but very significant. The sugar mill is in an overpopulated area of the capital. The discharge of waste is running into the already polluted Almendares River. Soot spread in all the western areas of Havana City	Smoke, soot, noise, and vibration. Discharge of molasses and technological wastes. Insects in large quantities; mephitic exhalations
"Arquímides Colina" sugar mill Torula yeast factory and brewery Bayamo, Granma Province	Bayamo area	Smoke, soot, noise, and vibration. Intense discharge of molasses (including torula), running into the hydric network of the rivers Bayamo and Cauto

Factory	Area affected	Pollution
"Guatemala" sugar mill Torula yeast and bagasse board factory, brewery Mayarí, Holguín Province	Mayarí area; town of Preston and Mayarí Bay (especially sea life)	Smoke and soot. Discharge of molasses and technological wastes (including alkaline catalyzers) into an open pond. Overirrigation, damaging the dynamic level of the underground water
"Antonio Guiteras" sugar mill Torula yeast and bagasse board factory Puerto Padre, Las Tunas Province	Puerto Padre area, but very intense; affects the town of Delicias and the village of Puerto Padre, as well as the sea life of Puerto Padre Bay	Smoke and soot. Discharge of molasses, alkaline catalyzers, and technological waste (including torula) in an open ditch, running to the sea. Danger of autocombustion of dry bagasse bricks. Insects
"Jesús Menéndez" sugar mill Torula yeast factory Chaparra, Las Tunas Province	Chaparra area, including town of Chaparra and sea life	Smoke, soot, and noise. Discharge of technological wastes (including torula wastes) and molasses directly into an open pond and a ditch running into the sea
"Ciro Redondo" sugar mill Torula yeast and bagasse board factory Morón, Ciego de Ávila Province	Morón area	Smoke, soot, and noise. Discharge of technological wastes and molasses directly into the Pozo Brujo River
"Bolivia" sugar mill Torula yeast factory Cunagua, Camagüey Province	Cunagua area, but very intensive in the town of Cunagua	Smoke, soot, noise, and vibration. Mephitic exhalations. Discharge of technological wastes into an open pond. Insects. Danger of fires from autocombustion of dry bagasse piles
"Primero de Enero" sugar mill Torula yeast factory Ciego de Ávila, Ciego de Ávila Province	Ciego de Ávila area, including the surface and subsurface hydric systems	Vibration, noise, smoke, and soot. Discharge of molasses and technological wastes (including torula wastes) in an open ditch
"Uruguay" sugar mill "Panchito Gómez Toro" Paper factory Jatibonico, Sancti Spíritus Province	Jatibonico area, including hydric network	Smoke and soot. Noise and low-frequency vibrations. Discharge of molasses and a great amount of technological waste from the sugar mill and paper factory (including oil and lubricants) into the hydric system of the Jatibonico del Sur River

Source: Oro 1992, 88, 90–102.

According to Oro (1992, 55), "the paper industry is one of the main causes of contamination of the Cuban surface water network because of the direct launch of wastes into ditches and ponds. The larger of the factories, the French-built 'Panchito Gómez Toro,' located near the 'Uruguay' sugar mill at Jatibonico, is emitting waste into the Jatibonico del Sur River."

Torula yeast

In 1965 Cuba built a pilot plant to produce torula yeast, a form of animal feed for cattle, poultry, and swine, made from molasses. On the basis of favorable pilot results, Cuba built ten additional plants with equipment imported from France and Austria. A Cuban scientist (Sánchez Hernández 1992) has identified torula yeast waste water as one of the most significant sources of stream and soil pollution in Cuba. According to his research, more than 30,000 cubic meters of waste water with an average 20 grams per liter of solids and very high acidity are produced daily by the nation's torula yeast plants. Lack of treatment of this waste water constitutes an important environmental problem.

Distilleries

The "Havana Club" rum distillery in Santa Cruz del Norte, La Habana Province, has been reported to discharge molasses and industrial waste products into the marine shelf near La Habana's eastern beaches, damaging sea life and water quality (Oro, 1992, 114).

Nickel Mining and Processing

With an estimated 5.3 percent of the world's land-based economically exploitable nickel resources and 13.3 percent of additional nickel resources (United Nations 1980, 7), Cuba has been an important player in world nickel supplies since the 1940s. In the 1980s and 1990s Cuba expanded its nickel-producing capacity in order to increase export revenues. These efforts have been supported first by assistance from the former Soviet Union and other socialist countries and more recently by joint ventures with Western nickel producers. Mining and processing of nickel ores has affected—and continues to affect—soil, air, and water resources in a large region in northeastern Cuba as well as the nation as a whole.

Cuban Nickel Industry

Land-based nickel deposits occur as two different types of ore: sulfides and oxides (or laterites). Sulfides are generally found deep underground and require deep mining, whereas laterites—which typically contain nickel mixed in complex fashion with other minerals—tend to occur near the surface and can be mined by open-pit methods. Refining methods for the two types of ore are also very different: sulfides can be concentrated by flotation, yielding product containing 15 percent nickel, which can then be smelted and refined to produce nickel metal; laterites cannot be concentrated and must be treated directly using either pyrometallurgical (smelting) or hydrometallurgical (leaching) processes to obtain refined nickel. Production costs for the two types of processing are significantly different, with energy costs much higher for laterites because the ores must be treated directly without concentration (United Nations 1980, 8).

According to industry experts, the smelting and refining necessary to extract the nickel content is eight to ten times as energy intensive for laterites as for sulfides (Moran 1976, 262). Cuba's nickel deposits are concentrated in the Nipe Mountain Range, on the north coast in Holguín Province (formerly part of Oriente Province), with other deposits in Camagüey and Pinar del Río Provinces (Marrero 1950, 271–75). Cuban nickel deposits are in the form of laterites, containing about 1.3 percent nickel as well as minor concentration of other metals, particularly cobalt.

The development of nickel mining and processing in Cuba was closely connected to the strategic defense needs of the United States during World War II. The first processing plant, at Nicaro, began operations in 1943. It was constructed by agencies of the U.S. government and operated by a subsidiary of the Freeport Sulphur Company. The production process used in the plant—ammonia leaching—permitted the recovery only of nickel and some cobalt, in the form of nickel-cobalt oxides and sinter, with iron and chromium disposed of as waste products. Production ceased in March 1947, after the end of World War II, and the plant was not reopened until January 1952, as a joint venture between the National Lead Company and Cuban investors, when mobilization associated with the Korean War again stimulated U.S. demand (Grupo Cubano 1963, 1081).

In 1953 the Freeport Sulphur Company began construction of a new nickel processing plant at Moa Bay, on the north coast of Oriente Province, east of Nicaro. The Moa Bay plant relied on a different technology—sulfuric acid leaching—to process the nickel ore. The output of the plant, in the form of nickel sulfide, would be shipped to facilities owned by Freeport Sulphur in Freeport, Louisiana, for smelting. The plant began to undergo testing in 1959; it was nationalized by the Cuban government in 1960 and began to operate commercially in 1961 (Pozo 1980, 20).

Production of nickel (nickel and cobalt mineral content) in 1970 was approximately 37,000 tons, split almost evenly between the Nicaro (renamed

"René Ramos Latour") and Moa ("Pedro Sotto Alba") plants. In the 1970s Cuba expanded production capacity of the two plants (to 30,000 tons per annum for the Nicaro plant and 24,000 tons per annum for the Moa plant) and, in the 1980s, began construction of two others: the "Comandante Ernesto Che Guevara" plant at Punta Gorda and a similar facility at Las Camariocas.[2] Both of these plants relied on the ammonia leaching process used at Nicaro and were rated as each being able to produce 30,000 tons of nickel-cobalt oxides per annum (Habashi 1993, 1170). The "Comandante Ernesto Che Guevara" plant began to produce nickel and cobalt oxides in 1987.

According to official Cuban statistics, nickel and cobalt (mineral content) production peaked in 1989 at nearly 46,600 tons and fell steadily through 1994: 40,700 metric tons in 1990, 33,300 tons in 1991, 32,400 tons in 1992, 30,200 tons in 1993, and 26,900 tons in 1994. Production rebounded in 1995, when it reached 42,900 tons, and a new production record of 55,800 tons was projected for 1996 (Banco Nacional 1995, 9).

Commercial ties between Cuba and Western mining companies have been influential in the recovery of the nickel industry. Canadian nickel-cobalt producer Sherritt, Inc., began to purchase nickel and cobalt sulfides produced by the Moa Bay plant in 1991 for further processing at its refinery in Fort Saskatchewan, Alberta, Canada. In 1994 Sherritt and Cuba's Compañía General del Níquel, the state-owned entity that operates the Cuban nickel industry, created a joint venture to develop and market Cuban nickel resources. According to the agreement, both the Moa Bay nickel production plant and the Fort Saskatchewan refinery became assets of the joint venture. The Cuban government granted the joint venture mining concessions sufficient to supply the Moa Bay plant for twenty-five years. Plans are for the joint venture to upgrade and expand the Moa Bay plant significantly and to create a jointly owned marketing corporation (Pérez-López 1995a, 10). Reportedly, Western Mining Corporation of Australia and the Cuban enterprise Commercial Caribbean Nickel, S.A., signed a letter of intent in 1994 to form a joint venture to assess developing a nickel deposit at Pinares de Mayarí. If the feasibility study is positive, a joint venture would build a new nickel production plant at the site and market the output (Pérez-López 1995a, 10).

Nickel Mining and the Environment

As discussed earlier, Cuban laterite deposits occur very close to the surface and can be exploited using strip-mining techniques. Typically, a thin strip of subsoil is removed from the mountainous area where the deposits are located, and then dragline shovels and heavy mining equipment extract huge volumes of raw material that is transported to the processing plants with trucks and/or conveyor belts.

Areas that have been mined have the appearance of "lunar landscapes," checkered with craterlike depressions where the mining equipment has dug deep (Núñez Jiménez 1980). The mined areas are denuded from any vegetation and unable to absorb rainwater, which runs off washing away fertile soil in surrounding areas and causing further erosion. Reportedly, heavy erosion from surface mining is filling Moa Bay with earth (Knox 1995). When exposed to moisture, the mined areas tend to give off acids that are carried by rainwater and pollute nearby streams and rivers.

During the July 1980 session of the National Assembly that took up a draft law "on the rehabilitation of soils and conservation of terrestrial waters," Vice President Raúl Castro, who grew up near the nickel-rich area of Nicaro, had harsh words for the nickel-mining industry. An article in the newspaper *Granma* reported his intervention as follows:

> Raúl added that perhaps because he was born near there, he is deeply disturbed by the nickel mines. He also fought in that region [during the revolution] twenty years ago. He has witnessed, particularly in the last twenty years, the destruction that is being wreaked. . . . The bulk of the damage has been done since the triumph of the revolution. . . . He recalled that while visiting this area some years ago with comrade Fidel [Castro], Fidel also made the comment: When are we going to stop this destruction? . . . Raúl commented that he has observed from the ground and from a helicopter how the mineral is extracted. Tractors and bulldozers come first and remove the vegetable cover, dumping it in gulleys and valleys and damaging the vegetation of these areas. They then make holes to dig up the nickel and when they are finished, they leave. (Resolución unánime 1980, 3)

The most significant environmental disruptions caused by strip mining can be mitigated by reclamation, that is, returning the land as closely as possible to its original condition by leveling the area, covering it with topsoil, and planting a vegetable cover. The Cuban press has reported on reclamation efforts in Ocujal, an area that was surface mined many years ago to supply the Nicaro plant, but reclamation does not appear to be the rule in the strip-mining operations. In Ocujal, mountainous areas that were surface mined using contour or terracing methods reportedly have been covered with topsoil prior to reforesting with pines, and small dams have been constructed to capture water runoff (Marrero 1985). Cuban officials have stated that in the Moa Bay region, strip-mined areas are not being reclaimed, "since there is still usable mineral, although its exploitation will require the construction of a plant with a different technological process" (Pozo 1980, 18). Ian Delaney, chairman of Sherritt, Inc., has stated that Moa Bay's worst pollution problem comes from erosion after ore is stripped from the hill sides: "[Cubans] have not reforested to the extent they should have" (Knox 1995). According to two Cuban journalists, in Moa Bay "a deep blue sea takes on a reddish color because of the soil that is carried by its waves and by the rivers that flow into its waters" (Castro and Ramos 1984, 22).

COMARNA Vice President Ferrer has stated that "some reforestation has been done, but at a slower pace than the mining itself. Reforestation is more complex here [in nickel strip-mining areas] since it involves bringing back the soil and guaranteeing proper fertilization as well as planting. Transportation of all these 'ingredients' becomes a major dilemma nowadays with the oil shortage" (Reed 1993, 33).

As was noted in Chapter 6, reforestation efforts have faltered in the 1990s while strip mining to support nickel production has increased.

Nickel Processing and the Environment

The Cuban nickel plants are significant sources of air, land, and water pollution (see Table 7.2). The ammonia-leaching technology used at Nicaro (and replicated without improvement at Punta Gorda and Las Camariocas) "proved to be one of the most polluting processes that ever went into operation due to extensive dust emission" (Habashi 1993, 1173). Meanwhile, the Moa Bay plant is prone to leaks of toxic hydrogen sulfide into the atmosphere (Habashi 1993, 1175). Journalist Fred Ward, who visited Nicaro and Moa in the late 1970s, described the following situation:

> Great sweeps of the coastal regions around the two towns look as if some supernatural blight had stricken the people and their environs. The pervasive red dust covers and penetrates everything. Three days later, after two baths a day, the color is still caught in hair and ears. The official joke is that it shows up on white shirts a week later in Havana. To the people who breathe it all day, it is no laughing matter. At Nicaro, so much of the dust is blown by the prevailing winds from gigantic smokestacks that a red streak is visible across the land for miles. Several of the roads in Moa receive such quantities of gases from the plant that newcomers gag and choke on the spot. (Ward 1978, 218)

According to residents of the Moa area, sulfur compounds used in the processing of nickel pollute the air and water, producing acid rain (Knox 1995). A

Table 7.2. Main sources of pollution—nickel processing plants

Facility	Range of ecological impact of pollution	Type and magnitude of pollution
"Comandante René Ramos Latour" Nicaro, Holguín Province	Nicaro area and national impact. Significant damage to the towns of Nicaro, Felton, and Levisa. Waste in the marine shelf is affecting the environment from the mouth of the river Mayarí to Tánamo Bay	Colossal quantities of smoke and soot, NH_3, noise, and vibrations. A huge amount of dust in the quarries. Two million tons of technological waste per year in an open pond. Potential NH_3 discharge
"Pedro Sotto Alba" Moa, Holguín Province	Moa area, national and potentially international impact. The amount of H_2S launched into the atmosphere is capable of generating acid rain in the eastern part of Cuba and its surroundings and spilling waste into Moa Bay. The coral reef barrier has suffered heavy damage and will be completely destroyed in ten years	Great quantities of H_2S launched into the atmosphere. A huge amount of dust in the quarries. Nearly 2 million tons of waste per year
"Comandante Ernesto Che Guevara" Punta Gorda, Holguín Province	Local area and national impact. Significant damage to the town of Punta Gorda. Much waste running through the Punta Gorda River and out to sea	Great amount of smoke and soot, NH_3, noise, and vibrations. A huge amount of dust in the quarries. Two and one half million tons of technological waste per year. Potential NH_3 discharge
CAME-I plant Las Camariocas	Most likely the same impact as the "Comandante Ernesto Che Guevara" plant	Most likely the same impact as the "Comandante Ernesto Che Guevara" plant

Source: Oro 1992, 80–82.

worker reported that "the rain in Moa is acidic enough to produce *picazón*—a stinging sensation—when it falls on bare skin. . . . He and other residents said children tend to suffer from respiratory problems such as asthma and persistent coughing" (Knox 1995). Residues, in the form of slurry, from the ammonia leaching process are disposed of in holding ponds. Because the bottoms of the ponds are not impermeable, water quickly drains underground, leaving a solid-ified mass that amounts to 2 million tons of material per year for each plant. The water that drains underground contaminates subterranean water (Habashi 1993, 1177).

The nickel plants also discharge large quantities of pollutants into streams, rivers, bays, and ultimately the ocean. In 1990 Rosa Elena Simeón, then president of the Cuban Academy of Sciences, "condemned nickel industries in the Moa region which discharge tons of chemical wastes into local rivers daily," a practice that "damages flora and fauna and poses serious dangers to human life" (Prieto 1990). A Cuban journalist reported in 1994 that a group of Cuban scientists were working on a method to recover some of the minerals and chemicals produced as waste by the Moa Bay plant. According to this report, liquid waste produced by the plant, "flows at the rate of 12,000 cubic meters a day and carries into the sea a wide range of light and heavy metals, such as sulfates and great amounts of sulfuric acid, which is used in the lixivating process of the laterictic mineral. Every day 72 tons of aluminum, 48 tons of chrome, 15 tons of magnesium, and 30 tons of the dangerous sulfuric acid get dumped into the sea. This harms the marine flora and fauna and, in the long term, could cause irreversible damage" (Nickel Plant 1994, 15).

The main environmental problem associated with the acid leaching process used at the Moa Bay plant, according to a nickel industry expert, is the large amount of sulfuric acid that must be neutralized before adding hydrogen sulfide and the large amount of waste acid that has to be disposed of (Habashi 1993, 1172). A Soviet shipping and marine environment specialist described the situation at Moa Bay as follows: "Each day, more than 450 cubic meters of waste products from the nickel enrichment process are dumped into the ocean. When a ship arrives at that port, it is as if it were subjected to a galvanic bath, with the hull being cleaned off from any adhesions. For this reason, it is imperative to do something to preserve nature at Moa" (Hernández 1982, 29).

A study conducted by Cuban scientists (González, Torres, and Ramírez 1995) of the bay of Levisa, the site of the Nicaro nickel-processing plant, found substantial levels of water pollution. The study identified extremely high levels of iron, cobalt, manganese, and nickel, all important components of laterites. The sources of the pollutants were runoff from strip-mining areas and industrial waste from the Nicaro nickel-processing plant.

Calcium carbonate is one of the chemical compounds used in the acid-leaching process at the Moa Bay plant to neutralize the acidic nickel and cobalt sulfides. According to Cuban sources, calcium carbonate—in the form of coral sand—is dredged from a large coral reef that runs near Cuba's coast at Moa. This coral reef is reportedly the second largest in the world after the Great Barrier Reef off the northeast coast of Australia (Pozo 1980, 21). Contradicting the statements by Cuban officials, a spokeswoman for Canadian company Sherritt, Inc., has stated that the calcium carbonate used is not from coral reefs, but "from coral mud that has long decayed by natural methods. . . . We don't do anything to wreck reefs" (Knox 1995).

Cement

Cuba is relatively well endowed with deposits of limestone, silica, and other raw materials necessary to make cement. Production of cement in Cuba began in 1918 at the "El Morro" plant in Mariel, Pinar del Río Province. In the 1950s two other plants were brought on line to meet growing domestic demand for cement: "Cementos Nacionales" or "Cementos Titán," in Santiago de Cuba, and "Santa Teresa" in Artemisa, Pinar del Río (Grupo Cubano 1963, 1118–19). Production capacity in 1958 was estimated at 940,000 metric tons per annum (Escobar Casas 1987, 16).

In the 1970s Cuba built two large cement plants: at Nuevitas, the "26 de Julio" plant, with East German technology, and capacity of 600,000 metric tons per annum, and at Sancti Spíritus, the "Siguaney" plant, with Czechoslovakian technology, and capacity of 670,000 metric tons per annum. In 1980 Cuba started up two even larger plants, both purchased from East Germany: the "René Arcay" plant, with annual capacity of 1.45 million metric tons, which replaced the "El Morro" plant in Mariel; and the "Karl Marx" plant in Guabairo, near Cienfuegos, with annual capacity of 1.65 million metric tons. Annual cement production capacity was estimated at 5 million tons in the late 1980s (Escobar Casas 1987, 17).

There are two different processes for making cement: dry and wet. To a large extent, the steps in both processes are the same—crushing and grinding of raw materials, burning, and finish grinding. In the wet process, however, the crushed and ground raw materials are blended into a slurry by adding water before introducing them into the kiln for burning. The product that emerges from the kilns, called clinker, has the size of marbles; some gypsum is added to the clinker before it is ground into cement in the form of a very fine powder. Technical studies have shown that the wet process is almost 40 percent more energy-intensive than the dry process, and that the larger the kiln, the greater the energy efficiency (Jankowski 1983, 151).

The crushing and grinding of raw materials and of clinker generate large amounts of dust that affect the areas surrounding the plants. The Cuban press has reported efforts to reduce air pollution during the grinding of clinker at the Mariel plant by installing new filters and fans to recover the dust particles before they are expelled into the air (Palazuelos Barrios 1990b). Dust particles also escape together with smoke from the kilns during the burning process. In the early 1980s Cuba contracted with an East German enterprise to provide electronic filters for the Nuevitas cement plant. According to a Cuban journalist (Norniella 1982a,b), the new filters would "solve the problem of the escape of cement dust from the plant, which affects the entire city of Nuevitas." Reportedly, the new equipment would reduce the release of cement dust fortyfold, from the then-current conservative estimate of 12 grams per cubic meter to 0.30 grams per cubic meter (Norniella 1982a).[3]

Of the six cement plants in operation, two (the two largest and newest facilities, "René Arcay" at Mariel and "Karl Marx" at Cienfuegos) use the dry manufacturing process, and the remaining four ("26 de Julio" at Nuevitas, "Siguaney" at Sancti Spíritus, "José Mercerón" at Santiago de Cuba, and "Mártires de Artemisa" at Artemisa) use the wet process. Cuban sources indicate that the dry technology was chosen for the newer plants because it was more energy efficient, consuming 60 kilograms of oil products per ton of cement less than the wet technology (Escobar Casas 1987, 19). All six of Cuba's cement plants are significant sources of air and water pollution, but the newest plants at Mariel and Cienfuegos are particularly heavy polluters. According to scientist Oro,

the dry system [used in the Mariel and Cienfuegos cement plants] is a low energy-consumption one, basically consisting of the clinker direct-heating-into-the-tunnel ovens. However, during this process a huge mass of dust is launched into the atmosphere, making it critical for the dry system's cement line of production to have an accurate electric filtering and cleaning of dust and smoke. Both Cuban dry-system cement factories have serious problems with the electrofiltering process, but in the Mariel factory, it is of catastrophic magnitude, with thick dust covering the entire village, harbor, and their vicinity. The dust column normally runs 10–15 miles windward, and often more. (1992, 49)

Pollution generated by the Mariel plant is affecting the village and harbor of Mariel, the coastline from the bay of Cabañas to the bay of La Habana, the marine shelf, and the Florida Straits (Oro 1992, 104). This assessment may also be also applicable to the Cienfuegos plant, which relies on similar technology. The "Mártires de Artemisa" plant in Artemisa, Havana Province, is a heavy emitter of dust and smoke, which affect the area from the town of Las Cañas to Artemisa, Candelaria, and Guanajay (Oro 1992, 78). These environmental problems probably also affect the other three cement plants—located at Santiago de Cuba, Nuevitas, and Sancti Spíritus—that use the wet production process.

Oro has also identified as heavy polluters two asbestos-cement plants located at Boyeros, in Ciudad de la Habana Province, and in Artemisa, La Habana Province (1992, 76–77). Both of these plants launch very large amounts of asbestos dust into the atmosphere, with significant adverse effects on the population and the environment given the high toxicity of asbestos.

Oil Production, Transportation, Refining, and Consumption

Cuba's output of primary energy products is modest, with the exception of bagasse, which is produced by and consumed almost exclusively by the sugar industry. No significant coal deposits have been found on the island. Hydroelectric resources are limited: Cuban rivers have low heads, carry relatively small volumes of water, and are subject to uneven rates of flow throughout the year. Commercial oil production has increased significantly in the 1980s and 1990s, but domestic output is still quite small relative to domestic needs.

According to official Cuban statistics on primary energy consumption for the period 1977–1988 (Comité Estatal 1989, 164), the share of total primary energy consumption represented by oil products ranged from 54.2 percent in 1983 to 61.3 percent in 1986, averaging about 60 percent in 1986–1988. Over the period 1977–1988, biomass (bagasse, fuelwood, charcoal, ethyl alcohol) accounted for 29.3–37.6 percent of total primary energy consumption, averaging about 30 percent for 1986–1988. Other forms of primary energy—hydroelectricity, coal and coke, manufactured gas—were relatively small contributors to overall consumption.

Thus, the Cuban energy balance is heavily tilted toward oil products: crude oil, diesel fuel, gasoline, kerosene, jet fuel. Cuba relies heavily on imports of crude oil and refined products to meet its energy needs. This requires transportation through Cuban ports of large volumes of imported crude oil and refined products that affect the quality of waters in Cuba's ports and bays. Cuban oil refineries are also significant sources of air and water pollution.

Oil Production

Commercial oil production in Cuba began in 1915 with the discovery of the Bacuranao field; another commercial field was discovered at Jarahueca in 1943. Oil output from these two fields was small, averaging about 4,000 metric tons per annum during 1950–1954. The discovery of the important Jatibonico field in 1954 shot up production to an average of about 30,000 metric tons per annum during 1955–1958 and gave rise to a flurry of concession applications and exploratory drilling activities by domestic and foreign companies. Small fields were discovered subsequently at Catalina, Cristales, and Guanabo. However, as most exploratory wells either turned up dry or found oil in quantities too small or of too low quality to justify commercial exploitation, the exploration boom subsided.

At the end of October 1959, the revolutionary leadership, convinced that Cuba had vast oil reserves that had not been exploited by the foreign oil companies operating in Cuba because the companies could reap larger profits from refining and marketing imported crude, seized the exploration records of the oil companies. With financial and technical assistance from the Soviet Union and Romania, Cuba undertook an ambitious program aimed at boosting oil production. Production for 1960–1967 averaged 50,000 metric tons per annum, rose to over 200,000 metric tons per annum in 1968–1969 when output peaked at the Guanabo field, and steadied at almost 140,000 metric tons per annum in 1970–1974 as production declined at mature fields (such as Jatibonico and Cristales). Production from new fields east of La Habana (Boca Jaruco and Varadero) pushed output above 250,000 metric tons per annum in 1975–1978.

According to a Cuban source, over the last three decades exploratory drilling exceeded 2 million meters and thirteen oil deposits were discovered; out of the 14,000 square kilometers that have been identified as likely to contain oil deposits, 50 percent are in the marine platform, primarily on the northern coast of Matanzas, La Habana, and Pinar del Río Provinces (Figueras 1994, 26). In the 1980s and 1990s domestic oil production increased rapidly as newly discovered fields were brought under production. Production reached nearly 940,000 metric tons in 1986 but slipped back to about 720,000 tons in 1988–1989.

Because of difficulties in financing oil imports, Cuba has singled out oil production as one of the priority industries during the special period. An important factor in Cuba's current rush to increase domestic oil production is the participation of foreign oil firms, which were banned from the country in the early 1960s. In October 1994 eleven foreign companies were participating in oil exploration and exploitation in Cuban territory, both on land and offshore. Firms from France, Canada, the United Kingdom, Sweden, and Germany were operating in eighteen of the thirty-three oil fields that had been opened for foreign investment through joint ventures. Most of the joint ventures seem to be production partnership agreements, or risk contracts, between the foreign companies and Cubapetróleo, S.A. (Cupet), a corporation created in 1991 to manage the domestic oil industry. These joint venture agreements include (Pérez-López 1995a, 9):

▶ A production partnership agreement between Cupet and a French consortium consisting of Total, the major French oil company, and Compagnie Européenne des Petroles (CEP) whereby the French companies were granted the right to search for oil offshore in a 2,000-square-kilometer block; the consortium would also carry out seismographic surveys and drill four exploratory wells in the bay of Santa Clara, along Cuba's northern coast, over a six-year period. (Total reportedly withdrew from Cuba at the end of 1994 when its offshore exploration activities in the bay of Cárdenas resulted in two dry holes.)

▶ Cupet and British Borneo Oil Corporation signed a risk agreement in November 1994 giving the British company the right to explore and produce oil in a 6,000-square-kilometer area in the central region of the country.

▶ Canada North West Energy Ltd. (a subsidiary of Canadian company Sherritt, Inc.), Fortuna Petroleum (subsidiary of Canadian company Talisman), and Swedish company Taurus have also signed agreements to explore for and produce oil in Cuba. Other foreign corporations involved in oil exploration in Cuba include Oil for Development (subsidiary of the French retailing company Bourgoin), Geopetrole (subsidiary of French company Geofinanciere), Geotel (France), Premier (United Kingdom), and Anglers (Canada).

▶ In February 1993 Cuba launched the first round of open bidding for oil exploration and production rights on the island. Eleven onshore and offshore tracts—ranging in size from 560 to 2,400 square miles—were offered to bidders.

Domestic oil production exceeded the 1 million metric ton per annum mark in 1993—when 1.1 million metric tons were produced—and increased to 1.3 million metric tons in 1994. In late 1996 Cuba was reportedly producing oil at the rate of 1.4–1.6 million metric tons per annum (Isla produce 1996). About 90 percent of crude oil is extracted from two fields: Boca Jaruco-Via Blanca and Cárdenas-Varadero, in the northern coast of La Habana and Matanzas Provinces

(Figueras 1994, 26). Crude oil obtained from the Varadero and Boca Jaruco fields tends to have high density, high viscosity, and high sulfur content (Petróleo crudo 1986). According to a Cuban journalist (Santiesteban 1997b, 50), Cuban crude is so dense that it has difficulty flowing by gravity; light crude oils produced by Algeria, for example, average 44.9 degrees in the density scale of the American Petroleum Institute (API), Angolan crude averages 34.7 degrees, Mexican and Venezuelan heavy crudes 28.0 and 23.4 degrees, respectively, and Cuban crude 15.2 degrees.

The rush to increase oil production reportedly has put pressure on the Cuban environment. Raúl Gómez de Molina, a member of the Cuban independent environmental group Grupo Ecopacifista Sendero Verde (Green Path Eco-Pacifist Movement) until his move to the United States in 1986, has reported extensive environmental damage to beaches located east of the city of La Habana, such as Santa María del Mar, Guanabo, Tarará, Jibacoa, and El Abra, from residues and waste products of oil wells drilled in the area since the early 1980s (Santiago 1990). More recently, Alberto Díaz-Masvidal, an expert on the Cuban oil industry, has expressed concern about deep offshore drilling by the French company Total in the vicinity of Cayo Blanco, near the Bahamas. According to Díaz-Masvidal, the equipment being used in drilling well "Cayo Blanco I" may not meet international standards; the use of substandard equipment could increase the risk of accidental discharge of crude oil, which water currents could carry to Varadero Beach or to the Florida coast (Remos 1994).

Oil Transportation

Meeting Cuba's energy consumption needs requires imports of large volumes of crude oil and refined products. Prior to 1960, most imports of oil and oil products originated from Venezuela and the U.S. Gulf region. In 1960 the Soviet Union became Cuba's main supplier of oil, establishing a 6,500-mile "oil bridge" between Black Sea and Cuban ports. According to official statistics, in the 1960s Cuban imports of oil and oil products (virtually all from the Soviet Union) amounted to 4–5 million metric tons per annum; imports rose to 6–10 million metric tons per annum in the 1970s, and reached as high as 13.5 million metric tons in 1985 and 1987–1988.

The very large volumes of oil and oil products Cuba imported, coupled with constraints on the capacity of port facilities to receive such imports, has meant a very large number of tankers and of tanker loads of liquid fuels transported to Cuba. According to a Cuban journalist, in an article published in the mid-1980s, tankers in the Cuba oil trade ranged from 32,000 to 45,000 metric tons (Abascal

López 1985, 189; Oramas 1986); eleven marine terminals were able to receive oil imports (Rodríguez Castellón 1987, 193). The main oil import terminals are located in the bays of La Habana, Matanzas, Cienfuegos, Santiago de Cuba, and Nuevitas; other facilities at Mariel, Batabanó, Nueva Gerona, Caibarién, Cárdenas, Manzanillo, Santa Cruz del Sur, Casilda, and Santa Lucía are used mainly for cabotage (Oro 1992, 57). Assuming an average tanker capacity of 40,000 metric tons, 230 trips would have been necessary to transport the 9.2 million tons of oil and oil products imported in 1983 and 275 trips for the 11.0 million tons imported in 1988. This heavy traffic of tankers carrying oil and oil products creates significant risk of oil spills. Even though no major accidental spills have occurred, the heavy traffic of tankers into bays to unload imports affected water quality, particularly oil spilled during unloading and expelled from ships while tanks were being cleaned.

Thus far, Cuba has been spared from the catastrophic environmental disruption that would be associated with a major oil spill, despite the fact that the archipelago is located at the crossroads of some of the most traveled international shipping lanes. The most serious incident occurred in January 1980, when the Greek-flag transshipment tanker *Princess Anne Marie* ran aground off Cuba's Cape San Antonio, on the southern coast of Pinar del Río Province, while en route from the Grand Cayman Islands to Port Neches, Texas, with a cargo of crude oil and fuel oil owned by Texaco (Cuba Holds 1980). Reportedly, 5,300 metric tons of crude oil and 700 metric tons of fuel oil escaped from the ship, creating an oil slick a mile long and 1.5 miles wide that affected fishing areas, coral reefs, beaches, and over 7 miles of coastline around the bay of Corrientes (Martí 1980; Unpredictable Consequences 1980). Smaller spills have occurred in the Cuban oil trade, but available reports are very circumspect about the amount of liquid fuel spilled and environmental impacts; for example:

▶ In June 1970 the Soviet-flag tanker *Mekhanik Afanasev,* described in the Cuban press as "one of the largest oil tankers used in transporting oil to Cuba," ran aground in the bay of La Habana, near El Morro Castle. There are no reports of the magnitude of the oil spill (Encalló 1970).

▶ In 1986, 1,200 metric tons of fuel oil were spilled in the bay of Cienfuegos from the Soviet tanker *Luknivitsy* when equipment failed and personnel in charge did not take prompt corrective action to stop unloading operations (Nota Informativa 1986; Griñán 1986).[4] The cleanup operations brought together a large number of volunteers and members of the armed forces working with a variety of makeshift devices (Griñán 1986). In a trial con-

ducted in May 1987 seven workers were found guilty of dereliction of duty
and condemned to prison sentences of one to two years plus fines to compensate for cleanup costs (Dictan sentencia 1987).

▶ A Greek-flag tanker under contract to the Cuban government to carry out
oil cabotage operations caught fire in January 1993 while unloading crude
oil in the bay of Matanzas. There are no reports of the quantity of oil that
might have been spilled (Se incendia 1993).

▶ A collision between the Panamanian-flag oil tanker *Shavadar,* carrying
imported oil and the *Bravo,* a smaller tanker registered in Saint Vincent and
the Grenadines, carrying domestic crude oil in the bay of Matanzas in
March 1998, resulted in a spill of 500 tons of oil, according to official sources
(Quinientas toneladas 1998).

In part to reduce environmental damage, but primarily for efficiency considerations, Cuba began in the 1980s to build a large oil terminal at Matanzas
and a network of pipelines to distribute crude oil and refined products. The
Matanzas oil terminal would accommodate tankers carrying up to 150,000 metric tons of oil, three times the capacity of Cuba's terminals (Oramas 1986, 3).
Pipelines would carry the imported oil from the ships to holding tanks, where it
would be mixed with domestic crude from Boca Jaruco and Varadero also coming through pipelines and eventually travel to oil refineries in Cienfuegos (220
kilometers) and La Habana (90 kilometers) (Oramas 1986; Santiesteban 1987a,
51). By the early 1990s the supertanker port and the pipeline from Matanzas to
Cienfuegos had been completed (CONAS 1994, 92), although it is not clear how
heavily they were being used given the drastic reduction in oil imports.

Oil Refining

Cuba's first oil refinery was built near the port of La Habana in the 1890s by
Cuban and U.S. investors. The refinery, usually referred to as the Belot refinery,
was modernized and expanded several times, including a major renovation in
the 1950s when it became the property of Esso. Also in the 1950s two other
major international oil companies built refineries in Cuba, Shell Oil Company in
La Habana and Texaco in Santiago de Cuba. A small, domestically owned refinery also operated in Cabaiguán.

In May 1960 the Cuban government ordered the three foreign-owned
refineries to purchase and process crude oil that the Cuban state had obtained
from the Soviet Union in barter for sugar under a bilateral trade agreement
signed by the two countries in February 1960.[5] The rationale for the government's order was that crude oil imported by the oil companies for refining, paid
for in dollars, was a burden to Cuba's dwindling dollar balances, whereas the

bartered Soviet crude did not have a negative effect on such balances. The three refineries balked, and on 29 June 1960 they were seized by the Cuban government. The two refineries in La Habana were joined and subsequently have operated as a single enterprise under the name "Ñico López"; the Santiago de Cuba and Cabaiguán refineries were renamed "Hermanos Díaz" and "Sergio Soto," respectively. It has been estimated that national refining capacity was 6.3 million metric tons of crude oil per annum, with the bulk of the capacity at the La Habana (3.9 million tons) and Santiago de Cuba (2.2 million tons) facilities (Rodríguez Castellón 1987, 193).

In the mid-1980s Cuba began construction of a large oil refinery in Cienfuegos using Soviet technology. As discussed earlier, a pipeline was built from the port of Matanzas to carry imported crude oil to be processed at the Cienfuegos refinery. Refining capacity of the new plant has been given as 3 million metric tons of crude oil per annum (CONAS 1994, 92). The Cienfuegos refinery was completed at about the time when Cuban oil imports and consumption declined sharply, and therefore the plant has not been made fully operational. Cuba and the Mexican enterprise Mexpetrol formed a joint venture in September 1994 to finish and upgrade the Cienfuegos refinery and to provide crude oil to be refined by the underutilized plant. Subsequently, it was announced by Cuban authorities that the plan to operate the facility with Mexpetrol had been put on hold (Pérez-López 1995a, 9).

The three oil refineries in operation are deemed by experts to be among the heaviest polluters in the country, contributing heavily to air, water, and ground pollution. According to Oro (1992, 58), from these refineries "highly contaminated wastes are launched into the bays, such as heavy metals, oil, and other toxic wastes in significant amounts. It is particularly severe in the heavily polluted bay of Havana where the impact of oil refining wastes is very high." According to a study conducted by Cuba's Ministry of Transportation under a grant from three UN agencies, 87.5 percent of the industrial wastes were being dumped directly into La Habana Bay (Ministerio de Transporte 1985, 22). Detailed information on air and land pollution caused by the oil refineries compiled by Oro are reproduced in Table 7.3.

Oro (1992, 72) also notes that although the Cienfuegos refinery is not yet operational and therefore it is premature to comment on its polluting potential, its prototype in Bashkiria, Soviet Union, was a heavy polluter, systematically emitting hydrogen sulfide (H_2S) and lead as waste products. However, according to Cuban sources (Morales 1985, 62):

> The refinery under construction in the Bay of Cienfuegos will be equipped with a large plant to treat wastes to ensure that the ecological

balance of that bay richly endowed with natural resources is maintained. The refinery can be considered as one of the most modern in the world. It will not emit oil products or other biological elements. "The water that will be dumped in the bay at the end of the industrial processes will be nearly potable," according to engineer Carlos Reyes. To protect the environment, very modern equipment to prevent gases from contaminating the atmosphere will be installed. Water for the industrial processes will come from a reservoir of about 800,000 cubic meters which is being constructed on the Damují River, so that the refinery will not affect the balance of water used by other industries and by the population of Cienfuegos.

Two Soviet experts who visited Cuba in 1982 to assess pollution of bays and coastal areas had a critical view of bays polluted by oil refineries and other industrial plants:

> Today, the bays of La Habana and Moa are practically dead. Not only can natural resources not be extracted from these areas, but they also pollute neighboring coastal areas. In Santa María del Mar [near La Habana] oil products can be seen floating on the surface of the water. In Santiago de Cuba, although intensive work is being done to reduce pollution, over 60% of the bay's water is heavily polluted, and in the rest it is almost impossible to swim .... With regard to the "Ñico López" refinery [in the bay of La Habana] ... it may be possible to reduce the pollution emitted by the plant by 70 to 80% ... by a system of floating barriers at the point where waste products are expelled. Oil concentrated by this system could be recovered using surface skimming equipment. (Hernández 1982, 29–30)

It is worth noting that the Cuban press reported in 1992 (Salazar 1992) about progress made by scientists at the Instituto de Oceanología (Oceanology Institute) in developing bacteria that speed up biodegradation of oil. It was anticipated that in the future these bacteria would be used to reclaim heavily polluted bays and in other industries.

Oil Consumption

Central planning and cheap oil energy supplies contributed significantly to the relatively low efficiency of energy use in socialist countries. In a study comparing energy consumption in Western and Eastern Europe, Moroney (1990,

Table 7.3. Main sources of pollution—oil refineries

Facility	Range of ecological impact of pollution	Type and magnitude of pollution
"Hermanos Díaz" oil refinery Santiago de Cuba, Santiago de Cuba Province	Santiago Bay	Lead, other heavy metals, potassium, sulfurs, oils, petroleum, and diverse technological wastes launched into Santiago Bay
"Sergio Soto" oil refinery Cabaiguán, Sancti Spíritus Province	Village of Cabaiguán and its vicinity	Lead, heavy metals, potassium, sulfurs, oil, petroleum, and diverse technological wastes in an open ditch, affecting both the surface and subsurface hydric system. Smoke and steam with high H_2S contents
"Ñico López" oil refinery Regla, Ciudad de la Habana Province	Havana Bay and national. Havana Bay is heavily polluted. In Tiscornia's Cove, the bottom mud contains twenty-five to forty times the normal amount of heavy metals. The polluted waters are going out of the bay during low tide, running northeasterly 40–50 miles	Lead, other heavy metals, potassium, sulfurs, oil, petroleum and diverse technological wastes launched into Havana Bay. Smoke and steam bearing H_2S
Cienfuegos oil refinery Cienfuegos Province	Jagua Bay and national	Although plant is not fully operational, its prototype in Bashkiria (Soviet Union) systematically had a high pollution index, including lead and H_2S

Source: Oro (1992, 72–74).

201) found that the European centrally planned economies consumed far more energy—approximately twice as much—relative to gross domestic product than the European market economies. Confirming this assessment, another analyst (Gwiazda 1991, 81) posited that "the economies of the Comecon countries [the Eastern European socialist countries] use, on average, at least 2.5 times the amount of primary energy in the production of one unit of gross domestic product, compared with the highly industrialized Western countries." According to Moroney (1990, 212), "this pattern of intensive energy use may well be a rational response by end users to historical energy subsidies and to central planning emphasizing energy-intensive industry." In the development strategy of centrally planned economies, "energy was conceived as a significant, abundant, and cheap factor for achieving extensive growth" (ECU 1991, 5).

Data on apparent consumption of oil and oil products in Cuba suggest that the pattern of high energy consumption observed in Eastern European socialist countries also applied. Apparent consumption of oil and oil products rose from about 2.3 million metric tons in 1959 to nearly 11.8 million metric tons in 1988, nearly a fivefold increase, or a 5.6 percent average annual growth rate.

The availability of cheap Soviet oil affected Cuba's development strategy, particularly in the industrial sector. Applying the heavy industrialization model pursued by the Soviet Union and Eastern Europe, Cuba invested in energy-intensive industries such as steel mills and capital goods production. Nickel-refining, cement, and fertilizer plants that have been built rely on technologies imported from the Soviet Union and Eastern Europe that have low energy efficiency. Domestic and foreign specialists estimated that Cuba could reduce energy consumption by 30 percent without affecting the level of economic activity through a combination of more efficient plants and equipment and elimination of wasteful practices (Hernández Cruz 1987, 36).

Miguel Alejandro Figueras (1994, 46–47) has made estimates of Cuba's energy intensity of gross national product—the ratio of energy consumption in kilograms of oil or equivalent to gross national product in million pesos—for the 1970s and 1980s. This ratio fell steadily from around 1,700 kilograms per million pesos in 1970 to about 750 kilograms per million pesos in 1985, or by about 55 percent, but climbed in 1985–1988. Cuba's energy intensity of gross national product in 1978 and 1987 was about three times that of Japan or France, and twice that of the United States.

Despite the increase in domestic oil production, imported oil and oil products still accounted for over 90 percent of apparent consumption of oil and oil products in the second half of the 1980s. In 1988 Cuba devoted 2.6 billion pesos to imports of oil and oil products, 34.2 percent of the value of total imports (Comité Estatal 1989, 263). Domestic oil's share of apparent consumption has probably increased in the 1990s as production rose and imports plummeted. Considering the very severe shortage of convertible currency and the draconian cuts in imports, the share of total imports accounted for by oil and oil products probably rose from the roughly one-third recorded in 1988 to at least one-half in the 1990s.

Nearly 60 percent of Cuba's consumption of oil and oil products in 1988 (total consumption was 11.1 million metric tons) was by the industrial sector of the economy. Other important consuming sectors, and their shares of total consumption, were: transportation, 11.9 percent; "other" activities (which might include use by the military), 10.2 percent; population, 7.9 percent; construction, 6.5 percent; and agriculture, 4.2 percent (Pérez-López 1992, 32).

Electricity

In 1988 the Cuban electrical system had a generation capacity of 3.853 megawatts; 98.7 percent of the generation capacity corresponded to thermoelectric plants (using fuel oil, diesel fuel, crude oil, natural gas) and the remaining 1.3 percent to a hydroelectrical plant (Pérez-López 1992, 36). In that same year the Cuban electrical system generated 14.5 gigawatt-hours of electricity; 99.5 percent of the electricity was generated by thermoelectric plants.[6] Thermoelectric plants used up over 3.3 million metric tons of liquid fuels in 1988, primarily fuel oil, or about one-half of total oil consumption by the industrial sector and 28 percent of the nation's apparent consumption (Pérez-López 1992, 36).

In the early 1990s Cuba's electricity production industry consisted of sixteen thermoelectric plants located throughout the island but primarily near the largest population (La Habana and Santiago de Cuba) or industrial centers (Cienfuegos and Nuevitas). These plants contributed significantly to air, water, and land pollution. According to Oro (1992):

▶ The "Este de la Habana" plant, a Soviet-built 200 MW power plant located in Santa Cruz del Norte, launches large amounts of smoke, soot, and hydrogen sulfide into the atmosphere. The power plant also has leakages of acid steam. Finally, the plant launches oil waste products into the marine shelf, near eastern La Habana's beaches (103).

▶ The "Antonio Maceo" plant, located in Regla, near the city of La Habana, emits large amounts of smoke, soot, and hydrogen sulfide into the atmosphere and launches oil waste products into the bay of La Habana. The plant is also the source of noise and vibrations that affect its vicinity (105).

▶ The "Otto Parellada" plant, located in Tallapiedra, within the city of La Habana, launches large amounts of smoke, soot, and hydrogen sulfide into the atmosphere and oil waste products into the bay of La Habana. The plant is also the source of noise and vibrations that affect its vicinity in the colonial area of La Habana (107).

In the mid-1980s Cuba began to experiment with the use of domestic crude oil—rather than imported fuel oil—to run electricity generation plants. Technical changes were made to the "Antonio Maceo" thermoelecric plant in Renté, Santiago de Cuba, to the "José Martí" plant in Matanzas, to the "Frank País" plant in La Habana, and to the "Máximo Gómez" plant in Mariel to run on domestic crude oil (Petróleo crudo 1986). The switch to high sulfur domestic crude exacerbated the air pollution problems associated with these plants (Petróleo nacional 1987).

Transportation

The transportation industry consumed 11.9 percent of total oil and oil products in Cuba in 1988. There is no breakdown of consumption by, for example, road, rail, and air transport, but presumably the largest consumer is road transport. Within the category of road transport, the bulk of consumption of oil products is by the state sector, since private automobiles, buses, and trucks in socialist Cuba are rare. Motor vehicles that run on gasoline emit carbon monoxide (CO), hydrocarbons, and nitrogen oxides as pollutants; hydrocarbons and nitrogen oxides combine in the atmosphere to form ground-level ozone, the principal ingredient in urban smog. Motor vehicles that run on gasoline containing lead also emit lead into the atmosphere, and those that run on diesel fuel emit hydrocarbons and sulfur dioxide (SO_2) (Walls 1994, 1–2).

Overall, the figures for motor vehicles per capita for Cuba are far below those for developed countries and advanced developing countries. According to data reported to the UN, Cuba's stock of passenger cars and commercial vehicles in 1988 was 241,300 units, of which 208,400 (86.4 percent) were commercial vehicles (United Nations 1993); data for more recent years are not available. Cuba's ratio of 38 motor vehicles per 1,000 population in 1985 (United Nations 1995b) was much lower than for developed countries (United States, 717; Canada, 567; Australia, 532; Germany, 451; France, 440; Italy, 434; Austria, 412) but also lower than for Eastern European countries for which data were available

(Bulgaria, 162; Hungary, 156; Poland, 122) and for the more advanced developing countries of Latin America (Argentina, 175; Brazil, 88; Chile, 73; Mexico, 94; Uruguay, 117; Venezuela, 117).

As has been observed for Eastern European countries, there were some ecological advantages to this "backwardness," since fewer automobiles per capita translate into lower levels of emissions (DeBardeleben 1991, 4). However, compared with motor vehicles in the United States and other developed countries, Eastern European motor vehicles tended to be more polluting for two reasons: first, pollution abatement technology in the socialist countries lagged behind the developed countries, and, second, the stock of motor vehicles in socialist countries tended to be much older and therefore lacked modern pollution control equipment, such as catalytic converters and electronic fuel injection, that newer cars have (Walls 1994, 3). About 25 percent of Hungary's stock of automobiles consisted of vehicles equipped with outmoded two-stroke engines that emit high levels of hydrocarbons, particles, and aldehydes (WRI 1992, 63).

Although statistics on pollution abatement technology and age of the Cuban stock of motor vehicles are not available, it is possible to make inferences about the origin of the stock of motor vehicles—and therefore the extent to which they might embody pollution control equipment—from data on Cuban imports. The Soviet Union, Poland, and Hungary were the main sources of Cuba's motor vehicle imports during the 1970s and 1980s. Around 1975, when the price of sugar in the world market skyrocketed and Cuban hard-currency import capacity boomed, Cuba did turn to Argentina (automobiles produced by a subsidiary of Ford Motor Company) and Japan (buses and trucks) to purchase motor vehicles, but this shift in sourcing was only temporary. The Soviet Union was by far the largest supplier of automobiles to Cuba during the entire period, followed by Poland. Over the five-year period 1982–1986, the Soviet Union and Poland provided 81.7 and 12.8 percent, respectively, of the nearly 59,700 automobiles Cuba imported; Hungary and the Soviet Union provided 54.0 and 35.5 percent, respectively, of the over 6,300 buses imported; and the Soviet Union provided 64.9 percent of the nearly 44,500 trucks imported.

Soviet and East European motor vehicles were heavy polluters, because they were not equipped with catalytic converters and other pollution abatement devices. The East German "Trabant," for example, was effectively banned in Germany after reunification (through a prohibitive tax) because it lacked catalytic converters to reduce pollutants (Tagliabue 1991). In a tirade against the low quality of Eastern European manufacturing products, Castro (Castro Discusses 1990, 15) had the following to say about the air pollution generated by Hungarian buses: "The Hungarian buses travel six kilometers on a gallon of fuel. They fill the city with smog. They poison everybody. We could get together some data. We could get statistics about the number of people killed by Hungarian buses." Unable to purchase spare parts from Hungary in hard currency to keep the bus fleet in operation, the Cuban mass transportation system essentially collapsed in the early 1990s. Cuba has accepted donations of used buses—removed from service after completing their useful life—from Canada, Spain, and the United Kingdom to attempt to rebuild its public transportation system. These units are several years old and most likely are not equipped with modern pollution abatement technologies. Ingenious Cubans have also created new forms of public transportation vehicles, for example, bus bodies pulled by truck cabs, called *camellos* (camels), as short-term solutions to the transportation crisis. These homemade vehicles, constructed with environment-unfriendly engines, are significant sources of air pollution.

Photographs taken by journalists and travelers of La Habana and other cities show numerous examples of U.S.-made automobiles from the 1950s still operating on the island.[7] These vehicles operate on leaded fuel and lack the pollution abatement technologies that have been developed in the West in the last thirty years.

Other Industries

Fertilizer

In 1958 Cuba had nearly twenty plants mixing organic fertilizers located throughout the island; most of the raw materials were imported (CERP 1965, 563). In the 1970s Cuba built two large nitrogen fertilizer plants at Cienfuegos and Nuevitas.

An aging fertilizer mixing plant at Regla, near the bay of La Habana has been identified over the years as a significant air and water polluter and as providing an unhealthy environment for its workers (La industria 1985). Oro (1992, 112) states that the Regla plant launches ammonia into the atmosphere and phosphates and nitrates into the atmosphere and the sea. The very toxic pollution emitted by this plant affects the village of Regla and the bay of La Habana.

The Cienfuegos and Nuevitas fertilizer production plants are also very significant polluters. Both plants launch great quantities of smoke and soot, as well as steam containing ammonia, that affect the surrounding areas; they also dump industrial wastes into the bays of Cienfuegos and Nuevitas (Oro 1992, 115–16).

The "Revolución de Octubre" plant at Nuevitas, in particular, has done significant damage to the bay of Nuevitas, which is heavily polluted because of the dumping of chemical fertilizers. According to two Soviet experts on marine pollution who visited the area, a contributing problem is that "significant amounts of chemical fertilizer are stored [adjacent to the bay], which are washed into the bay by the rain" (Hernández 1982, 30). Independent journalist and environmentalist Endel Cepero reported in July 1997 that the bay of Nuevitas was heavily polluted with sulfur, smoke, lead, acids, and untreated water. The principal sources of pollution of Nuevitas Bay were a wire and electrodes factory, the local fertilizer plant, a thermoelectric plant, a cement factory, and the "Potrerillo" marine fuel depot (Cepero 1997).

Mining

Among the mining operations that have been identified as major sources of pollution are

▶ the Delita Gold Mine in the Isle of Pines. Although the polluted area is small, it is affected by a very high concentration of toxic substances, including arsenic, mercury, and cyanide wastes, dust composed of arsenic and sulfur, and mercury steams (Oro 1992, 52).

▶ copper and polymetallic mines such as Júcaro (in Bahía Honda), Matahambre, Santa Lucía (in Matahambre), and Mella (Matahambre) in Pinar del Río Province and El Cobre in Santiago de Cuba Province, which emit large quantities of toxic wastes into open ditches and ponds. Pollution from the Júcaro mine has affected a nearby pine forest as well as surface and subsurface water resources. Hydrogen sulfide discharges from the Mella mine have affected 600 acres of forests (Oro 1992, 52, 83–86).

▶ quarries mined to obtain raw materials for the construction materials industry that affect nearby areas through deforestation, explosions, and dust. Examples of these problems are evident in the areas surrounding the Camoa-I and Camoa-II limestone quarries near San José de las Lajas and the quarries supplying the Mariel cement factory (Oro 1992, 52).

Chemicals

Several plants producing basic chemicals are heavy polluters. They include the "Patricio Lumumba" chemical plant, located in Matahambre, which produces mainly sulfuric acid and recovers some lead, zinc, copper, and gold. This plant reportedly launches enormous amounts of hydrogen sulfide (containing heavy metals) into the atmosphere and industrial waste onto the marine shelf; the air pollutants have affected the villages of Santa Lucía and La Sabana, where asthma and pulmonary diseases are rampant, and technological waste has affected the coastal and shallow-water marine life (Oro 1992, 87). The Rayonitro plant in Matanzas and the chlor-alkali complex in Sagua la Grande expel acid steams into the atmosphere (Oro 1992, 53); the latter facility also pollutes the Sagua la Grande River and the estuary at Isabela de Sagua with heavy metals, principally mercury (González, Lodenius, and Otero 1989; González 1991). Two plants located within the city of La Habana producing detergents—Suchel and Sabatés—launch into the atmosphere significant amounts of smoke and steam. The pollution emitted by these plants is particularly significant because of the heavy population density in nearby areas and the large amount of pollution discharges by other sources (Oro 1992, 79, 107).

Important sources of pollution within the metalworking industry are the large steel mill complex "Antillana de Acero," located about 20 kilometers from the city of La Habana, and the "26 de Julio" combine plant in Holguín. The latter plant, which produces the KTP sugarcane harvesters, emits significant quantities of smoke, soot, and noise, which affect the city of Holguín (Oro 1992, 54, 113).

Other

Other industrial plants identified by Oro (1992) as significant sources of pollution are

▶ the "Melones" manufactured gas plant, in the city of La Habana. It emits large amounts of smoke, soot, and mephitic steams into the atmosphere (117).

▶ tallow processing factories in Regla and San Miguel del Padrón, both in the city of La Habana. The two plants emit strong unpleasant odors, smoke, and acid steams; waste products from the San Miguel del Padrón plant are run into a ditch, and those from the Regla plant run directly into the bay of La Habana (106, 109).

▶ a feed plant in Boyeros, city of La Habana, that emits unpleasant smells and huge amounts of dust into the atmosphere and also attracts large numbers of insects. Emissions from the plant are harmful to the Boyeros areas, including the urban areas of Río Cristal, María del Carmen, and Capdevila (89).

▶ a marble polishing factory in Marianao, city of La Habana, that launches carbonated and silicated abrasive dust into the atmosphere and is also the source of intensive noise. It affects the area of La Ceiba in the municipality of Marianao (130).

▶ a plant producing tires in the municipality of Cotorro, near the city of La Habana, that launches soot, including black carbon particles, which runs windward about 2–3 miles and affects air quality of a very heavily populated area (131).

▶ the ice cream factories "Guarina" and "Coppelia," located in the 10 de Octubre and Boyeros municipalities of the city of La Habana, respectively, and a milk complex operated by the Ministry of the Food Industry in La Habana Vieja.

▶ The "Coppelia" plant discharges milk wastes into the already-polluted Almendares River; the "Guarina" plant and the milk complex do the same into the bay of La Habana (110–11, 132).

A recent report (Nogueras Rofes 1997) indicates that a corn fermentation plant in the industrial zone adjacent to Cienfuegos Bay is releasing large amounts of industrial discharges into the bay's waters, where bacteria counts are alarmingly high. Pressure to develop the industrial zone and government inaction regarding environmental controls have combined to make the situation so critical that a researcher from the provincial Hygiene and Epidemiological Center considers the damage to the bay to be irreversible. According to a joint study by the National Institute of Hydraulic Resources and the Public Health Ministry, the situation has worsened because of poorly developed pollution abatement infrastructure.

Summary and Implications

Over three decades of socialist industrial development policies appear to have aggravated Cuba's environmental situation. Although many of the sources of industrial pollution being discharged into the soil, the atmosphere, and the waters in and around Cuba were already present before the revolution (e.g., sugar mills, nickel mines and refineries, oil refineries, cement plants), the levels of discharges increased greatly as the industrial sector expanded and because Cuba incorporated into its industrial infrastructure inferior and more polluting Soviet and Eastern European technology.

The evidence is convincing that Cuban planners and plant managers have been driven by bigness and maximization of physical production goals, with very little attention paid to protecting the environment. Ian Delaney, chairman of Canadian-based Sherritt, Inc., a company that has entered into joint ventures with Cuba to mine and process nickel, summed up the attitude of Cuban authorities as follows: "They should take the pollution problems more seriously" (Knox 1995).

The pollution associated with the production of sugar and derivatives increased in proportion with sugar output during the 1980s, when annual sugar production exceeded by about one-third average production during the 1950s (8 versus 6 million tons per annum). Air and water pollution levels also rose as the installed capacity of oil refineries, cement plants, and other industrial facilities increased severalfold between the 1960s and the late 1980s. No less environmentally damaging was the expansion of the country's mining industry, particularly due to the widespread use of strip mining techniques and of ammonia and acid leaching processes for the production of nickel. Inattention to the environmental consequences of open-pit mining, ineffective land reclamation efforts, and inadequate systems to abate industrial discharges from nickel processing have caused severe air, soil, water, and coastal pollution. Because of the general neglect of pollution abatement technology, several of Cuba's excellent bays—La Habana, Moa, Nipe, Cienfuegos, which have small openings to the ocean and therefore are subject to low water flows—are highly contaminated, with many reports suggesting that the damage has been so great as to be irreversible.

On the positive side, the level of industrial pollution in Cuba, though not insignificant, was limited by the fact that under the international socialist division of labor, Cuba was assigned the role of an agricultural and raw material producer, with industry relegated to a secondary role. Further, many of the worst consequences of air and water pollution were mitigated by Cuba's insular and elongated geography and by the presence of marine winds that help dissipate air contamination. The evidence is insufficient at this time to assess the long-term consequences of industrial and agricultural pollution on bodies of inland water, although it has been documented that many of Cuba's coastal areas have been decimated by industrial, mining, and agricultural discharges. A geologist who worked for more than twenty years at Cuba's Instituto de Geografía (Geography Institute) claims that of the country's 32 cubic kilometers of water reserves, 13 cubic kilometers are contaminated with chemical residues, including carcinogens such as nitrates (García Azuero 1994, 6A). Although most of these chemical residues can be traced to agriculture, many others originated from the industrial sector.

❧ 8 ❧

Nuclear Energy and the Environment

REPORTING FROM the city of Cienfuegos, on Cuba's southern coast, independent Cuban journalist Olance Nogueras Rofes wrote in 1995 about the nuclear power plant under construction at nearby Juraguá. Nogueras Rofes' column, published in a U.S. newspaper, began as follows:

> Juraguá dances on a string of incredulity. Majestic, sacred, untamed, the Cuban nuclear power plant awaits the start-up of its first reactor, product of collaboration between Russian strategies and the interests of multinational corporations, responding to the much-used slogan, "The Project of the Century." . . . Within the mountains of cement of the plant grows the worst nuclear catastrophe of the Hemisphere. Therein lie horror, desolation, and suffering for millions of human beings and the genetic future of an entire region. Juraguá is a time bomb. (Nogueras Rofes 1996)

Based on interviews with Cuban experts associated with the Juraguá project—not identified by name in the article—Nogueras Rofes went on to discuss several specific areas of concern: defective components supplied by the Soviet Union for the project, poor quality of construction work, inadequate training of nuclear workers.[1]

This chapter examines environmental issues associated with the use of nuclear power in Cuba. During the pre-revolutionary period, the use of nuclear energy for scientific, medical, and economic purposes in Cuba was limited, as would be expected to be the case for such a new technology. After the revolutionary takeover of 1959, nuclear technologies assumed a higher profile, in part because of Fidel Castro's naive enthusiasm for nuclear energy to solve Cuba's chronic fuel shortages and the influence of Soviet thinking on overcoming problems through technological interventions. We focus on the largest—and potentially most environmentally harmful—undertaking by the Cuban government in the nuclear field: the construction of nuclear power reactors for electric power generation. This is followed by a discussion of safety issues associated with Soviet-built nuclear power reactors. Finally, in the last section of the chapter we discuss specific safety concerns about the much-delayed and currently suspended Juraguá nuclear power plant based on analyses by numerous Western experts and testimony of scientists associated with the design and construction of the plant.

Nuclear Energy in Pre-Socialist Cuba

In November 1947 the Cuban government established the Comisión Nacional de Aplicaciones de la Energía Atómica a Usos Civiles (National Commission for the Use of Atomic Energy for Civilian Purposes) under the aegis of the Ministry of Public Health and Social Welfare. Among the functions of the commission were to encourage research on atomic energy and its peaceful applications, particularly in medicine and industry; to seek to identify natural deposits of radioactive mineral ores; to promote the use of nuclear techniques and materials in the health sector; to offer scholarships and opportunities to travel abroad to specialists to enhance their education; and to control the handling and use of radioactive materials (Castro Díaz-Balart 1990a, 332).

In 1949 radioisotopes began to be used in Cuba for the diagnosis and treatment of certain illnesses, particularly malignant tumors. As the practice of nuclear medicine expanded during the 1950s, several medical centers acquired trained personnel and equipment to use these techniques. Around 1956 Cuba hosted the Second Latin American School on the Uses of Radioisotopes, sponsored by UNESCO and attended by physicians and scientists from Cuba and the rest of the hemisphere. Cuban representatives also attended several professional gatherings of nuclear scientists sponsored by UNESCO and other organizations, such as the first two UN Conferences on the Peaceful Application of Nuclear Energy held in Geneva in 1955 and 1958. In 1957 Cuba became a member of the

IAEA and participated regularly thereafter in its activities (Pérez-López 1979, 3).

In June 1955 the Comisión de Energía Nuclear de Cuba (CEN, Cuban Nuclear Energy Commission) was created to promote and coordinate peaceful uses of nuclear techniques in industry and agriculture. At that time, two modest nuclear physics laboratories were operating in Cuba, one at the University of La Habana and the other at the private university Santo Tomás de Villanueva. In 1957 CEN sponsored the First National Symposium on Civilian Uses of Atomic Energy, attended by leading nuclear scientists from around the world.

Pursuant to the Atoms for Peace Program, the United States entered into thirty-seven "peaceful bilaterals" with individual countries between 1956 and 1962, covering the provision of nuclear research and power reactors as well as technical advice and training (*Atoms for Peace* 1990, 6–7). Cuba and the United States negotiated such an agreement in June 1956; the agreement was approved by the Cuban Senate in November 1957 (Convenio Sobre Cooperación 1958). Pursuant to the agreement, the United States pledged, among other things, to assist Cuba with the design, construction, and operation of nuclear research reactors and to supply enriched uranium to fuel them.

In the late 1950s CEN planned the installation of a nuclear research center equipped with a nuclear research reactor to be supplied by a U.S. manufacturer. The process of obtaining the requisite approval for the export of the equipment and fuel from the U.S. Atomic Energy Commission had not been completed at the time of the overthrow of the Batista regime and therefore the installation of the nuclear research reactor did not occur.

At least three projects involving the installation of commercial nuclear power plants were on the drawing board during the late 1950s. These projects were either abandoned because they were not economically feasible or aborted by the 1959 revolution. The first project, with financing from the Banco Nacional para el Desarrollo Económico y Social (BANDES, National Bank for Economic and Social Development), involved plans to build a small nuclear power plant (20.5 megawatt generating capacity) to provide electricity to an agroindustrial complex near the Matahambre copper mines in Pinar del Río Province. CEN approved construction of the plant, which was to be carried out by a British–U.S. consortium consisting of Mitchell Engineering Company of England, the American Machine and Foundry Company of New York, and General Nuclear Engineering of Florida. In September 1958 Cuba and the United States signed an atomic power agreement that authorized the sale or lease to Cuba of up to 700 kilograms of uranium enriched to 20 percent and 8 kilograms of uranium enriched to 90 percent for this facility (Status Report 1958). The second project, at an earlier stage of planning, was the establishment of a steel mill in Oriente Province equipped with two nuclear power reactors that would supply electricity (Status Report 1958). The third project involved plans by Compañía Cubana de Electricidad (Cuban Electric Company) to construct a small boiling-water nuclear power reactor (11.5 megawatt generating capacity) in the center of the island, near the Zapata Swamp. Reportedly, the plans were abandoned because the plant was not economically feasible (Castro Díaz-Balart 1990a, 335).

Nuclear Energy in Socialist Cuba

That Cuba's socialist government has devoted substantial resources to nuclear energy is not surprising given Fidel Castro's long-standing enthusiasm regarding the ability of nuclear science to solve one of the country's most chronic economic problems: a severe shortage of commercial energy resources.

More than forty-five years ago, in his defense at the trial where he was accused of leading the July 1953 attack on the Moncada Barracks, Castro laid out a vision for a revolutionary government in Cuba that would resolve the nation's economic, social, and political problems. Among many proposals, Castro stated that "today possibilities of taking electricity to the most isolated spots of Cuba are greater than ever. The use of nuclear energy in this field is now a reality and will greatly reduce the cost of producing electricity" (Castro 1953, 52). The prescience of the Cuban leader has been noted by his son, Fidel Castro Díaz-Balart (1986, 5), the erstwhile Cuban nuclear energy czar, who has reminded us that this pronouncement by his father was made in October 1953, "nearly a year before the Soviet Union put into operation the first nuclear power plant in the world."

Elements of the Cuban Nuclear Program

In the second half of the 1980s the Cuban nuclear program consisted of four elements: (1) construction of the first of several nuclear power plants to generate electricity, (2) introduction of nuclear technologies in other areas of the national economy, (3) promotion of basic and applied research, and (4) creation of a system of nuclear safety and radiological protection (Castro Díaz-Balart 1990b, 50–52).

Electricity generation

As will be discussed, Cuba harbored ambitious plans to build several commercial nuclear power plants to generate electricity. This aspect of the Cuban nuclear program was under the direction of the Nuclear Directorate of the Ministry of Basic Industry.

Use of nuclear power in other areas of the national economy

In addition to electricity generation, Cuba focused on the use of nuclear technologies and ionizing radiation in various areas, for example,

▶ *Medicine:* Labeling compounds for diagnostic methods, radiation sterilization of medical and pharmaceutical products, and radiotherapy for treating malignant tumors (a linear electron accelerator for treating malignant tumors was in operation);
▶ *Agriculture:* Irradiation of agricultural products such as potatoes, onions, garlic, cocoa, and spices; labeling of nitrogen-based fertilizers to determine the precise timing and dosage to be delivered; experimentation with the sterile insect technique to control pests that attack sugarcane and corn; and use of radioimmunoassay techniques for early diagnosis of pregnancy in cattle.
▶ *Industry:* Use of radioactive tracers and nuclear analytical techniques in hydrological studies in the sugar industry.

Basic and applied research

Development of basic and applied research as essential parts of the infrastructure was needed to carry out plans to operate nuclear power plants and use nuclear techniques in the economy. Particularly important in this regard was the work of the Center for Applied Nuclear Development Studies and the Nuclear Research Center; the latter institution, created jointly with the Soviet Union, was to be equipped with a 10-megawatt research reactor, several laboratories, an Isotope Center (aimed at meeting growing domestic demands for radio-pharmaceutical products and labeled compounds), and a Nuclear Instrument Application and Development Center.

Nuclear safety and radiation protection

With over 1,200 instruments and facilities run by some 800 specialists and more than 1,400 workers subject to occupational exposure to radiation, Cuba required a regulatory framework that would ensure human and environmental protection against radiation. A Center for Radiation Protection and Hygiene was created to monitor and operate the system of radiation protection—in conjunction with an environmental radiological surveillance network covering laboratories located throughout the country—and to manage low-level radioactive wastes.

Environmental Concerns

Independent of the special environmental problems associated with commercial nuclear power plants for electricity generation, which are discussed in more detail later, several other elements of Cuba's nuclear power program raise environmental concerns because of the risk of radioactive contamination they pose to air, land, and soil. Scientist José Oro (1992, 25–32, 68–69) has identified the following projects as being particularly capable of harming the environment:

▶ "La Quebrada" nuclear research center, near the town of Pedro Pi, 25 kilometers southeast of the city of La Habana. The research center, located within a 5-square-kilometer compound, consists of several facilities, including an operational 10-megawatt Soviet-designed research reactor, a zero-power Hungarian-designed research reactor, and an underground nuclear fuel deposit.

▶ Radioisotopes Production Center, located at "La Quebrada" nuclear research center. The basic equipment for the center was purchased from Argentine suppliers; equipment also originated from Japan, Germany, and other nations.

▶ "IV Workshop," a secret experimental installation for processing nuclear waste. The workshop was built by the Ministerio de las Fuerzas Armadas Revolucionarias (MINFAR, Ministry of the Revolutionary Armed Forces) and SEAN in an underground bunker on the Managua Air Force Base, in the municipality of Arroyo Naranjo, 15 kilometers southeast from the center of the city of La Habana. The storage and processing of nuclear waste at a military air force base where there are large stores of jet fuel and ammunition and frequent takeoffs and landings of military airplanes is very risky. Also of concern is the facility's location, which is very near the heavily populated city of La Habana.

▶ "IX Workshop," a projected depository for nuclear and highly toxic wastes to be built near Manicaragua, in the Escambray Mountains, in central Cuba. The purpose of the facility would be to deep-bury radioactive waste from nuclear power plants as well as other highly toxic wastes. Such waste storage facilities raise concerns about contamination of underground waters as well as air and land from chemical explosions.

Nuclear Power for Electricity Generation in Cuba

During an address to Cuban construction workers on Builders Day, 5 December 1974, President Fidel Castro announced plans to build a commercial nuclear power plant in Cuba: "In 1977–78 we shall begin construction of a nuclear power plant with an initial capacity of 440 MW, and then a second unit. They shall begin to operate in the next five-year period, that is, during 1981–85" (Castro 1974, 2). Subsequently it was announced that the first nuclear power plant would be built at Juraguá,[2] near Cienfuegos, in the central region of the island, and would consist of four nuclear reactors, each capable of producing 440 megawatts of electricity; construction of two other plants was contemplated, the first near Holguín in the eastern region of the island and the second in the western region (Castro Díaz-Balart 1986, 9–10).[3] Cuban sources estimated that each 440-megawatt nuclear reactor would save 600,000 metric tons of oil and oil products per annum (Castro Díaz-Balart 1985, 47), thereby reducing substantially the dependence on imported oil and oil products and improving the overall trade balance.

Construction of two nuclear reactors at Juraguá began in the first half of the 1980s. The Juraguá reactors are Soviet-designed pressurized-water reactors, each with 440 megawatts of electrical generation capacity, generally known as VVER-440. The prototype for the VVER-440 was built in Novovoronezh, in the Ukraine, and began commercial operation in 1972. The Soviet Union built and operated several VVER-440 reactors and also exported them to Bulgaria, Czechoslovakia, Finland, East Germany, Hungary, Poland, and Romania (Pérez-López 1987b, 82).

Ground was broken at the Juraguá site in the late 1970s. Actual construction of the facilities to house the first two reactors began in 1983 and 1985, respectively. The slowdown and eventual near-break of traditional commercial relations between Cuba and the former Soviet Union that occurred in 1989–1990 affected deliveries of Soviet materials and equipment for the plant. In late September 1990, as Cuba experienced a severe economic crisis and electric power shortages were common because of shortages of oil imports, Castro (1990, 3) lamented: "We are building a nuclear power plant with the USSR. It was going to have four 400,000-kilowatt reactors. The construction of the first two is at an advanced stage. How much we could use those reactors now! Those reactors have not fallen behind schedule through our own fault. When will they be ready? We do not know. Will they ever be completed? We do not even know that."

In September 1992, speaking before an assembly of construction workers at

Juraguá, Castro announced that work on the nuclear power plant would be suspended indefinitely because of the lack of resources to pay for hardware and services to complete the plant and the inability to reach a satisfactory financing agreement with Russian authorities (Castro 1992). According to Cuban experts (see Table 8.1), at the time the Juraguá project was suspended 75 percent of construction work and 20 percent of installation of equipment for the first nuclear reactor had been completed (Serradet Acosta 1995).

In October 1995 Russia and Cuba entered into an arrangement to complete the Juraguá reactors. Russia pledged $349 million for the project. Cuba and Russia also agreed to create a joint venture that would seek to attract one or more foreign partners to assist with financing, construction, and operation of the reactors. The joint venture would operate the reactor and sell the electricity to the Cuban government; the new partner or partners would be the first to recoup their investment, the Russian partner second, and the Cuban state third. To date, new partners have not been found—despite overtures to Siemens AG of Germany, Ansaldo SpA of Italy, and Electricité of France (Rosett and de Córdoba 1995; Benjamin-Alvarado 1996a, 5). Thus, the Juraguá facilities remain "mothballed" until such time as a suitable deal to complete them can be consummated.

Since 1995, the IAEA has been providing Cuba with supervision and advice in the conservation, or "mothballing," of the Juraguá nuclear power plant during the suspension of construction in order to facilitate the resumption of the power

Table 8.1. Construction status of nuclear reactor no. 1, Juraguá nuclear power plant, as of May 1996

Activity	Projected	Completed	Completed/projected (%)
Construction			
Earth moving (thousands of cubic meters)	4,321	3,600	83
Concrete (thousands of cubic meters)	354	270	76
Reinforced steel (tons)	27,023	23,396	86
Steel inserts (tons)	4,870	4,461	92
Metal structures (tons)	9,967	7,799	78
Equipment			
Thermomechanical (tons)	2,031	1,174	44
Pressurized steam (tons)	1,039	655	63
Ventilation (tons)	990	241	24
Electrical (tons)	14,742	2,066	14

Source: Benjamin-Alvarado 1996b, 10.

plant's activities (U.S. General Accounting Office 1997, 17). In response to criticism by members of Congress regarding IAEA assistance to Cuba, a U.S. State Department spokesperson stated that the objective of the assistance is to maintain continuous inspection of the facility and that no funds are being used to help Cuba in restarting construction of the plant (EU defiende 1997).

Cuban Minister of Foreign Investment and International Cooperation Ibrahim Ferradaz stated in early January 1997 that Cuba continued to seek one or more foreign partners to complete the Juraguá plant and attributed the lack of success to the U.S. Helms-Burton Act, which specifically sanctions foreign participation in the project.[4] According to Cuban officials, Russia and Cuba invested the equivalent of $1.1 billion in the construction of the plant before work ceased, and an additional $750 million would be needed to put the first reactor into operation (Frank 1997).

However, at a major address marking the Cuban Day of Science in mid-January 1997, President Castro announced that the Juraguá plant was "indefinitely postponed" as efforts to recruit a foreign partner had not succeeded. Castro stated that "there was no hope for now" to solve the country's energy problems through nuclear power and that completion of the Juraguá facility would have to be "postponed for some time." Instead, Castro proposed modernizing existing thermoelectric plants to improve their efficiency and reduce oil consumption (Postergada 1997). Nevertheless, in mid-1997 Russian and Cuban sources suggested that plans to complete the Juraguá plant were still alive. Russia's vice minister of Atomic Energy told journalists in June 1997 that several potential partners in a multinational consortium were studying necessary legal documents and the consortium would be formalized by January 1998 (Rusia reanudará 1997). In July, Danilo Alonso, director of Cuba's Nuclear Energy Information Center, stated that low-cost nuclear energy continued to be an attractive alternative for Cuba and "there is no motive to renounce that option"(Cuba intent 1997). In February 1998 Russia's Minister of Civil Defense Serghei Shoisu told journalists in La Habana that his country was pursuing discussions with prospective joint venture partners to complete the Juraguá plant (Rusia desea 1998).

Safety of Socialist Nuclear Power Plants

On 26 April 1984 two explosions rocked reactor no. 4 at the Chernobyl nuclear power station, 50 miles north of Kiev, in the Ukraine. The explosions partly destroyed the roof of the building housing reactor no. 4, which quickly became engulfed in flames. Wind from the southeast initially carried radioactive discharges spewed out from the damaged plant to the Ukraine, Lithuania, Latvia, Poland, Sweden, and Norway, and later, when the winds shifted, to Germany, the Netherlands, and Belgium (Oberg 1988, 253–54).

The nuclear reactors at Chernobyl were graphite-moderated boiling-water reactors generally known as RBMK. These reactors, designed and built in the Soviet Union, were extensively used for commercial power generation in that nation. Reactor no. 4 at Chernobyl, an RBMK-1000, lacked the pressure-tight containment structure common in nuclear plants in the West. Had this reactor been so equipped, the containment structure might have confined much—perhaps most—of the radioactive materials that were actually spewed into the atmosphere. During the most serious nuclear accident in the United States, the 1979 Three Mile Island accident, a containment structure played a crucial role in preventing the escape of radioactive material, although the enormous amount of energy released at Chernobyl might well have breached the strongest containment structure.

The Chernobyl accident brought into the open the significant differences in the approach to nuclear power safety between the West and the Soviet Union. The Western approach traditionally has been to build in redundant systems to counteract a range of uncertainties, including low-probability occurrences such as loss-of-coolant accidents and core meltdowns. The Soviet approach has been to emphasize care in the design, manufacture, and installation of equipment (engineered safety) and to minimize redundant systems by limiting them only to "credible" occurrences, which exclude low-probability events (Lewin 1977; Pryde and Pryde 1974; O'Toole 1978). In the pre-Chernobyl era Soviet nuclear experts argued that their nuclear power plants, built in a system where the profit motive that drives capitalists to cut corners and to produce substandard goods and services did not exist, were completely safe and that Western-style safety systems were both unnecessary and expensive (MacLachlan 1979; Fialka 1978). Reportedly, after a visit to the United States in the late 1970s, a Soviet nuclear science delegation came away convinced that containment structures "were placebos to 'placate the people,' necessary because of 'negative dramatization' by the American press" (Babbitt 1980).

Soviet nuclear scientists and government officials also boasted about the accident-free record of their nuclear program, despite persistent reports in the West about the occurrence of major accidents. One such accident occurred in the Ural mountains around 1957–1958 presumably involving stored nuclear waste at a weapons production complex (Medvedev 1978; Trabalka et al. 1979 and 1980). It was not until 1979 that Soviet officials confirmed that nuclear accidents had occurred in their country but downplayed their seriousness (Soviet Official 1979; Top Soviet 1979).

With the lifting of the information veil over the former Soviet Union and the Eastern European socialist countries, the seriousness of the public safety problems associated with Soviet-built nuclear power plants became evident. Murray Feshbach (1995, 38–39) has compiled a list of the most widely reported accidents, serious and minor, at such plants in recent years; the following are among them:

▶ Between 1986 and 1992, there were 118 fires at Soviet nuclear power plants, 60 percent of which occurred in the machine and reactor halls; most of Russia's nuclear power plants are still inadequately protected against fire.

▶ A breakdown occurred in a regulatory valve shutter on one of the water cooling pipes at Sosnovy Bor on 24 March 1992, resulting in the release of radioactivity into the atmosphere.

▶ An oil switch explosion occurred on 29 April 1992 requiring the partial shutdown of the Balakovo nuclear facility.

▶ A short circuit occurred in the system for protection against earthquakes in the Khmelnitskiy nuclear station in the Ukraine in May 1992, and a fault occurred in the measuring systems at the South Ukrainian power plant in September 1992, which resulted in the temporary shutdown of the plants.

▶ A fire occurred in the backup electrical system at the Kozloduy nuclear power plant in Bulgaria in September 1992.

▶ Radioactive leaks developed at the Ignalina nuclear power plant in Lithuania in October 1992 and at the Kursk plant in Russia in December 1992; the latter accident resulted in casualties among the staff.

Soviet Nuclear Reactors

Within the socialist community, the Soviet Union controlled the design and production of nuclear reactors, with other socialist countries typically supplying components of nuclear power plants (e.g., instrumentation, steam turbines, pumps). The Soviet Union produced four basic commercial reactors: (1) graphite-moderated reactors (RBMKs), such as the one involved in the Chernobyl accident; (2) first-generation pressurized-water reactors VVER-440 Model 230; (3) second-generation pressurized-water reactors VVER-440 Model 213 or Model V318 reactors; and (4) large pressurized-water reactors VVER-1000. RBMKs were only constructed within the Soviet Union, whereas VVER reactors were constructed domestically and also exported.

Generally speaking, all Soviet reactors have been judged by Western experts to suffer from safety deficiencies arising from inadequate design, substandard fabrication, and faulty construction practices, although the severity of the deficiencies varies by type of reactor (Chakraborty 1995, 47; Wilson 1995). The most severe problems are those of the RBMKs. Design and construction inadequacies result from a lack of consistent design basis, failure to account for adequate separation and system redundancies, and insufficient fire and seismic support systems. RBMKs lack containment structures (generally, a steel-lined concrete, domelike structure that serves as the ultimate barrier to the release of radioactive material in the case of an accident); a core meltdown is likely to be catastrophic in these reactors. VVER-440 Model 230s also have serious design deficiencies, chief among them are the lack of a containment structure, absence of containment for pipe breaks, no substantial emergency core-cooling system, failure to separate critical electric and fuel systems, and inadequate fire protection. The more modern VVER-440 Model 213s remedied the most serious problems in the Model 230 series. They have "some degree of containment with a bubbler suppression and a large volume inside the partial containment" (Wilson 1995, 40). The VVER-1000s are the most modern Soviet-designed nuclear reactors, similar in concept to Western designs. The VVER-1000s feature a reactor surrounded by

Table 8.2. Nuclear power plants in operation or under construction worldwide using the Soviet VVER-440 reactor (as of 31 December 1994)

Country and plant	Status	Date of operation or scheduled completion
Bulgaria		
Kozloduy 1	Unit in commercial operation	December 1974
Kozloduy 2	Unit in commercial operation	December 1975
Kozloduy 3	Unit in commercial operation	January 1981
Kozloduy 4	Unit in commercial operation	August 1982
Cuba		
Juraguá 1	Unit under construction	N/A
Juraguá 2	Unit under construction	N/A
Czech Republic		
Dukovany 1	Unit in commercial operation	August 1985
Dukovany 2	Unit in commercial operation	September 1986
Dukovany 3	Unit in commercial operation	May 1987
Dukovany 4	Unit in commercial operation	December 1987
Finland		
Loviisa 1	Unit in commercial operation	May 1977
Loviisa 2	Unit in commercial operation	January 1981

a full containment structure similar to that used in Western nuclear reactors. AlthoughVVER-1000s are superior to the rest of the Soviet reactors, concerns have been raised about their instrumentation and control systems, performance of steam generators, and reactor power stability.

Table 8.2 lists all of the VVER-440 units under construction or in operation worldwide as of the end of 1994.[5] This type of Soviet-designed reactor is the most relevant for this discussion since Cuban officials have stated that the Juraguá plant will be equipped with a variant of the VVER-440s. According to Table 8.2, at the end of 1994 there were twenty-six VVER-440s operating commercially in seven countries (Bulgaria, Czech Republic, Finland, Hungary, Russia, Slovakia, and Ukraine) and two under construction (both in Cuba). The older VVER-440s—Model 230—were built in the 1960s and 1970s in Russia (Novovoronezh), Bulgaria (Kozloduy), and Slovakia (Bohunice) in the 1960s and 1970s. The remaining VVER-440s in operation correspond to the newer, and relatively safer, Model 213 or its variant Model V318 (see later).

Table 8.2. *(continued)*

Hungary		
Paks 1	Unit in commercial operation	August 1983
Paks 2	Unit in commercial operation	November 1984
Paks 3	Unit in commercial operation	December 1986
Paks 4	Unit in commercial operation	November 1987
Russia		
Novovoronezh 3	Unit in commercial operation	June 1972
Novovoronezh 4	Unit in commercial operation	March 1973
Kola 1	Unit in commercial operation	December 1973
Kola 2	Unit in commercial operation	February 1975
Kola 3	Unit in commercial operation	December 1982
Kola 4	Unit in commercial operation	December 1984
Slovakia		
Bohunice 1	Unit in commercial operation	April 1979
Bohunice 2	Unit in commercial operation	January 1981
Bohunice 3	Unit in commercial operation	May 1985
Bohunice 4	Unit in commercial operation	March 1986
Ukraine		
Rovno 1	Unit in commercial operation	September 1981
Rovno 2	Unit in commercial operation	July 1982

Source: World List 1995.

The Loviisa nuclear power plant in Finland has two Soviet-design VVER-440 Model 213 reactors. The two reactors were modified in order to meet Western safety standards. The Finnish customer constructed a steel and concrete containment structure around the reactors and contracted with the U.S. corporation Westinghouse for an ice-condenser emergency cooling system and with the German corporation Siemens for modern instrumentation. Reflecting the mix of Soviet reactors and Western safety hardware and engineering, the Loviisa reactors are often referred to as "Eastinghouse."

It is interesting to note that with German reunification all eight of the Soviet-designed VVERs in operation or under construction in the former East Germany were shut down because of safety concerns: "West German experts made recommendations for safety improvements at the four East German VVER-440/230 reactors at Lubmin Nord. Because at that particular moment [1991] there was an excess of generating capacity in West Germany, it was decided to shut them down (together with a VVER-440/213) rather than to upgrade them. Three other VVER-440/213 reactors under construction at Lubmin Nord and two VVER-1000 at Stendal were canceled at the same time" (Wilson 1995, 40). Two VVER-440 Model 230 reactors in Mezamor, in the then-Soviet Republic of Armenia, were shut down in January 1989 "for security reasons" after a major earthquake struck near the plant. The government of the now independent Republic of Armenia stated that it intended to restart the reactors in the second half of the 1990s after making them "more earthquake resistant" (Bonner 1993) and did so in late 1995, despite serious safety reservations by Western nations (Hiatt 1995; LeVine 1995).

Safety Review of VVER-440 Reactors

In 1990–1991 a group of specialists from twenty-one countries and the IAEA conducted safety evaluations of ten Soviet-designed VVER-440 Model 230 reactors operating commercially in Bulgaria (Kozloduy), Czechoslovakia (Bohunice), and the Russian Federation (Novovoronezh and Kola). The overall result of the evaluation was that "these plants have serious shortcomings in safety in comparison with international practice. At the same time, however, they have some features that make them less sensitive to disturbances than plants of other types" (Niehaus and Lederman 1992, 24; Ripon 1992, 69). The scope of the reviews included core design, system analysis, mechanical and component integrity, instrumentation and control, electric power, accident analysis, fire protection, plant management and organization, quality assurance, operator training and qualification, conduct of operations, maintenance, technical support, and accident planning.

The IAEA distinguished between design and operational problems, and ranked issues according to their safety significance from Category I ("departure from recognized international practices; it may be appropriate to address them as part of actions to resolve higher priority issues") to Category IV ("issues . . . of the highest safety concern; immediate action is required to resolve the issue") (Niehaus and Lederman 1992, 28). Over 1,300 specific safety problems were identified in the four plant complexes, 56 percent related to design and 44 percent to operational experience; 7 percent of the problems were deemed to be in Category I, 34 percent in Category II, 46 percent in Category III, and 13 percent in Category IV (Ripon 1992, 72–73).

While some of the results of the review with regard to design issues are specific to the Model 230 reactors, some also apply to the newer-generation reactors such as those under construction in Cuba. The operational concerns also apply to Cuba, since they are deemed to "emanate from sociological, economic, and political problems of the closed, state-controlled societies" (Chakraborty 1995, 46).

Design issues

The most critical design-related safety issue regarding the VVER-440 Model 230 reactors was the potential for common-cause failures affecting the core decay heat removal system because of the lack of segregation of systems. In these reactors the systems that operate the turbines and provide emergency cooling are located very close together. A common-cause accident—such as a fire in a cable gallery—could affect the emergency cooling system. Similarly, a natural event such as an earthquake could affect the source of service water as well as water for the emergency cooling system. The IAEA report identified over two dozen design-related safety issues involving the VVER-440 Model 230.[6]

Operational issues

The IAEA found significant divergences from normal international practices in the operation of VVER-440s.[7] It concluded that they resulted "largely because the WWER [VVER] operators were isolated from the international community until recently. The safety review missions concluded that immediate attention is needed at the plants to improve the approach to operations, to improve the standards of maintenance, and to instill a higher sense of safety awareness in their staff. In a number of instances, the key elements necessary to establish a safety culture were missing" (Niehaus and Lederman 1995, 30).

Resource constraints were evident in all aspects of power plant management and support; plant layouts and piping and flow diagrams were not always available and consistent with the as-built facility; safety studies were not comprehensive and lacked adequate and full documentation; basic and spare parts were in short supply; trained personnel were leaving the industry to seek better financial rewards elsewhere; and the overall economic stagnation was affecting the morale of the workers. Even worse were the conditions in the nuclear regulatory side of the industry, where rampant inflation, the government's inability to meet basic needs of workers, and the lack of adequate controls, a mandate, a clear sense of direction, and adequate oversight powers, coupled with the lack of a clear government policy for nuclear regulation, significantly eroded the effectiveness of regulatory functions (Chakraborty 1995, 47).

The lack of a nuclear safety culture was a significant long-run problem, with implications well beyond the time interval when some of the design and mechanical concerns of the power plants could be addressed. According to a nuclear safety expert,

> Little attention has been paid to nuclear safety culture in the former Eastern Bloc countries. Inadequacies in nuclear safety culture seriously affect the safe operation of the nuclear power plants in these countries. Recent reviews reveal that inadequacies exist in all aspects of nuclear power development, including system design practices, selection of sites, construction habits, operating procedures, training, maintenance practices, and accident response procedures.... Lack of adequate attention to the details of industrial safety practices by nuclear plant operators, together with inadequate regulatory requirements, resource constraints, and a clear national nuclear safety policy, is hindering implementation of many of the safety improvements that have been proposed through many international programs. This is perhaps the most significant hindrance to nuclear safety within the former Eastern Bloc nuclear industries. Even if design deficiencies, regulatory shortcomings, and operating procedures are corrected, without very serious attention to safety culture, a measured improvement in the nuclear power safety in these countries could not be achieved. (Chakraborty 1995, 47)

International Response

Concerns about the devastating effect of a major nuclear accident at any of the fifty-eight Soviet-built commercial nuclear power reactors located in the former Soviet Union and Eastern Europe on the populations and resources of the host country and of the entire community of nations prompted a strong response by Western nations. In July 1992 the G7 countries[8] announced a multimillion dollar emergency plan to improve the safety of Soviet-built reactors, which considered the shutting down of the most dangerous plants. The plan consisted of (1) operational safety improvements such as training of plant personnel, (2) near-term technical measures (e.g., new equipment or enhancements of existing equipment) that will produce early benefits, and (3) enhancement of independent regulatory agencies in the countries operating the reactors. With regard to near-term technical measures, the first priority was the twenty-five reactors that were considered to be the least safe by Western safety experts: fifteen RBMKs (two located in the Ukraine, two in Lithuania, and eleven in Russia) and ten VVER-440 Model 230s (four in Bulgaria, four in Russia, and two in Slovakia). The donors hoped that these plants would be shut down as soon as practicable. The second priority was the remaining thirty-three Soviet-built reactors, fourteen VVER-440 Model 213s and nineteen VVER-1000s, deemed to be less risky that the older generations of reactors. The improvements proposed by the donors would enhance safety for longer periods of operation.

In January 1993 the G7 countries agreed to final details on a fund to finance the plan, to be managed by the European Bank for Reconstruction and Development (Simons 1993). Bulgaria and Slovakia were expected to be the first countries to receive assistance.

As of May 1994, about $785 million had been pledged to implement the G7 emergency safety assistance plan, but only about 7 percent of the projects—worth about $57 million—had been completed (U.S. General Accounting Office 1994b). Among the reasons for the delays in the implementation has been the unwillingness of U.S. and European contractors to undertake work without protection from liability in the event of a nuclear accident. As an IAEA official put it, "if Western manufacturers go in there and people get hurt by something that has a Western piece bolted to it, this can have all sorts of repercussions.... Western companies are very eager to go and get the business, but their legal departments are very nervous" (Simons 1993). Another problem is the lack of commitment on the part of the countries where the nuclear reactors are located to shut them down, giving rise to the perverse result that assistance being given may in fact be encouraging the continued operation of these reactors.

Cuban Nuclear Reactors

As discussed earlier, construction of the Cuban nuclear power plant at Juraguá has been formally suspended since September 1992. Cuba's attempts to secure foreign financing to permit completion of the plant have not borne fruit to date. The consensus about serious design inadequacies and operational problems with Soviet-built reactors has strengthened opposition to the construction of nuclear power plants in Cuba. Interestingly, the opposition to nuclear power plants in Cuba does not come from environmental organizations within the island—since the authoritarian regime does not allow opposition groups and has a monopoly over the domestic media—but from private groups and elected officials in the United States, primarily those in the neighboring state of Florida.

Cuban officials remain adamant in their view that the Juraguá reactors meet international safety standards and do not represent a threat to the environment or to the health and well-being of the population of the island and of neighboring states. In a well-publicized 1985 article, then energy czar Fidel Castro Díaz-Balart (1985, 81–82) stated that "the Soviet VVER reactors, which will be installed at Juraguá, have been demonstrated to have superior safety levels than similar commercial power reactors supplied by Western producers."

Cuba sees criticisms of the safety of its nuclear power program as politically motivated: inappropriate external interference aimed at destabilizing the regime. Nevertheless, testimony of several Cuban nuclear experts who were involved in the planning and construction of the Juraguá plant and have defected to the West have raised concerns about the quality of the nuclear hardware and of the construction work, the siting of the facility, and Cuba's ability to respond to a nuclear accident.

Cuba's Views on the Safety of the Juraguá Plant

Cuban officials have stated that the two nuclear reactors being constructed at Juraguá are VVER-440s Model V318, a "more advanced version of similar units now in operation in several European countries . . . [whose] containment structure and other safety and operational parameters place them well within the present requirements of nuclear power generation technology" (Castro Díaz-Balart 1990b, 49–50). One of the distinguishing features of the Model V318 reactor seems to be antiseismic protection (Castro Díaz-Balart 1990a, 356) in the form of a containment structure. At a hearing before the U.S. Senate Subcommittee on Energy in June 1986, a U.S. Department of Energy expert stated that based on a model of the Juraguá plant and on aerial photographs of construction activities at the site (1) the plant appeared to be laid out in the same fashion as the Loviisa plant in Finland, with reactors located inside a containment structure; and (2) construction activities were consistent with Western-style containment structures, since the structures around the reactors were cylindrical rather than rectangular as is characteristic of VVER-440s built in the Soviet Union without U.S.-style containment structures (U.S. Senate 1986, 16–17).

Over the years, Cuban officials have said the following regarding the construction of the Juraguá plant:

▶ In 1981 the director of the state enterprise responsible for the construction of the Juraguá plant stated that the plant will have "extremely thick walls" and "will be able to withstand earthquakes of 8 degrees on the [Richter] scale, 30-meter high waves, and the impact of 20-ton planes flying at 750 kilometers per hour" (Copa 1981, 13–14).

▶ In mid-1984, Castro said about construction at Juraguá:

> This colossal project requires digging up millions of cubic meters of rock, hundreds of thousands of cubic meters of concrete in situ, dozens of thousands of tons of steel. The plant has been designed, and is being constructed, with all safety measures. It has been designed to withstand earthquakes, since, although they are not frequent and of low intensity, the region of Cienfuegos is within the seismic areas of the nation. This nuclear power plant is being constructed to withstand a large earthquake. . . .
>
> It is protected against the possibility—which they say occurs or may occur every 10,000 years—that a 30-meter wave may reach our shores. . . . It is also protected against the unlikely event that there may be an airplane accident, that is, that a large jet plane would collide with one of the reactors. Thus all theoretical risks have been prevented, and that requires, of course, more concrete and more steel. (1984, 2)

▶ In 1985 Fidel Castro Díaz-Balart wrote (1985, 77): "The radiological safety [of the reactors] is guaranteed through a series of very strict measures which include design, installation techniques, and operation of the reactors, so that it can be assured that the active zone is cooled and hermetically sealed under any circumstance, even in the face of risks as improbable as the crash of an airplane against the plant's buildings, an earthquake, or a giant wave."

▶ In May 1986 the Cuban magazine *Bohemia* published a roundtable discussion among nuclear experts about the Chernobyl accident in which a Cuban technician described some of the safety features of the Juraguá reactors:

> Another important safety feature [of the reactor under construction at Juraguá] is that it will have a system of multiple barriers, generally known as a containment system, whose objective is to limit the escape of radioactive materials into the atmosphere. . . . The first barrier consists of the shells that contain the fuel. The second is [the reactor pressure vessel] . . . which prevents the escape of radioactivity to work

areas [within the power plant]. Finally, our nuclear power plant will have, for each reactor, an enclosure around all components [and the reactor pressure vessel] . . . ; its principal objective will be to limit the escape of radioactivity into the environment even in the unlikely event that there occurred the most serious foreseeable accident in that installation. (Petinaud Martínez and González Quintana, 1986, 65)

▶ In August 1986, writing to Florida Congressman Michael Bilirakis, who had expressed concerns about the safety of the Juraguá reactors, President Castro (1986) said: "our nuclear power plant contemplates, for each reactor, [the construction of] a building or containment structure which will include all components and equipment associated with the primary system of the reactor; its principal objective will be to limit the escape of radioactivity into the environment even in the unlikely event that there occurred the most serious foreseeable accident in that installation."

▶ In 1991 a SEAN official wrote in response to challenges to the safety of the Juraguá plant by elected representatives from the state of Florida:

The reactors [being built in Cuba]—the VVERs—have operated in the Soviet Union since 1964 without a single accident. Since 1977 they have been built and operated successfully in the Soviet Union, Czechoslovakia, Bulgaria, Germany, Hungary, and Finland. The design of the reactor that will operate at Cienfuegos (the so-called VVER-440, Model 318) includes triple safety systems, with independent electric power and control systems and a steel and concrete containment system similar to the one at the Loviisa plant in Finland, one of the safest plants in the world. . . . Pursuant to domestic legislation, and in accord with international norms, there is a licensing process for all significant steps regarding siting, construction, start-up, operation, and decommissioning of the plants. A strict system of supplier certification, on-site monitoring, and laboratory testing guarantees the quality of materials and equipment. The execution of any activity with a significant impact on safety requires preparatory steps, including permits and approvals by state supervisory organs. With regard to personnel, which has been meticulously selected, Cuba has trained more than 1,000 professionals in universities in the Soviet Union, Czechoslovakia, Hungary, Germany, and in the island; more than 700 specialists have received training in operation and maintenance of power plants abroad. Cubans have undergone more than 1,000 training sessions with an average length of 6 months in the Soviet Union, Czechoslovakia, Bulgaria, Poland, Hungary, Germany, and Spain, and have participated in programs sponsored by the IAEA in Europe and Latin America. (Castañeda Gómez 1991, 26–27)

▶ In mid-1995 Vice Minister of Basic Industry Rodrigo Ortíz told the newspaper *Granma:*

> Our nuclear reactor is safe. We have trained personnel who have operated this type of reactor in other plants similar to the one at Juraguá. . . . Quality controls have been implemented at this site." He further stated that the World Association of Nuclear Power Operators had proposed 48 modifications to the reactor under construction in Cuba and that 38 of those have been incorporated in the design of the Cuban reactor, with the rest to be made before the plant is operational. Finally, according to Ortíz, the plant will be equipped with instrument and control systems supplied by a Western company. (Assessment 1995, 9)

Safety Concerns About Cuban Nuclear Reactors

The U.S. General Accounting Office (1992, 4–5), in a report requested by U.S. Senator Bob Graham, had the following to say:

> Cuba's nuclear power reactors are the newest model (known as the VVER440 model) of the Soviet-designed 440-megawatt pressurized water reactors (PWR) and are the first Soviet-designed reactors to be built in the Western Hemisphere and in a tropical environment. The Cuban model, called the VVER440 V318, is the model that the Soviet Union planned to export to other countries. The most notable difference between the Cuban model and other Soviet-designed reactors is that the Cuban reactors will have a full containment. The containment, a steel-lined concrete domelike structure, serves as the ultimate barrier to a release of radioactive material in the event of a severe accident.

Independent confirmation that the Cuban reactors will have containment structures is indeed welcome news. Nevertheless, several issues regarding the safety of the Juraguá nuclear reactors remain: (1) adequacy of Cuba's nuclear regulatory infrastructure; (2) adequacy and number of trained regulatory and operational personnel; (3) questions about the quality of components and of construction;

(4) concerns about the design of the reactors; (5) potential for natural disasters; and (6) the probability that radioactive pollutants accidentally released into the atmosphere from the Cuban nuclear reactors could affect neighboring countries, particularly the United States. Finally, another matter of concern is deterioration of the site and equipment during the lengthy period that work on the plant has been suspended.

Nuclear regulatory structure

One of the keystones of the Western approach to nuclear power safety is a structure where regulators are independent of promoters. This was not the case in the socialist countries, including Cuba, where nuclear regulatory, promotional, and operational activities were included within the same institution. Regulators are inhibited in doing their work—which often delays actions and creates conflicts between them and operators—when they are in a chain of command in which they must respond to authorities also responsible for operational aspects of a technology.

Decree No. 52 of October 1979, which created the Cuban Atomic Energy Commission (CEAC) and its operational arm, SEAN, endowed these organizations with both promotional and regulatory functions (Decreto No. 52 1979). Decree No. 56 of 1982, "On Regulation of the Peaceful Use of Nuclear Energy," explicitly devolved on CEAC–SEAN the issuance of licenses authorizing the "importation, acquisition, use, operation, processing, transportation, transfer, removal, storage, or export of radioactive substances or other sources of ionizing radiation or nuclear materials" (Decreto No. 56 1982, 96). Pursuant to Decree No. 56, CEAC/SEAN was designated as the organization responsible for directing (1) the National System of Accounting and Control of Nuclear Materials, (2) the System of Physical Protection of Nuclear Materials, Radioactive Substances, and Other Sources of Ionizing Radiation, (3) the System of Safety of Nuclear Installations, (4) the National System of Radiation Protection.

The subordination of regulators to promoters—evident in the construction of the Juraguá power plant—contributes to quality control problems. It also gives rise to conflicts between operational organizations, focused narrowly on completing work within deadlines and budget, and regulators who have safety as their objective. Quality control activities are often seen by construction officials as unnecessarily slowing down operations and reducing their profit margins, particularly when quality control experts find unacceptable work that has to be redone. When quality control functions are placed under the control of those carrying out the project, quality can be compromised for expediency and to meet operational goals.

Vladimir Cervera Cruz, a quality control technician at the Juraguá facility who defected in 1991, has stated that personnel responsible for quality control of construction worked directly for the Ministry of Construction, the organization building the plant (U.S. House of Representatives 1991, 57). This was confirmed by Pelayo Calante García, a former quality inspector at the plant, who told a U.S. congressional committee in 1995 that "the main problem is that the construction of the power plant is divided into two components: the construction company and the investment company. The investment company has to do all of the quality control for the construction work. But the department of quality control is in the construction company. For this reason, after they made some mistakes, it is very uncomfortable for the quality control [personnel] and they do not say anything to the investment division" (U.S. House of Representatives 1995, 31).

Trained regulatory and operational personnel

In order to operate a nuclear power plant safely—particularly in emergency situations—abundant trained human resources must be available. Some questions regarding Cuba's capabilities in this area have surfaced. Cuba's plans for the Juraguá plant had Soviet experts and advisors playing a significant role in construction and operation. Approximately 350–400 Soviet advisors were involved in the construction of the plant; moreover, Cuba and the Soviet Union had agreed that Soviet technicians would operate the plant for two years after startup (Benjamin-Alvarado and Belkin 1994, 22). With the breakup of the Soviet Union and financing difficulties of the Juraguá project, it is reasonable to expect a larger Cuban role in construction and operation. According to Cuban technician Vladimir Cervera Cruz, by 1991 the number of Soviet technicians at the Juraguá plant had declined to 100 plus (U.S. House of Representatives 1991, 56).

A former Cuban official associated with the Juraguá program has alleged that prospective nuclear plant operators have not received adequate training. Thus, according to this expert,

> individuals trained to be reactor operators have received 5 months of instruction from the Russians on a VVER440-megawatt model V230 reactor simulator at the Novovoronezh nuclear power plant in Russia. However, he said that the value of this training is questionable because this simulator does not resemble the reactor under construction in Cuba. In addition, he said that some Cuban reactor operator trainees had asked for training on a VVER1000-megawatt reactor simulator because it was similar to the reactor in Cuba, but he did not know why they had not been trained on it. Furthermore, according to an NRC

[U.S. Nuclear Regulatory Commission] official, Soviet-designed simulators are slow-response simulators and are considered deficient by U.S. standards because they do not simulate an accident as it would actually happen. (U.S. General Accounting Office 1992, 8)

Quality of components and construction

Experts who worked at the Juraguá plant state that there is no system in place to check imported reactor components to ensure that they meet quality standards. For example, two of these experts stated that "advisers from the Soviet Union could not guarantee that valves installed in the emergency core cooling system of the first reactor would function under certain conditions. Although the Soviet advisers told these officials that the valves had been tested, the advisers did not provide any documentation showing test results. Emergency core cooling systems are an important part of the reactor because they help ensure that, in the event of an accident in which coolant is lost, radioactive material does not escape into the environment" (U.S. General Accounting Office, 1992, 7).

A Cuban expert working at the Juraguá plant told journalist Nogueras Rofes (1996) that more than 50 percent of the components received from the former Soviet Union during 1982–1992 were "defective" and that Soviet nuclear experts warned Cuban officials that the quality of the valves installed in the reactor's emergency cooling system could not be guaranteed. Cuban nuclear scientist Vladimir Cervera Cruz confirmed this lack of quality control, stating that 60 percent of the components received from the former Soviet Union for the Juraguá plant were "defective" (Prevén un Chernobyl 1991, 1A).

Regarding Soviet nuclear components, Western expert Edward Warman has stated that "documentation has been a major problem and no matter what colleagues I've met with, in what country, they have very little in the way of documentation to support what it is they receive from the Soviets and again I am sure that the problem is no different in Cuba" (U.S. Senate 1991a, 48). U.S. expert Nils Díaz has stated the following regarding components for the Juraguá plant:

There is very little control over what comes out to Cuba as parts and components. Cuba is accepting components from seven countries located thousands of miles away. These components come sometimes with a page, a tag, sometimes with good documents. There is no quality assurance on the control of the components. Cuba has no testing devices to verify the quality of these components. These are being bought on the fact that they are good whether they have documentation or not. There is no design control, no procurement documentation control, no control of purchases and materials, no control of equipment. There is no identification and control of components. The tagging on devices that are installed has been reported as being lost rapidly. They disappear so you no longer will be able to go to an elbow valve or a welding to determine which elbow valve or welding it was. (U.S. Senate 1991a, 42)

Concerns regarding the quality of construction have also surfaced. Cuban nuclear scientist Cervera Cruz, who was in charge of checking the quality of welds in the Juraguá reactors, has stated that he and a Soviet colleague examined X-rays of about 5,000 weld sites that had already been approved by Cuban government inspectors; when subjected to radiographic inspection, 10–15 percent of the welds were defective (Defector 1991, 5). Cervera Cruz told a U.S. congressional committee in 1991 that the violations of the Soviet construction quality code he observed, "put in doubt the quality assurance of the plant, which is something that we require in all nuclear power plants. It would make it difficult to operate the plant according to the international standards if the plants were to be completed. These deficiencies . . . have now reached a point in which it would be very difficult to correct" (U.S. House of Representatives 1991, 54; Berg 1991).

Finally, another Cuban nuclear scientist who left the island in 1992 reported that the Juraguá reactor's support structure "was built poorly but was approved after repairs." He stated that "there are serious problems inside the first reactor building, including faulty seals and flaws in the support structure that holds the reactor vessel in place" (Benesch 1992).

Design questions

An official of the U.S. Nuclear Regulatory Commission (NRC) knowledgeable about the Juraguá reactor has expressed concern about the design of the plant's containment system. The NRC official stated that the design of the pressure suppression system was based on analytical models and had not been tested; the NRC would not allow such a system in a U.S. nuclear power reactor unless it had undergone extensive testing (U.S. General Accounting Office 1992, 11–12). Corroborating this concern, nuclear expert R. Wilson (1995, 40) has stated that "The containment [unit of the reactors being built in Cuba] has never been tested, however, so it is unclear what range of accidents will be contained. The Czechs and Hungarians argue that they will run the reactors no matter what the outcome of such a test, so why bother about the test?"

Another open question is how the system to handle the excess steam that would be generated by a severe accident would operate. In the Cuban reactor, the steam is condensed to water in a bubbler-condenser system so that pressure is reduced in the containment structure. In a worst-case scenario, if the steam bypassed the bubbler-condenser system and reached the upper portion of the containment in pressures greater than the 7 pounds per square inch retention capability that this section of the containment is designed to withhold (other portions of the containment structure are designed to withstand pressure of 32 pounds per square inch), the containment could be breached and a radioactive release could occur. By comparison, containment structures in U.S. pressurized water reactors are designed to withstand pressures of about 50 pounds per square inch throughout the entire containment structures (U.S. General Accounting Office 1994b, 5).

Finally, because the reactors being built in Cuba have been modified to include additional safety features, they are "a plumbing nightmare," according to Western nuclear expert Nils Díaz; there is "a proliferation of systems and plumbing that goes against present safety criteria of minimizing the number of systems, simplifying the plant, trying to avoid the many, many, many types of components that pile up on top of each other and do complicate the safety picture" (U.S. Senate 1991a, 40).

Natural disasters

Two forms of natural disasters that could threaten the integrity of the Juraguá reactors and potentially bring about accidental escapes of radioactivity are hurricanes and earthquakes. Although Cuban authorities have stated that the Juraguá reactors have been designed to withstand these natural disasters, nevertheless some concerns remain.

Nuclear scientist Nils Díaz (U.S. Senate 1991a, 48) has expressed the following about the indirect effect of a hurricane on Cuban nuclear power plants: "Cuba is uniquely susceptible to hurricane damage. I'm not suggesting for a moment that containment wouldn't be able to withstand 100 mile an hour winds, but what concerns me is the extended loss of off-site power. I would suspect that it might be a real problem for many hours at a plant like that.... [L]oss of off-site power is among the most serious things that worry you." The approach to deal with this potential problem is to engineer additional back-up systems, anticipating lack of power for a day or so. There is no evidence that the Juraguá nuclear power plant is thus equipped.

Cuban geologist José Oro, former director of Natural Resources of the Ministry of Basic Industries who left the island in 1991, claims that the seismological conditions at Juraguá are adequate. The statistical probability of an earthquake in the area—as calculated by Soviet scientists—was low: 6.0 on the Richter scale in a period of 100 years and 7.0 in a period of 1,000 years (Oro 1992, 21). Oro further states that some Western scientists have questioned the Soviet calculations. Experts at the U.S. Geological Survey (USGS) have requested information from Cuba that would allow making an analysis of the risk of an earthquake in the Juraguá area, but such information has not been provided to them. An assessment of seismic potential of the Cienfuegos area by an international insurance group[9] estimated that it could produce an earthquake with a probable magnitude of 5.0 on the Richter scale in a period of fifty years (Fultz 1995, 7).

According to USGS officials, the Caribbean plate, a geological formation near the southern coast of Cuba in the area of Guantánamo, is active and may pose large to moderate seismic risks to Cuba and to the nuclear reactor site (Fultz 1995, 7). On 25 May 1992 an earthquake whose epicenter was located at Pilón, in southeastern Cuba, caused moderate damage in the Manzanillo and Niquero areas. The Cuban National Seismological Center reported the intensity of the earthquake as 5.2 on the Richter scale (Sismo causa 1992); meanwhile, the Geological Institute, located in Golden, Colorado, reported it as having an intensity of 6.9 on the Richter scale (Sismo causa 1992; Sismo pudo 1992). The discrepancy has not been satisfactorily resolved.

Geologist Oro has charged that the Cuban government understated the true intensity of the Pilón earthquake and avoided reporting on its impact on the central area of the country in order not to draw attention to the danger that earthquakes could pose for the Juraguá nuclear power plant. According to Oro, nuclear power plants built in areas susceptible to earthquakes of intensity 6.5 or more on the Richter scale must be constructed using special antiseismic techniques; for this reason, Cuban authorities reported a much lower intensity (only 5.2 on the Richter scale) for the 1992 Pilón earthquake (Remos 1992).

Cross-border radioactive pollution

An accident at the Juraguá nuclear power plant that released into the atmosphere radioactive pollutants would affect areas distant from the release location. Radioactive pollutants would form a plume that could be transported and dispersed by air currents. The area affected by the release, and the concentration of pollutants affecting each location, would be a function of many variables, among them the magnitude of the release and the time of the year when the release occurred.

In 1992 analysts at the Air Resources Laboratory of the U.S. National Oceanic and Atmospheric Administration conducted a simulation of the probability of a plume impact, average time of plume arrival, and relative plume concentration from a single pollutant release from the Juraguá plant using climatological data for summer 1991 and winter 1991–1992 (Heffter and Stunder 1992). According to the simulation study, areas within Cuba near Juraguá would be most severely affected, but the pollution would also affect other Caribbean islands, Mexico, and the United States. A discharge during the summer months, when east-to-west trade winds prevail, could carry radioactive pollutants over the entire state of Florida and also affect the states of Texas, Louisiana, Arkansas, Mississippi, Alabama, and Georgia in about four days; after a discharge during the winter months, when the trade winds are weaker and would meet resistance from westerly winds, pollutants would be likely to move toward the east, affecting areas as far north as Virginia and Washington, D.C., in about four days (Heffter and Stunder 1992; Fultz 1995, 7).

Deterioration of plant and equipment

As discussed earlier, work on the construction of the Juraguá plant formally ceased in September 1992. Russia reportedly granted Cuba a $30-million credit to shut down and mothball the facilities to prevent deterioration and permit restart and completion of construction at a later date. According to Vice Minister of Basic Industries Ortíz, Cuban and Russian experts developed a work program for the plant that "has allowed us to preserve what has been built and to maintain the technological equipment under optimum conditions" (Assessment 1995, 9).

A sharply different assessment of the condition of the Juraguá plant has been made by experts outside Cuba. Assistant Comptroller General Kenneth Fultz of the U.S. General Accounting Office told a U.S. congressional committee in August 1995 that film footage of the plant taken by a U.S. news organization in July 1995 "graphically portrays a nuclear plant that is in a serious state of deterioration, and appears to be abandoned rather than properly maintained" (U.S. House of Representatives 1995, 19). Among the examples of deterioration cited by Fultz were extensive rusting in several locations, including in the pressure suppression system and in the inside wall of the reactor; what appeared to be moss growing out of the containment vessel; overall exposure of reactor no. 2 to the environment, the weather, salt vapor, and so on; haphazard storage of equipment; and deterioration of the reactor vessel for reactor no. 1, which is stored in a metal shed that is not sealed and where corrosive salt air vapor can freely enter (U.S. House of Representatives 1995, 20–21). Former Juraguá quality inspector Calante García concurred with these views; he told the same U.S. congressional committee that "since 1987, the exposure to salt air has corroded safety equipment that has already been assembled such as pumps, valves, pipes, heat exchanger, and water tanks. Moreover, the nuclear reactor vessel should be protected in a nitrogen atmosphere in order to avoid rusting. However, the reactor vessel has been exposed to the salt air in open form, without any protection, since 1990. Rust on the reactor vessel can affect the integrity of the welds and can cause weakening of the reactor vessel" (U.S. House of Representatives 1995, 65).

The deterioration of the plant and equipment has been so severe that it led U.S. General Accounting Office official Fultz to conclude that, "on the basis of our review of this film, and I think that this is probably the most current information that we have to examine, that if we had concerns in 1992 [about the safety of the Juraguá plant] and we did, we have much more serious concerns today about the quality of those reactors" (U.S. House of Representatives 1995, 18).

Summary and Implications

Cuba's socialist regime assigned high priority to its nuclear power program. Very ambitious plans unveiled in the 1970s foresaw the construction of three nuclear power complexes spanning the entire national territory. The island's poor endowment with domestic energy resources was the official rationale for turning to nuclear power, but the decision was also driven by the Cuban government's desire to demonstrate the technological superiority of the socialist system and to enhance Cuba's leadership role among developing countries.

Consistent with Soviet thinking and practice, Cuban officials in charge of the nuclear power program exhibited great confidence in the safety of socialist nuclear power plants. The socialist approach to safety of nuclear power plants discounted the West's reliance on redundancy of systems and instead focused on engineered safeguards and a belief that socialist man, not driven by the profit motive and free from pressures to skimp on quality to cut costs, would produce equipment and build structures that were accident free. The Cuban leadership and nuclear power officials bought into this philosophy and uncritically accepted the myth that the use of nuclear power in the socialist countries had an unblemished safety record. This view changed with the 1984 Chernobyl accident and the lifting of the information veil that had prevented information about nuclear accidents in the former Soviet Union and the Eastern European socialist countries from reaching the West.

Cuba's very ambitious plans to build a dozen nuclear reactors was scaled down soon after it was announced. In the early 1980s construction of two reactors began at Juraguá, in Cuba's south-central area. Construction has been affected by cost overruns and numerous delays. In September 1992 Cuba announced suspension of construction because of lack of financing to acquire the necessary equipment and services from Russia. The Juraguá plant has been mothballed since then. Cuba and Russia continue to seek one or more foreign partners who would be willing to finance the completion of the plant.

The nuclear reactors for the Juraguá plant are Soviet-designed pressurized water reactors generally known as VVER-440. More pointedly, the Juraguá reactors are VVER-440 Model V318, a variant of the VVER-440 Model 213; the Model 213 reactor has been built in the former Soviet Union and also in other Eastern European nations. The safety of the VVER-440 family of reactors has been a matter of concern for the international community, and a multilateral plan financed by Western countries to shut down or improve the safety of nuclear reactors in operation in Europe has been put into place. The distinguishing feature of the Model V318 reactor seems to be a containment structure around the reactor vessel. Containment structures are required in Western nuclear power plants but generally have not been part of the design of power plants built in the former Soviet Union and Eastern Europe.

The potential operation of the Juraguá nuclear power plant raises two sets of safety concerns: (1) those associated with the safety of Soviet-designed nuclear power plants in general, and of untested equipment in the Model V318 reactor in particular; and (2) those arising from reports from experts who were formerly associated with the planning and construction of the Juraguá plant who have provided specific examples of instances of quality of construction problems, lack of quality assurance for imported parts and components, and inadequate training of personnel. The indefinite suspension of construction of the Juraguá plant because of difficulties in obtaining financing announced by Fidel Castro in early 1997 is a positive development for Cuba's environment because the risk of an environmental disaster associated with a nuclear accident is avoided.

❧ 9 ❧

Regional Development and the Neglect of La Habana

ONE OF the most important development objectives pursued by Cuba's socialist government since the onset of the revolution was to fundamentally transform the country's pattern of urban growth and regional development. Policies implemented over several decades to accomplish this objective had various goals, but all were essentially intended to alter the traditional primacy of La Habana over the rest of the country by neglecting the capital city so as to bring about a more equitable and balanced pattern of regional socioeconomic development. Over time, however, this decision gave rise to serious environmental stresses because the basic sanitary infrastructure of the city was not expanded to keep up with population growth, and the upkeep of existing facilities was woefully inadequate.

For centuries, La Habana had been the country's preeminent city. It was the largest urban agglomeration, as well as the country's center of economic, cultural, and administrative life. In the late 1950s La Habana was a typical primate city, its population being 3.5 times larger than the combined populations of the next two largest cities (Santiago de Cuba and Camagüey). Educational levels, employment prospects, sanitary standards, access to social and medical services, and overall urban amenities were vastly superior in La Habana than in any other region of the country, whether urban or rural. In the late 1950s La Habana accounted for 75 percent of the country's nonsugar industrial output and for 90 percent of its shipping activities (Acosta León and Hardoy 1972, 55).

In 1953, the year of the last pre-socialist population census, La Habana, with 1.23 million inhabitants, housed 21 percent of the country's total population. During the 1943–1953 intercensal period, the city's population expanded at an annual growth rate of 2.6 percent. By the late 1950s La Habana accounted for 26 percent of the country's total population growth (Acosta León and Hardoy 1972, 51). Between 1959 and 1981 (the date of the last Cuban census) La Habana's population growth rate was curtailed, in part because of policies designed with that effect in mind, but primarily due to demographic developments that adversely affected its continuous growth. Between 1953 and 1970 (the date of the first census enumeration under socialist rule) La Habana's annual population growth rate declined to 2.2 percent, and during the 1970–1981 intercensal period it further declined to 0.7 percent. By 1970 the share of the country's total population residing in La Habana had been reduced to 20.8 percent, a percentage slightly lower than in 1953. A little over a decade later, in 1981, 19.8 percent of the country's total urban population resided in La Habana.

By the end of the 1970–1981 intercensal period, most of the largest Cuban cities were experiencing a faster pace of demographic growth than La Habana, especially cities in the eastern provinces of the country, such as Holguín, Bayamo, and Las Tunas (see Table 9.1). Urban population growth rates were lower than in the previous intercensal period (1953–1970), however, mostly due to the ceiling effects produced by higher urbanization rates and a rapid decline in the birth rate that followed a period of rising fertility during the 1960s. Nevertheless, La Habana still retains its primate character, the city's population being over three times as large as the combined population of the country's two next largest cities, which are still, as in the 1950s, Santiago de Cuba and Camagüey.

La Habana's Demographic Growth and Emigration

Permanent emigration has been an important demographic variable behind the slowdown of La Habana's growth. Particularly intense periods of emigration occurred during the early 1960s, especially between 1960 and 1962, when close to 200,000 emigrants left the country. An even larger emigration surge followed between 1966 and 1972, during the so-called freedom flights, when emigration from Cuba was regulated by a "memorandum of understanding" between the Cuban and U.S. governments. During this period, 344,000 Cubans emigrated to the United States. Another emigration watershed occurred in 1980 when approx-

Table 9.1. Population size and growth rate of Cuba's principal cities, 1953–1981

City	Population			Growth rate		
	1953	1970	1981	1943–1953	1953–1970	1970–1981
La Habana	1,223,899	1,786,522	1,924,432	2.6	2.2	0.7
Santiago de Cuba	163,237	277,600	347,279	3.3	3.2	2.1
Camagüey	110,388	197,720	245,525	3.2	3.5	2.0
Holguín	57,573	131,656	186,943	4.8	5.0	3.2
Santa Clara	77,398	130,241	172,223	3.7	3.1	2.6
Guantánamo	64,671	129,005	167,255	4.3	4.1	2.4
Cienfuegos	57,991	80,758	102,791	0.9	2.0	2.2
Matanzas	63,916	86,596	100,813	1.5	1.8	1.4
Bayamo	20,178	71,484	100,622	2.2	7.7	3.2
Pinar del Río	38,885	75,485	96,660	4.0	4.0	2.3
Manzanillo	42,252	75,565	87,423	1.5	3.5	1.3
Las Tunas	20,431	53,734	84,857	4.8	5.8	4.2
Ciego de Avila	35,178	57,869	74,280	4.0	3.0	2.3
Sancti Spíritus	37,741	57,818	71,949	2.9	2.5	2.0
Cárdenas	43,750	54,913	59,626	1.7	1.3	0.7
Palma Soriano	25,421	42,380	56,389	4.9	3.0	2.6
Sagua la Grande	26,187	35,514	42,730	0.9	1.8	1.7
Güines	29,226	39,173	41,625	2.6	1.7	0.5
Florida	21,159	34,043	39,728	11.6	2.8	1.4
Baracoa	11,459	20,926	31,119	0.9	10.4	5.6
Colón	15,755	26,472	35,217	3.2	3.1	2.6
Nuevitas	12,390	20,737	35,108	1.0	3.1	4.9

Source: Luzón 1987, 218, 257.

imately 125,000 Cubans departed through the port of Mariel. In all, some 793,000 Cubans entered the United States between 1959 and 1980 (Díaz-Briquets and Pérez 1981, 26). Many others—in the tens of thousands—found their way to other countries.

The 1980 international round of national censuses enumerated close to 54,000 Cubans in seventeen Western Hemisphere countries (including Canada), and four European countries (CELADE 1989). The vast majority of the emigrants had roots in the greater La Habana region (Eckstein 1982, 121; Díaz-Briquets and Pérez 1981, 27). Cuban emigration to the United States continues to this date, now regulated by the 1995 U.S.–Cuba Migration Accord, which assures Cuba a minimum of 20,000 U.S. immigration visas annually. As in the past, a disproportionate number of current emigrants continues to be drawn from La Habana.

Another factor contributing to the long-term trend of slower demographic growth for La Habana was the rapid decline in fertility rates that occurred in Cuba during the 1970s, following a brief baby boom during the 1960s. Between 1970 and 1979, the total fertility rate (or the number of children the average woman has over her reproductive lifetime) declined from 3.7 to 1.8, or by 51 percent (Hollerbach and Díaz-Briquets 1983, 37). Although the decline in fertility rates was national in scope, it was most marked in urban areas, and most of all in La Habana. In the former province of La Habana, home to the capital city, the birth rate declined by 63 percent, from 31.3 births per 1,000 population in 1963 to 11.5 in 1980, as opposed to 14.1 for the country as a whole (Hollerbach and Díaz-Briquets 1983, 30, 40–43).

Internal Migration to La Habana

Large-scale in-migration to La Habana from other parts of the country compensated for the demographic loss associated with permanent emigration. As shown in Table 9.2, the province of Ciudad de La Habana, together with the province of Camagüey, exhibited the highest internal migration rates of any province (the rates are expressed as net migrants per 1,000 population of the provinces). Of particular note in the table are the high outmigration rates from Cuba's eastern provinces throughout the 1971–1981 period and into 1989. The provincial internal migration rates obscure the fact, however, that some of the fastest growing cities up to 1981 (such as Holguín, Las Tunas, and Ciego de Avila, see Table 9.1) are located in the provinces with some of the highest outmigration rates.

Not surprisingly, periods of high internal migration from the interior of the country to La Habana have coincided with periods of high emigration. During the early 1960s, for example, many of the La Habana houses vacated by emigrants were occupied by in-migrants from eastern Cuba, who came to the city as part of the new political and administrative elite, or in search of expanded educational opportunities (Gugler 1980, 32). A comparable surge in migration to La Habana in 1981 can be noted in Table 9.2: the highest internal migration rates to Ciudad de la Habana Province were recorded immediately after the Mariel emigration outflow.

Some analysts have noted that among the factors contributing to the relative slowdown in La Habana's demographic growth were "rigidities of a largely administered labor market," where many regulations impeded the free flow of labor (Lehmann 1982, 326). Another credible explanation posits that with reduc-

Table 9.2. Internal migration rates, by province, 1970–1989 (per thousand)

Province	1971	1973	1975	1979	1981	1976–1981	1989
Pinar del Río	−1.8	−0.3	2.2	−1.6	−1.3	−0.4	−3.1
La Habana	0.7	2.9	2.1	3.4	3.3	3.9	0.7
Ciudad de la Habana	5.8	5.1	5.5	4.2	6.3	3.7	1.8
Matanzas	0.6	1.7	3.6	0.2	1.7	1.5	0.8
Villaclara	−1.6	−1.3	2.7	−1.6	0.3	−1.6	−2.0
Cienfuegos	−4.2	−2.6	−2.3	0.6	2.0	0.0	4.0
Sancti Spíritus	−2.8	−4.5	6.9	−2.9	1.6	−2.5	−0.2
Ciego de Avila	2.1	−0.6	1.1	5.4	3.9	4.7	4.3
Camagüey	6.2	5.5	10.9	5.4	4.7	5.2	−0.5
Las Tunas	−6.8	−5.9	−9.2	−2.4	0.0	−3.3	−0.8
Holguín	−2.3	−3.4	−3.5	2.5	−3.1	−2.6	−2.7
Granma	−4.3	−4.4	−5.7	−6.1	−9.7	−5.7	−5.6
Santiago	−2.9	−1.3	−4.5	−3.0	−5.1	−3.3	−2.7
Guantánamo	−6.4	−6.1	−8.4	−8.3	−9.8	−6.5	−6.1

Source: Luzón 1987, 178. For 1989 data, Comité Estatal 1989, 73.

tions in differentials in living conditions between La Habana and the rest of Cuba, the attractiveness of the internal migration option was reduced (Gugler 1980; Eckstein 1982; Lehmann 1982).

Policies Promoting Balanced Regional Development

The rapid demographic growth rates observed for many of Cuba's secondary cities during the 1960s and 1970s can be explained by examining the policies pursued to redress past patterns of unequal regional development. These policies were in part predicated on the desire to reduce overall rural-urban differences in living standards, and in particular differences in living conditions between La Habana and other parts of Cuba, but were also based on the policy goal of promoting the accelerated development of regions of the country that were considered to have lagged in their economic growth, including promoting the growth of secondary cities as agricultural, industrial, and transportation hubs. Such cities included Holguín, Bayamo, Las Tunas, Cienfuegos, Nuevitas and Pinar del Río, as well as smaller cities and towns (Acosta León and Hardoy 1972; Luzón 1987).

Regional development policies also rested on the maxim of "a minimum of urbanism and a maximum of ruralism," articulated by President Fidel Castro in 1966 (Gugler 1980). This was partly to be achieved by relocating some of the urban infrastructure traditionally located in cities (such as schools and health posts) to rural areas and by allocating a major share of national infrastructure investments to the countryside. This strategy also included the establishment throughout rural Cuba of 335 new towns, where the formerly dispersed rural population was to be concentrated. These settlements were designed to provide improved housing and better health, educational, and social services to the population. They were also part of a greater design to alter Cuba's rural landscape (see Chapter 4) by concentrating the rural population in small urbanized areas to permit the establishment of large-scale state farms reliant on high levels of mechanization and a more intensive use of agricultural land. The strategy was predicated as well on the accelerated development of a denser and more efficient transportation infrastructure (Acosta León and Hardoy 1972, 55; Centro de Estudios 1976, 168).

The reallocation of resources to the development of the rural infrastructure and to the acceleration of the development of the economic infrastructure of secondary cities meant a shifting of resources away from La Habana. The goal of the regional development strategy largely succeeded since "between 1953 and 1981 a number of socioeconomic provincial discrepancies [in access to health and educational services, sanitary facilities, etc.] were substantially reduced," although differentials remained over time (Díaz-Briquets 1988, 59).

The industrial infrastructure of the country also experienced a notable transformation mostly due to a more geographically dispersed location of new facilities. There is reason to believe, however, that the maintenance and upkeep of the infrastructure and industrial facilities has been as lax in the rest of Cuba as in La Habana, and that many of the same maladies affecting their proper functioning and environmental disruptions also plague the country's secondary cities. In cities like Santiago de Cuba and Holguín, for example, water distribution systems appear to be as poorly maintained as those of La Habana (Batista 1997).

The Neglect of La Habana

Achieving balanced regional development entailed minimizing infrastructure investments in the city of La Habana, which in the past had consumed a disproportionate share of national resources. A concomitant of this strategy was reducing as well the rate of housing construction in the city, a policy made more acceptable by the considerable number of dwelling units in the city being vacated by emigration and taken over by La Habana residents or internal migrants to the city. The long-term consequences of the "neglect by design" of the city were fast in coming: numerous commentators over the years have noted the rundown conditions of the city's infrastructure, despite occasional and feeble attempts to reverse decades of neglect (Segre, Coyula, and Scarpaci 1997; see also Luzón 1988).

Among the more environmentally damaging features of this neglect was the decision not to expand water distribution and sewerage systems originally designed to serve a population much smaller than that of La Habana today to keep pace with growth, as well as the failure to properly maintain these systems. The neglect of the housing infrastructure has led to the loss of thousands of residential units, while shoddy construction practices and the limited number of new housing units built over the last several decades has severely aggravated the city's housing deficit.

Failure to modernize the city's industrial infrastructure has also contributed to environmental stress, most notably in La Habana Bay, which is among the most polluted bodies of water in the world. The result is a gargantuan restoration bill. A recent estimate assumes that to address the infrastructural, housing, and environmental problems facing the city of La Habana at the close of the century it would be necessary to invest between 10 and 14 billion dollars (Segre, Coyula, and Scarpaci 1997, 326).

With the collapse of the socialist world, and Cuba's ensuing economic crisis, migration from the interior of the country to La Habana intensified again in the 1990s. Between 1993 and 1995, 45,000 internal migrants are said to have moved to La Habana from other parts of Cuba (Cuban Radio 1996, 13). In 1996 alone it is estimated that 55,000 internal migrants, most of them originating from the country's eastern provinces, settled in La Habana (Entra en vigor 1997). This surge in migration has been attributed to a severe worsening of overall living conditions as a result of the national economic crisis and the belief by the migrants that conditions are better in La Habana, where dollars circulate more freely. With the arrival of the migrants, many new shanty towns have emerged. About 76,000 migrants, dispersed in 188 shanty towns, occupy 23,000 poorly constructed dwellings. There are also reports of increases in the population density of already-crowded urban housing and the occupation of abandoned buildings (Capital en ruinas 1997). An interesting aside is that in La Habana today almost as many people live in shanty towns as they did in 1959 (Centro de Estudios 1976, 165).

The migration to La Habana has occurred despite a growing and acute housing shortage, aggravated in no small measure by the decision made in the 1960s and sustained over several decades not to allocate sufficient financial resources to maintain the city's "historic" housing stock and sanitary and water distribution infrastructure. According to government estimates, half of the 560,000 dwellings in La Habana are in a poor state of maintenance, and an additional 60,000 cannot be repaired and must be demolished. A further 75,000 dwellings are being propped by wooden beams, and 7,800 other dwellings are slated to eventually be equipped with supporting beams (Capital en ruinas 1997).

In 1997 the government responded to what was perceived as an internal migration crisis by tightening the administrative rules determining who could reside in La Habana, even forcibly deporting unauthorized migrants to the city to their home communities (Darling 1997; Plan para frenar 1997). According to Decree 217 of May 1997, prospective migrants to La Habana must receive prior approval from the municipal authorities of the locality within the city where the prospective migrant intends to reside. Those found in violation of this decree could be fined from 200 to 1,000 pesos and "would be under the obligation of returning immediately to their original place of residency" (Entra en vigor 1997).

Given the deteriorated status of most dwellings, the limited amount of new housing construction, and the demographic growth of the city in the last several decades, it can be concluded that, on the whole, La Habana's housing situation today is far worse than it was in the late 1950s, when the socialist government came to power. The same conclusion applies to the water distribution and sewerage systems, for much the same reasons. The sections that follow review the evidence that lead us to that assessment and that highlight the environmental stresses confronting La Habana at the end of the 1990s.

Water and Sanitation

By international standards, socialist Cuba's record in providing drinking water supply and sanitation services to its citizens is quite remarkable. According to statistics provided by Cuba to the Pan American Health Organization (PAHO), in 1992 Cuba provided drinking water to 100 percent of its urban population and 91 percent of the rural population and sewage and excreta disposal to 100 percent of the urban population and 68 percent of the rural population (PAHO 1994, vol. 1, 275, 278).[1] Cuba's record of providing drinking water to 98 percent of its population was topped in the Latin American and Caribbean area only by Barbados (100 percent) and exceeded by a large margin the overall area average (80 percent) as well as the record of advanced countries in Latin America such as Argentina (64 percent), Brazil (92 percent), Chile (87 percent), Mexico (83 percent), Uruguay (83 percent), and Venezuela (68 percent). With regard to sanitation services, Cuba's record of serving 92 percent of the population was topped only by Barbados (100 percent) and the Bahamas (98 percent) and exceeded the area's average (67 percent) as well as the records of Argentina (89 percent), Brazil (73 percent), Chile (83 percent), Mexico (66 percent), Uruguay (82 percent), and Venezuela (55 percent). Official estimates of the percentage of the population served by sanitary facilities, however, appear to grossly exaggerate the true figures. Other estimates suggest that only about half of the population, rather than 92 percent, really has access to these facilities (see Chapter 5).

Cuba's socialist government has not made the necessary investments to maintain and improve the city of La Habana's water and sewer systems. This neglect manifests itself particularly in three ways: (1) water is available only at certain times of the day, (2) water is lost because of leaks in the distribution system, and (3) the sewage system often backs up and raw sewage flows onto the streets (Cereijo 1992, 88). La Habana's neglected water and sanitation infrastructure constitutes a serious environmental and public health problem. As a Cuban analyst has put it,

> Among the most serious problems [of La Habana] is poor environmental quality, particularly regarding water, including water supply deficiencies and the high contamination of surface waters—such as the Bay of La Habana and rivers that flow into it; air pollution as a result of the poor condition of transportation equipment, and sound pollution. The ocean frequently floods low-laying coastal areas. Moreover, trees in green areas of the city are increasingly being cut down because they are in the way of overhead electric lines or because their roots interfere with sewer lines, but more importantly because of the lack of an effective environmental culture. (Fernández Soriano 1997, 29)

La Habana's water and sewer infrastructure is very old. Very little maintenance has been given to plants, distribution, collection, and transmission systems, with government interventions focused almost exclusively on repairing breaks and leaks (Cereijo 1992, 86). La Habana's main water supply lines date to the 1870s, when the Isabel II or Canal de Vento aqueduct was built (Grupo Cubano 1963, 255–56). In the 1950s the Cuenca Sur (Southern Watershed) and Marianao aqueducts were constructed and investments also made in rebuilding some of the main water distribution lines in the city (Grupo Cubano 1963, 1177).

Cuban experts have estimated that, nationally, fewer than 8 percent of the population receiving drinking water supply in the mid-1990s had service twenty-four hours a day, with the average being between eight and fifteen hours of service per day (Atienza Ambou et al. 1995, 12). The condition of the water mains is so poor that an estimated 55 percent (Segre, Coyula, and Scarpaci 1997, 6) or 50 percent (Batista 1997) of the water pumped through La Habana is lost through leakage. This is a chronic problem, as in 1964 it was already estimated that 30 million gallons of water were lost daily because of leaks in the distribution lines (Area Handbook 1971, 117). In the Habana Vieja section of the city, water has not been available through faucets and pipes for several years, and tank trucks deliver water to the neighborhoods each day (Batista 1997).

The city's sewer system, inaugurated in 1913, was built for a population of 600,000; its service area was the present-day areas of Habana Vieja, Centro Habana, Cerro, Diez de Octubre, Plaza, and the oldest section of Miramar. Other areas of the metropolitan area rely on septic tanks or discharge runoff and raw sewage directly into rivers, streams, and canals (Segre, Coyula, and Scarpaci 1997, 102). Under 50 percent of the population of the city of La Habana is currently served by the sewer system (Lezcano 1994b).

Up to 90 percent of the sewer system of La Habana is not functioning adequately or at all (Lezcano 1994b). Frequent sewer backups experienced in La Habana may be the result of lift stations that are not functioning, sewage lines that have ruptured, sewage lines in which flow has been impeded because of the accumulation of debris, and treatment plants that are unable to accept flows (Cereijo 1992, 89). Overburdened storm sewers are unable to flush out rainwater, and the areas prone to flooding have been gradually increasing (Lezcano 1994b).

According to experts, the useful life of water distribution and sewer lines is fifty years for facilities that are properly maintained; useful life is lower for non-maintained lines (Cereijo 1992, 88). In 1959 the bulk of water and sewer facilities in Cuba's urban areas were already over forty years old (Cereijo 1992, 88). Out of the estimated 3,800 kilometers of water lines in La Habana, over 2,000 kilometers are deemed to be in substandard condition and only 400 kilometers are deemed to be in acceptable condition (Lezcano 1994b).

The combination of water main leaks and backup of sewer lines and storm drains creates a potentially very serious public health hazard, since sewage may be sucked into water lines by the zero or negative pressure that develops from the lack of flow (Cereijo 1992, 89) or may be mixed in with running water in flooded areas (Lezcano 1994b), thereby contaminating public water supplies. The most common outcome of the consumption of drinking water contaminated by organic wastes—such as can result from leaks in water mains—is acute diarrhea, although it is also possible to contract more serious illnesses such as typhoid fever and viral hepatitis (Atienza Ambou et al. 1995, 9).

Disposal of solid waste in La Habana has been a significant problem for many years. La Habana's main solid waste disposal site was Cayo Cruz, a small cay within La Habana Bay on the edge of Guasabacoa inlet. In the 1950s Cayo Cruz was described as "a great heap of garbage"; trade winds carried the stench of rotten garbage to the nearby poor communities of Luyanó, Atarés, and the southern portion of Habana Vieja (Segre, Coyula, and Scarpaci 1997, 96). In the 1970s the garbage dump at Cayo Cruz was decommissioned and replaced with a sanitary landfill at a new location near Calle 100 (Segre, Coyula, and Scarpaci 1997, 147). By now, both the Calle 100 and Guasabacoa landfills have exceeded their capacity, although they continue to be used despite the serious health problems this entails (Lezcano 1994b).

Air and Water Pollution

Despite the fact that socialist Cuba has a very small stock of road transportation vehicles relative to its population, air pollution in La Habana has been a growing problem because of the high level of particulates emitted by the transportation sector as well as by industry. To be sure, air pollution problems are far less severe than in Los Angeles or Mexico City, but nevertheless, smog is also a problem in La Habana.

As has been discussed in Chapter 7, Cuba's stock of automobiles and other road transportation vehicles is small but highly polluting. Vehicles in private hands tend to be very old and poorly maintained. The bulk of vehicles imported by the state were of Soviet or Eastern European vintage and were not equipped with catalytic converters and other pollution abatement devices. The public transportation system consists largely of Eastern European buses and trucks that are notorious for their low energy efficiency and high pollution levels.

La Habana differs from other Latin American capitals such as Mexico City, Buenos Aires, Santiago, and Lima-Callao in that it does not concentrate the bulk of the nation's industrial output (Segre, Coyula, and Scarpaci 1997, 227). It has been estimated that in 1990 about 23 percent of La Habana's labor force worked in industrial activities—light manufacturing, petrochemicals, and warehousing—clustered in three geographical areas. First, the back-bay areas of Regla, Luyanó, and Guanabacoa, which host oil refinery and petrochemical plants, an electric power plant, a fertilizer plant, grain towers, and extensive dockyards. Second, a corridor running along the southeastern railroad through San Francisco de Paula, Cotorro, and Cuatro Caminos, clustering light industry such as food processing and pharmaceutical packing houses. Third, a less clearly defined concentration of light industrial plants located in a southwestern and western axis near the international airport (Segre, Coyula, and Scarpaci 1997, 226-227). Although La Habana's overall concentration of industrial plants is low, industrial pollution is high because of the old age of the facilities (most of the heavy industry plants date from before 1959), extensive use of polluting Soviet and Eastern European equipment in the newer facilities, and overall poor maintenance.

Some 100 industrial plants located within the environs of La Habana are heavy air polluters; 53 percent of them are not capable of being retrofitted to abate air pollution and would have to be shut down to reduce pollution (Lezcano 1994b). The heaviest polluters are electric power plants, metalworking shops, gas production plants, and oil refineries. Air pollution in the form of smoke, dust, and industrial gases is most prevalent in the municipalities of 10 de Octubre, Marianao, Cerro, La Lisa, Regla, Habana Vieja, and Centro Habana (Lezcano 1994b). Industries within the city environs emitting high levels of pollutants into the atmosphere (see Chapter 7) include the following:

▶ In Regla, the "Ñico López" oil refinery (smoke and steam with high acidic content), the electric power plant "Antonio Maceo" (smoke, soot, and steam with high acidic content), a fertilizer plant (ammonia, phosphates, nitrates), and a tallow processing plant (strong unpleasant odors, smoke, acidic steam).

▶ In Marianao, the "Manuel Martínez Prieto" sugar mill (smoke and soot) and a marble finishing plant (carbonated and silicated abrasive dust).

▶ In Boyeros, an animal feed plant (unpleasant smells, dust) and an asbestos-cement plant (asbestos dust).

▶ In the municipality of El Cerro, twenty-one enterprises were identified in 1992 as polluting the atmosphere. Most of these enterprises emitted dust, smoke, and soot, but others also emitted asbestos particles. Among these enterprises were two foundries; five factories producing cement, cement blocks, and pipes; two factories producing soap; several factories producing processed foods; and a bus terminal (Cabrera Trimiño 1997, Annex 5).

Other heavy polluters within La Habana's environs are the "Melones" manufactured gas plant (smoke, soot, and mephitic steams); the Suchel and Sabatés plant, producing detergents (smoke and steam); the "Otto Parellada" electric power plant, located in Tallapiedra; a tallow processing plant in San Miguel del Padrón (strong unpleasant odors, smoke, acid steam); and a tire plant in Cotorro (soot and black carbon particles). Moreover, the huge dust plume of the "René Arcay" cement plant in Mariel, under certain weather conditions, can also affect air quality in the city.

An estimated 202 industrial plants in the metropolitan La Habana area pollute surface and underground waters; 100 plants require sizable investments to make "profound" changes in their production technology or treatment of residues, 40 have to be relocated, 30 require local abatement technologies, and 32 require extensive maintenance of existing waste treatment systems (Lezcano 1994b). Most rivers in the city of La Habana environs—Luyanó, Almendares, Cojimar, and Quibú—are heavily polluted by organic and chemical wastes and have lost 90 percent of their animal and plant life in the last twenty years (Lezcano 1994b).

Among the leading industrial plants in the La Habana environs polluting surface or underground water (see Chapter 7) are the "Manuel Martínez Prieto" sugar mill in Marianao (molasses and technological waste); the "Ñico López" oil refinery in Regla (lead, heavy metals, potassium, sulfur, oil, oil products, technological wastes); the electric power plants "Antonio Maceo," in Regla, and "Otto Parellada," in Tallapiedra, (metals and other waste products); a fertilizer plant located in Regla (phosphates, nitrates); tallow processing plants in Regla and San Miguel del Padrón (waste products); and the ice cream factories "Guarina," in the municipality of 10 de Octubre, and "Coppelia," in the municipality of Boyeros; and a milk complex in Habana Vieja (milk waste products).

La Habana Bay

La Habana Bay is one of the several protected bays shaped like a bag or pocket *(bahías de bolsa)* with which Cuba is endowed (Marrero 1950, 47). It is divided into three inlets—Marimelena to the northwest and Guasabacoa and Atarés to the southeast—which are reached by a deep channel. Although pocket bays are highly valued because of their safety, they are prone to contamination because of the relatively long period of time that it takes for water within the bay to be flushed out by sea currents, tides, and fresh water flowing into it. A study conducted in the 1970s found that the renewal time for the waters of La Habana Bay was five to six days (Ministerio de Transporte 1985, 25).

La Habana Bay has the dubious distinction of having been designated by the UN Development Program (UNDP) as one of the ten most polluted bodies of water in the world (Segre, Coyula, and Scarpaci 1997, 276). It is also one of the most severely polluted bays and coastal areas of the Caribbean region, eligible for assistance in environmental management from the Global Environmental Facility of the UN Environment Program (Riera 1998). An estimated 300 tons of organic matter and over 40 tons of oil and oil products are dumped into it daily. The concentration of waste products in the bay is such that at certain times of the day, when the winds blow in a certain direction, the stench is unbearable (Lezcano 1994b).

Since the 1970s the UNDP has been financing a project to address the pollution of the bay, starting with the identification of sources of pollution, measurement of pollution levels, and development of policies to stop and reverse the pollution. A preliminary report emerging from this project issued in the mid-1980s offers a snapshot of the pollution of this important bay:

▶ Pollution entering La Habana Bay consists primarily of industrial wastes and organic materials associated with sewage. Main pollution sources include the sewer systems of several residential locations, three electric power stations, an oil refinery, a paint manufacturing plant, two fertilizer plants, two slaughter houses, a bus depot, an alcohol and a rum distillery, a leather tannery, a yeast manufacturing plant, two fishing ports, and several flour mills.

▶ Over 104 metric tons of organic matter were dumped daily into the bay, an amount of organic matter associated with a population of 2 million people. The discharges were delivered by (1) the rivers that flow into the bay (about 40 percent), carrying mostly waste products from the alcohol and rum distilleries and the yeast production plant; (2) industrial plants that dumped their wastes directly into the bay (30 percent); and (3) sewer lines (Ministerio de Transporte 1985, 22).

▶ Significant concentrations of fecal matter were found within the bay and along the coastline to the west of the bay; during some time periods, concentrations were higher than those acceptable for public swimming (Ministerio de Transporte 1985, 37).

▶ Nearly 88 percent of the 32 tons of waste products dumped daily by industries originated with the "Ñico López" oil refinery (Ministerio de Transporte 1985, 22).

▶ Oil products flow into the bay primarily through storm drains, although the oil refinery and the gas manufacturing plant are also significant sources of such products (Ministerio de Transporte 1985, 23).

▶ Electric power plants dumped significant amounts of metals such as iron, vanadium, manganese, copper, zinc, chromium, lead, and cadmium into the bay. Sewer lines also contributed iron and lead (Ministerio de Transporte 1985, 24).

To summarize, two Soviet experts who visited Cuba in 1982 to assess pollution of bays and coastal areas concluded that La Habana Bay was "practically dead"; not only can natural resources not be extracted from the bay, but its waters also pollute neighboring coastal areas (Hernández 1982, 29–30).

Some efforts have been reported by Cuban officials in cleaning up La Habana Bay, but it is not clear what impact these have had in reversing decades of neglect. According to a CITMA official at a workshop held in early 1998, there is less hydrocarbon pollution in the harbor as a result of a decrease in maritime and port traffic and the construction of a permanent barrier at the "Ñico López" refinery, which prevents the flow of oil products into the Marimelena inlet; other forms of pollution have also been reduced with the introduction of new technology at the Evelio Prieto gas plant, the decision to stop fermenting sugarcane to make alcohol at the Habana distillery in Lawton, and the shutdown of the "Ciro Redondo" slaughter house in La Virgen del Camino. Very little progress seems to have been made in reducing pollution associated with discharge of organic matter, 70 percent of which flows through the storm (rainwater) drainage system and the three rivers that flow directly into the bay: the Luyanó, Martín Pérez and Arroyo Tadeo (Riera 1998).

Summary and Implications

La Habana, more than any other location in Cuba, shows the effects of several decades of misguided and centrally orchestrated development policies. Well-intentioned, but ineffectively implemented policies to reduce regional levels of social and economic development, while leading to relative improvements in living conditions in Cuba's secondary cities and in the countryside, resulted in a near catastrophic neglect of the country's capital city. The collapse of the Soviet bloc and the end of foreign economic subsidies have made La Habana's decay more poignant. Its housing and physical infrastructure is crumbling. Particularly serious problems have been documented regarding its water distribution and sanitary infrastructure. Pollution is severe in the streams running through the metropolitan area and in La Habana Bay and surrounding coastal areas. Although in some respects its environmental situation may not be worse than in other major capital cities in the developing world, La Habana is unique in that its current deteriorated condition resulted from the decisions made by an unchallenged political leadership with complete sway over the country's destiny for nearly four decades.

The seriousness of the situation has alarmed many Cubans, some in positions of authority, who had previously remained largely oblivious to the many environmental challenges confronting La Habana, as well as Cuba (Cuba: La Habana promueve 1997). Tragically, the country's current economic and political situation does not augur well for the inevitable restoration task. Cuba currently does not have, nor is it likely to gain access to in the near future, the many billions of dollars necessary to modernize La Habana's physical infrastructure and to begin to reverse its environmental deterioration.

๛ 10 ๛

The Special Period and the Environment

THE ECONOMIC crisis that has swept Cuba in the 1990s—during the "special period in peacetime"—has affected every facet of Cuban life. Analysts have focused a great deal of attention on the effects of the crisis on overall levels of economic activity, on population standards of living, and on the performance of specific sectors of the economy (such as sugarcane agriculture, electricity, or transportation). Relatively unstudied, however, are its consequences for other aspects of Cuban life, such as the environment.

In this chapter we review the effects of special period policies and outcomes on the Cuban environmental situation. The analysis is broad, examining the interactions between the special period and the environment across the economy at large, including agriculture, industry, mining, tourism, nutrition, and public health. In the first section we examine aspects of special period policies and outcomes that have had a positive impact on the environment. In the second section we examine policies and outcomes that have had a negative impact on the environment. We conclude with a critical discussion of arguments claiming that Cuba has embraced a more environmentally benign agricultural model and a description of the difficulties faced by independent environmentalists in Cuba.

Positive Interactions Between the Special Period and the Environment

The special period has had some positive effects on the Cuban environmental situation. As President Castro stated in the document presented by Cuba to the June 1992 Earth Summit, "[The special period] is a period of readjustment ... requiring maximum economizing and austerity in economic and social policies, along with many creative initiatives, a large number of which have come directly from the people. Many of the steps taken as a result of the special period fit in with the strategic lines prepared by the Revolution. Some of them have helped accelerate the policies put into effect by the country in defense of the environment" (Castro 1993, 49). The sector of the economy receiving the greatest attention because of the positive interaction between the special period and the environment is agriculture (Carney 1993). However, the special period has also positively affected the environmental situation in other economic sectors, such as industry, transportation, and public health.

Economic Growth and Environmental Degradation

To the extent that overall economic activity in Cuba has declined during the special period, so has the degradation of the environment associated with the emission of greenhouse gases from burning fossil fuels, the generation of industrial pollutants, and the contamination of water by runoff of chemical fertilizers and pesticides. The contraction in economic activity no doubt has had some positive effects on the environmental situation in Cuba, particularly since the economic contraction has been very sharp. As discussed in Chapter 2, Cuba's national output shrank by 35–50 percent between 1989 and 1993, with a slight recovery recorded during 1994–1997.

Perhaps the most tangible aspect of special period adjustment policies has been the sharp reduction in Cuba's ability to import. Between 1989 and 1993, Cuban imports fell from 8.1 billion pesos to 2.0 billion pesos, or by 75 percent. Although Cuba has not published foreign trade statistics by commodity since 1989, a reconstruction of such statistics carried out by the U.S. Central Intelligence Agency based on partner country statistics shows that reductions in imports over the period 1989–1993 affected all categories of imports. Particularly relevant for purposes of this discussion were the reductions in imports of raw materials (89 percent), transportation equipment (86 percent), semi-finished goods (79 percent), chemicals (72 percent), and fuels (71 percent). According to Cuban official Carlos Lage, in 1993 Cuba imported 5.7 million tons of oil and oil products ("Carlos Lage" 1993, 4), and domestic crude production reached 1.1 million tons, for a total apparent supply of 6.8 million tons; this compares to the approximately 11 million tons per annum that Cuba consumed during 1984–1988. Although imports rose in 1994 (3.3 percent) and 1995 (37.5 percent), in the latter year they amounted to about one-third of their 1989 value.

Agriculture

At the end of 1989 the Cuban state controlled the vast majority of the country's agricultural land, most of it organized as state farms. Agricultural machinery (equipped with internal combustion engines), irrigation, chemical fertilizers, herbicides, and pesticides were extensively used. As discussed in Chapter 4, by 1989 soil preparation for major agricultural crops— sugarcane, rice, tomatoes, potatoes, beans, plantains, citrus—was carried out using mechanized equipment. Mechanization was also heavily used in fertilizing, weeding, and harvesting of these crops. In sugarcane agriculture, for example, 100 percent of soil preparation and fertilization was carried out using machinery; so were 62 percent of weeding, 70 percent of sugarcane cutting, and 100 percent of sugarcane lifting and transportation activities. As a farmer who supervised production at a cooperative outside La Habana told a U.S. researcher (Simon 1997), one of his main tasks before the special period was to help dump tons of chemical fertilizers on fields and spray the ripening crops with drums full of pesticides; under the watchful eye of agronomists from the University of La Habana, "even if you didn't have a pest infestation, you applied pesticides, because that's what the central planners had decreed." Cuban agriculture also demanded high levels of chemicals inputs such as fertilizers and herbicides. Burning of oil products in agricultural machinery contributed to air pollution, while the use of fertilizers, herbicides, and pesticides did the same for soil and water resources.

According to official statistics, the agricultural sector consumed nearly 468,000 metric tons of oil and oil products in 1988, roughly 4 percent of the 11.1 million tons of such products consumed nationally (Comité Estatal 1989). Imports of fertilizers and of herbicides and pesticides amounted to 158 million and 80 million pesos, respectively, in 1989, representing roughly 3 percent of overall imports in that year. Imports represented 94 percent of fertilizers and 97 percent of herbicides used on the island (Rosset and Benjamin 1994b, 18). Two specialists have highlighted the heavy dependence of the Cuban agricultural sector on imported inputs (Díaz González and Muñoz 1995, 44–45): "[The] 'modern' development of the Cuban farming sector depended on its economic integration into the commercial trade agreements of the former socialist countries

(CMEA). Oil for farm machinery, machines, and fertilizers (as well as for steel and chemical raw materials which were manufactured locally) and pesticides had to be imported."

Vice President Carlos Lage, in a November 1992 interview with the Cuban press, gave some parameters of the magnitude of the reduction in imports of agricultural inputs during the special period (Carlos Lage 1992, 5). Lage stated that overall fuel imports for 1992 would amount to about 6.1 million tons, 53 percent lower than the 13 million tons that were imported annually prior to the special period.[1] Although the agricultural sector was assigned a high priority, Lage noted that fuel consumption by the agricultural sector in 1992 would be under 1,000 tons per day, 33 percent lower than the 1,500 tons per day consumed by the sector before the special period. The reductions were even sharper with regard to other agricultural inputs; for example:

▶ *Fertilizers:* Cuba formerly imported 1.3 million tons of nitrogen, phosphorus, and potassium fertilizers; 1992 imports amounted to 250,000 tons, an 80.1 percent reduction.

▶ *Herbicides:* Imports of herbicides, which formerly amounted to about $80 million per annum, were reduced to under $30 million, or by 62.5 percent.

▶ *Animal feed:* Imports of grains and fodder for animal consumption were reduced from 1.6 million tons to 450,000 tons, or by 71.9 percent.

These sharp reductions in imports of fuels, fertilizers, and other agricultural inputs during the special period forced changes in agricultural techniques that, although disastrous in terms of agricultural output, have had positive environmental consequences. By November 1991 about 12 percent of Cuba's agricultural tractors were idle because of lack of fuel, and 100,000 oxen had been trained for duty in animal traction (Roca 1994, 105). Fuel shortages also idled pumps used to draw irrigation water from aquifers, easing environmental pressures associated with the overextraction of underground water. In 1992 Cuba applied chemical fertilizers to 817,000 hectares of sugarcane, compared to 2,625,000 hectares in 1989; for herbicides application, the corresponding areas were 1.7 million hectares in 1992 compared to 2.2 million hectares in 1989 (Alvarez and Peña Castellanos 1995, 29). According to statistics gathered by the FAO (1993), Cuban consumption of fertilizers in agricultural year 1992–1993 amounted to 249,000 tons, 62.3 percent lower than the 661,000 tons consumed in agricultural year 1989–1990. Particularly sharp was the reduction in consumption of potash, which fell by 82.6 percent from 212,000 tons in 1989–1990 to 37,000 tons in 1992–1993.

Decree-law No. 142, enacted by the Cuban Council of State in September

1993, brought about an important change in land tenancy in Cuba. The statute created Unidades Básicas de Producción Cooperativa (UBPC, Basic Units of Cooperative Production) on state lands formerly organized as state farms. In essence, Decree-law No. 142 broke up state farms into cooperatives, giving cooperative members the right to farm their land in perpetuity and some degree of autonomy over their activities, for example, the crops they plant and the farming methods they use (Alvarez and Messina 1996). Compared to the situation that prevailed in state farms, where workers were essentially wage employees subject to a rigid decision-making process controlled from above, UBPC members are more likely to behave as land owners—for example, engaging in soil conservation and improvement activities, seeking to use cultivation methods that are less reliant on imported inputs—to the benefit of the overall environmental situation.

Recent writings on Cuban agriculture by foreign experts (e.g., Dlott et al. 1993; Gersper et al. 1993; Levins 1993; Vandermeer et al. 1993; Rosset and Benjamin 1993 and 1994a,b; Perfecto 1994) posit that in the 1990s Cuba has adopted an environmentally-friendly "alternative model" of agricultural development. Table 10.1 describes the main elements of this alternative model and contrasts them with those of the "classical model" of modern agriculture, practiced in California, the former Soviet Union, and in Cuba prior to the special period. Peter Rosset and Medea Benjamin (1993, 22) describe the "classical" agricultural model that prevailed in Cuba in the 1960s, 1970s, and 1980s as follows:

> It relies on intensive use of chemical fertilizers, pesticides, mechanization, feedlots, petroleum and petroleum by-products, hybrid corn varieties, and capital in the form of credit. It is based on crop monocultures planted on large holdings to take advantage of economies of scale. It has led to soil erosion, compaction, salinization and waterlogging, environmental contamination, pest resistance to pesticides and uncontrollable pest outbreaks. It has also led to an exodus of the population from rural areas as mechanization replaces human labor. It is also a model that has become increasingly expensive as input prices rise and farmers use increasing quantities to compensate for eroding soil fertility and the loss of natural pest controls. For developing countries it is even worse, because it creates a dependence on imports that use up scarce foreign exchange.

In contrast, according to these same experts (Rosset and Benjamin 1993, 22), the "alternative model," as put forth in Cuba, "promotes crop diversity rather than monoculture, organic fertilizers and 'biofertilizers' instead of chemical ones, and biological control and 'biopesticides' instead of synthetic pesticides. Further-more, animal traction is substituted for tractors, and planting is planned to take advantage of seasonal rainfall patterns in order to reduce reliance on irrigation. Local communities are to be more intimately involved in the production process, hopefully putting brakes on the exodus to the cities. . . . In essence, this [the 'alter-native model'] is a return to the importance of the family farmer."

Foreign experts have described in considerable detail advances made in Cuba with regard to certain elements of this alternative model, including man-agement of insect pests and weeds with minimal use of chemical pesticides and herbicides (Dlott et al. 1993; Perfecto 1994) and soil conservation techniques (Vandermeer et al. 1993). The literature that argues that there has occurred an organic farming revolution in Cuba has a "back to the future" element to it. It is ironic that the Cuban alternative agricultural model that has generated so much enthusiasm among foreign experts is essentially pre-revolutionary Cuba's agri-cultural model. Brian Pollitt (1997, 197) refers to this as a process of "technical regression" seeking to reestablish practices widespread in Cuban farming prior to what he calls the "industrialization of agriculture" of the 1970s and 1980s.

Notwithstanding its weaknesses, pre-revolutionary Cuba's agricultural model scored highly, for example, with respect to "diversification of crops and autochthonous production systems based on accumulated knowledge" as indi-vidual farmers made planting decisions—including intercropping—based on experiences accumulated over generations. Similarly, "introduction of scientific practices that correspond to the particulars of each zone" was unnecessary since decisions were decentralized. Corrective measures embodied in the alternative model are, to a large extent, aimed at blunting the distortions created by the imposition by the Cuban government since the 1960s of a socialist agricultural model with centralized decision making (Pollitt 1997).

In the same spirit, a report prepared by Cuba's Ministry of the Sugar Indus-try for the 1992 Earth Summit stated that "harvesting sugar cane without burn-ing the excess foliage is an integral part of the new Cuban harvesting system. Burning foliage, a common practice in most other countries, not only harms the environment but rules out the production of numerous valuable cane by-products" (Rose 1992, 39). What the report did not mention is that green sugar-cane was traditionally harvested in Cuba, and introduction of the so-called Aus-tralian system (which burns sugarcane prior to cutting) began in the early 1970s in the context of the mechanization of the harvests. The rationale for the shift was that burning the tops and leaves of the cane, as well as weeds, reduced the amount of extraneous matter picked up by mechanical harvesters. The environ-mental drawbacks of the Australian system—for example, that tops and leaves that could be used for other purposes (such as cattle feed) were lost; that trash

Table 10.1. Comparison of classical and alternative agricultural models

Classical model	Alternative model

External dependence of
    the country on other countries
    the provinces on the country
    localities on the province and the country
	Maximum advantage of
	    the land
	    human resources of the zone or locality
	    broad community participation
	    cutting edge technology, but appropriate to
	        the zone where it is used

Cutting-edge technology
    imported raw materials for animal feed
    widespread utilization of chemical pesticides
        and fertilizers
    use of modern irrigation systems
    consumption of fuels and lubricants
	organic fertilizers and crop rotation
	    biological control of pests
	    biological cycles and seasonality of crops
	        and animals
	    natural energy sources: hydro (river, dams,
	        etc.), wind, solar, biomass, etc.

Tight relationship between bank credit and
    production; high interest rates
	animal traction
	    rational use of pastures and forage for both
	        grazing and feedlots, search for locally
	        supplied animal nutrition

Priority given to mechanization as a production technology	

Introduction of new crops at the expense of
    autochthonous crops and production
    systems
	Diversification and autochthonous production
	    systems based on accumulated knowledge

Search for efficiency through intensification and
    mechanization
	Introduction of scientific practices that
	    correspond to the particulars of each zone;
	    new varieties of crops and animals, planting
	    densities, seed treatments, post-harvest
	    storage, and so on

Real possibility of investing in production and
    commercialization
	Preservation of the environment and the
	    ecosystem

Accelerated rural exodus	Need for systematic training (management, nutritional, technical)

Satisfying ever-increasing needs has ever-
    increasing ecological or environmental
    consequences, such as soil erosion,
    salinization, and waterlogging
	Systematic technical assistance

	Promote cooperation among producers, within and between communities

	Obstacles to overcome
	    difficulties in the commercialization of
	        agricultural products because of the
	        number of intermediaries
	    control over the market
	    poverty among the peasantry
	    distances to markets and urban centers (lack
	        of sufficient roads and means of
	        transport)

Source: Cuban Ministry of Agriculture, as reproduced in Rosset and Benjamin 1993, 23.

generated by green cane harvesting, which provides ground cover and helps preserve moisture in nonirrigated areas, was also lost; and that burnt fields have to
be replanted more frequently—were contemporaneously discussed in the
Cuban technical literature, but nevertheless the decision to burn fields prior to
harvesting was made from above and forced upon sugarcane growers (Pérez-
López 1991, 67).

Most foreign experts who have lauded the virtues of Cuba's organic farming
revolution have failed to note the low productivity associated with this model
and its implications for meeting the food demands of a population of 11 million.
An exception is Carmen Diana Deere (1993, 47), who has correctly pointed out
that lack of fuel and fertilizers has contributed to declining rice and sugarcane
yields and low yields for staple agricultural commodities and vegetables. Statistics on physical production of key agricultural commodities in 1989 and 1993
(CONAS 1995, 41) demonstrate the sharp reduction in production over this
period, including large declines for fruits (68.8 percent), rice (67.0 percent), milk
(65.7 percent), grains (61.3 percent), tobacco (52.2 percent), and tomatoes (50.8
percent). Reportedly, agricultural production declined by 35 percent in 1994
alone (Simon 1997). The low productivity of the new agricultural model puts
into question its permanence, an issue to which we will return later. Similarly,
discussing the productivity and costs of animal power versus tractors in the
sugar industry, sugar expert Pollitt (1997, 195) states: "Pro-ox propagandists
argued that in Mexico—a large oil producer—more than 60 percent of agriculture used animal power. What they overlooked was that large-scale use of
draught animals was generally associated with very low labor productivity and
widespread rural poverty."

Industry

Reductions in imports of fuels, raw materials, machinery, and spare parts
have reduced industrial activity and therefore air, soil, and water pollution associated with such activity during the special period. Since most industrial plants
are located at or near medium to large urban areas, the principal beneficiaries of
reduced pollution levels have been the cities. In 1988 the industrial sector used
6.6 million metric tons of oil products, or nearly 60 percent of overall consumption (Comité Estatal 1989). According to estimates, up to 80 percent of industrial
facilities on the island were idle at the end of 1993 due to a dearth of fuel, raw
materials, and spare parts (Mesa-Lago 1994, 11). This means that most industrial
plants were not operating boilers that generate greenhouse gases and that generation of industrial effluents and wastes was also significantly reduced.

The electricity generation industry, in particular, has been heavy hit by shortages of fuel and spare parts during the special period. As discussed in Chapter 7, 98.7 percent of Cuba's electricity generation capacity was embodied in thermoelectric plants that used fuel oil, diesel fuel, crude oil, and natural gas. Electricity generation in that year used up 3.3 million metric tons of liquid fuels, or half of the total amount of oil and oil products consumed by the industrial sector. Considering that overall fuel availability in 1993 was under 7 million tons, it is clear that consumption by the electricity generation system also had to be severely cut back.

According to official Cuban statistics, electricity generation in 1993 amounted to 11,004 gigawatt-hours, 27.8 percent lower than the 15,240 gigawatt-hours generated in 1989. To ration available electricity, the government instituted a system of rotating blackouts throughout the island. In the summer of 1993, electricity outages stretched for sixteen to eighteen hours, with some cities in the interior of the nation facing up to twenty hours of power outages per day. Such power outages not only inconvenienced consumers but also had the effect of shutting down factories that relied on electricity for power and affected other public services (Whitefield 1993).

The result was a sharp contraction of economic activity across all components of the industrial sector, including some of the industries deemed to be the largest polluters. Thus, during the period 1989–1993, nickel-cobalt production declined by 35.2 percent, cement by 72.1 percent, complete fertilizers by 89.5 percent, raw sugar by 44.3 percent, and steel by 68.7 percent (CONAS 1995, 41). As discussed in Chapter 7, however, import stringencies associated with the special period have prevented the importation of pollution abatement equipment, for example, for the cement industry.

Public Health

Special period policies have had some unintended positive effects on public health. For example, with regard to the widespread use of bicycles to provide means of transportation for the population, President Castro (1993, 49) has said: "To make cuts in transportation usage, a solution was introduced which is innovative because of its mass scale: the use of bicycles. Hundreds of thousands of bicycles were imported, several factories were modified to manufacture bicycles and almost a half a million bicycles have been distributed to workers and students. The proliferation of cyclists of all ages is perfectly compatible with the policies promoted for several years to guarantee health for all, including exercise programs for senior citizens. In this way, the current shortages of fuel, although they negatively affect daily life, also have a positive effect on the environment." Other "ecologically valuable" outcomes of the special period, according to President Castro (1993, 49–50), are "the intensified use of herbal medicine, the promotion of local fruit and vegetable gardens (even in residential areas in yards and terraces), the gradual utilization of animal traction in agriculture, the development of composting, and much more."

As the average daily intake of calories, proteins, and other nutrients has declined, the composition of the Cuban diet has changed. Cubans, accustomed to a diet high in fats and carbohydrates, have not only seen their food intake decline drastically, but have also been forced to attempt to supplement their limited diets with domestically grown fruits and vegetables, many of which are produced in home gardens. While this shift away from fatty foods may be beneficial, the weight of the evidence seems to point to a decline in health standards as a consequence of the deterioration in food intake (see later discussion).

Another positive effect of the special period is that it has reduced waste, since recycling has intensified. Policy changes instituted in 1993 that legalized certain forms of self-employment coupled with the reestablishment of artisan markets in December 1994 (Jatar-Hausmann 1996, 214) have spawned a significant light manufacturing industry that relies largely on waste products, since markets for intermediate goods do not exist. Recycling makes a positive environmental contribution by limiting the need for landfills and other forms of waste disposal systems. The amount of city waste has also declined as residents use their food leftovers to feed barnyard animals (e.g., chickens, pigs) in their yards, although these practices raise sanitation concerns.

Negative Interactions Between the Special Period and the Environment

The special period has also presented serious challenges for Cuba's environment. In an effort to overcome the economic crisis, the Cuban government adopted emergency measures in a number of areas. Because of the severity of the economic crisis, and the urgency in addressing it, "environmental considerations have been basically absent" from the analysis of policy measures (Atienza Ambou 1996, 63).

Among other policy actions, the Cuban government has aggressively sought foreign investment to stimulate the stagnant mining, tourism, and industrial sectors, putting additional pressure on the environment. Thus, the economic crisis "tempts the Cuban government to search for petroleum and mining resources, or to develop tourist facilities, on a scale and at a speed which also increase the likelihood of environmental damage, especially to Cuba's smaller cays and its marine resources" (Cole and Domínguez 1995, 5). According to Cuban government figures, 212 joint ventures with foreign investors had been established through 1995 in a broad range of sectors of the economy.[2] Shortages of imported food and medications threaten public health achievements of the last three decades. The economic crisis has also affected the overall scientific structure of the country, with potential adverse long-term implications.

David Collis (1996a, 451) argues that the economic crisis of the 1990s has increased pressure to sacrifice environmental protection for economic profit at a time when resources to remedy existing problems are scarce. He further points out that the economic policy changes that have been made to address the crisis (e.g., stimulation of foreign investment, legalization of the use of hard currency, authorization of self-employment) have "triggered a decentralized and semi-capitalist development that is incompatible with the existing environmental regulation structure designed for a centralized, socialist economy." Whether environmental authorities in Cuba will be able to influence development patterns in the second half of the 1990s remains an open question.

Mining

Through 1995, Cuba and foreign investors established twenty-eight joint ventures in the mining sector; twenty-seven of these partnerships were established in 1993–1994. The most significant foreign joint ventures are in the nickel industry, but there also has been some joint venturing in the production of gold, copper, zinc, and other minerals.

Joint ventures have expanded operations in nickel-producing areas in the eastern region of the island, placing severe pressures on an already serious environmental situation. A joint venture with Canada's Sherritt, Inc., operates the aging nickel ore processing plant at Moa Bay. According to a Canadian journalist, "Because of leaky equipment and other factors, the sulphur compounds used in the process pollute the air and water, producing what residents say is acid rain. Heavy erosion from surface mining is also filling Moa Bay with earth" (Knox 1995). Residents of Moa told the same journalist that they took it for granted that "one of the reasons a foreign mining company would be interested in operating in Cuba was that environmental standards would be lower" (Knox 1995). As discussed in Chapter 7, the Moa Bay plant dumps substantial volumes of liquid wastes—containing a wide range of light and heavy metals and sulfuric acid—each day into the ocean. It has also been alleged that the joint venture is using calcium carbonate from coral reefs in Moa Bay to neutralize sulfuric acid in waste materials, a charge Sherritt officials have denied.

Strip mining of the nickel ores has been an environmental concern for years, since remediation and reforestation have not kept pace with mining activity. This is a major source of environmental damage at Moa Bay, according to Sherritt chairman Ian Delaney (Knox 1995). The construction of new nickel production plants with the assistance of foreign corporations in the northeastern region of Cuba is on the drawing board. In particular, Western Mining Corporation of Australia and the Cuban enterprise Commercial Caribbean Nickel, S.A., signed a letter of intent in October 1994 forming a joint venture to assess developing a nickel deposit at Pinares de Mayarí, near where the other nickel production plants are located. Other joint ventures exploiting copper, gold, and silver deposits are also putting pressure on the environment.

To ease the bottleneck caused by shortages in imported fuels, Cuba has gone all out during the special period to increase domestic production of crude oil. The drive to increase crude production has meant that drilling and production have been permitted in certain areas that were formerly considered environmentally fragile, such as coastal areas. Cuba has also approved joint ventures with foreign firms to explore offshore, risking the possibility of an offshore oil spill that could harm the marine environment in Cuba and Florida (Rosendahl 1991).

Cuba and foreign investors had entered into twenty-five joint ventures in the oil sector through 1995. Most of these joint ventures (nineteen) were concluded in 1992–1993. According to a government broadcast in October 1994, eleven foreign companies were participating in oil exploration and exploitation in Cuban territory, both on land and offshore, and ten others were working in oil production services. Firms from France, Canada, the United Kingdom, Sweden, and Germany were operating eighteen oil fields on the island. The basis for most of these operations seems to be production partnership agreements, or risk contracts, between the foreign oil companies and Cubapetróleo, S.A. (Cupet), a corporation created by the Cuban government in 1991 to manage the oil industry.

Cuba began to lease offshore tracts to foreign oil corporations since 1990. In that year, a 2,000-square kilometer offshore block was leased by the Cuban government to a French consortium consisting of Total and Compagnie Européenne des Petroles. The consortium was also authorized to drill four exploratory wells in the bay of Santa Clara, along Cuba's northern coast; it also drilled in the bay of Cárdenas. In February 1993 Cuba launched a round of open bidding for oil exploration and production rights on the island, offering eleven onshore and offshore tracts, ranging in size from 560 to 2,400 square miles, to foreign bidders.

The location of Cuba's oil deposits near areas that are suitable to tourism development has caused conflicts within the leadership. According for President Castro, he had to intervene personally to stop efforts to produce oil in the Hicacos Peninsula (where Varadero Beach is located) that would be harmful to the development of the tourism industry. He said in 1990 (Castro 1993, 102): "of course we need oil! But when we did the analysis and the calculations, no matter how much oil might be deposited under the [Hicacos] peninsula, it would not generate as much revenue for the country as tourism. I personally had to do the calculations, showing them the value of the 100,000 tons of oil that they planned to extract from this area. Moreover, oil is an exhaustible resource while resources such as sun, air, and sea are inexhaustible. Finally, the oil [under the Hicacos Peninsula] is a heavy oil, with high sulfur content, and a low value in the world market."

Tourism

One of the leading sectors in Cuba's adjustment policies during the special period is international tourism. The Cuban leadership has identified the international tourism industry as one that can contribute significantly to the country's hard currency balances and has aggressively sought foreign capital to develop additional tourist facilities, particularly seashore resorts. Thus, tourism has shifted from being a complementary area of economic activity to a dominant one, with significant implications for the environment (Cabrera Trimiño 1997, 209). Through 1995, Cuba had entered into thirty-four joint ventures with foreign enterprises in the tourism sector. Particularly active in the Cuban tourism sector have been investors from Spain, Canada, Mexico, Netherlands, Jamaica, and Germany.

In the early years of the revolution the Cuban regime shunned international tourism for ideological reasons. As María Dolores Espino (1994, 148) puts it: "Tourism was perceived as too closely associated with the capitalist evils of prostitution, drugs, gambling, and organized crime. The revolutionary government discounted tourism as a vehicle for economic growth and development. During the 1960s and early 1970s, no major investment in tourism was undertaken. The vast tourism infrastructure built up during the pre-revolutionary years was left for the use of Cuban citizens and international guests from socialist and other friendly countries or simply abandoned. Some sixteen hotels were closed down, and hotel capacity was reduced by 50 percent." This began to change in the mid-1970s. In 1976, recognizing the potential economic benefits to be garnered from international tourism, the Cuban government created the Instituto Nacional de Turismo (INTUR, National Tourism Institute). In 1987 Cuba established the corporation Cubanacán, S.A., to develop joint ventures with foreign investors in the international tourism sector; Gaviota, S.A., a corporation reportedly connected with the Cuban armed forces, is also active in this area, as are other tourist enterprises. According to government statistics, Cuba received 620,000 international tourists in 1994, compared to about 158,000 in 1984. In 1994 there were 159 hotels suitable for international tourism with 23,000 rooms; 3,000 rooms were to be added in 1995, and plans called for a total of 50,000 rooms by the year 2000. In 1995 over 738,000 foreign tourists visited Cuba, 19.6 percent higher than in 1994, generating 1.1 billion pesos in gross revenues (BNC 1996, 10). In 1996 the Cuban tourism industry had a stock of 26,900 rooms and attracted over 1 million foreign visitors to the island, generating nearly 1.4 billion pesos in hard currency and making the tourism industry the largest source of foreign exchange (CEPAL 1997, tab. A23). Cuban analysts forecast that up to 2.5 million international tourists could be vacationing in Cuba by the year 2000 even if U.S. economic sanctions continue to prohibit U.S. tourists from doing so (CONAS 1995, 20).

The Cuban government claims to be sensitive to the fact that growth in the tourist industry depends not only on the availability of more and better facilities, but also on the implementation of cautious development plans to preserve the country's natural tourist attractions. However, in expanding its tourism infrastructure—hotels, recreation facilities, roads, airports, and so on—Cuba has often emphasized speed and low cost to the detriment of the environment. Cuban environmental protection official Helenio Ferrer has been quoted as saying that "tourism, like nuclear power, is like chemotherapy for us. It is something we need, yet it is something bad for us." Ferrer told a foreign visitor that "we built a seaside resort and the beach disappeared," reinforcing the dependence of tourism on the preservation of natural resources (Kaufman 1993, 33). Varadero, the premier beach resort area on the northern coast of Matanzas Province, was not developed in the most environmentally sound manner, according to the Cuban Institute of Physical Planning. While general legislation protecting coastal areas existed, no specific laws were in place. Thus, too many hotels were built, many were badly constructed and without sufficient space between them. In addition, the introduction of nonnative trees and plants to the area had an adverse ecological impact (Collis 1996, 453–54). Foreign observers note that while the Cuban government "talks a good line," in practice it is willing to sacrifice the environment in the name of economic survival (National Public Radio 1995).

The introduction of tourism into the country's pristine outer cays and islets and forest preserves has alarmed the conservation community. The northern cays of Ciego de Avila Province, for example, did not have a single hotel room in 1993: by 1997, they had over 1,500 rooms, an airport, more than 200 kilometers of roads—including 17 kilometers of roads over the sea—electricity generating units, and 100 kilometers of pipes to supply water from the mainland; 6,000 rooms are expected by the year 2000 (Bohemia Article 1997). Promotional materials from the Cuban government advertise the tourism facilities of Cayo Coco and Cayo Guillermo, in the Sabana-Camagüey sub-archipelago, as the largest tourism complex in the Caribbean, with over 1,700 hotel rooms, managed through joint ventures between Spanish corporations Tryp Hotels, Sol-Meliá, and RIU, the Italian corporation Viaggi del Ventaglio, and Cuban enterprises Cubanacán, S.A., and Gran Caribe (Visión real 1998). Some members of the conservation community note that while the country was sheltered from market forces (due to Soviet subsidies), it did not have to face the same difficult growth-environment tradeoffs that other island countries in the Caribbean did. As a result Cuba remains as one of the world's "richest storehouses of unique animals" (Dewar 1993), with some experts estimating that about 40 percent of Cuba's species still remain to be discovered (National Public Radio 1995).

Ecotourism, or "responsible tourism," is a relatively recent phenomenon in the global tourism industry. Ecotourists differ from the traditional "sun and sand" tourists who flock to the Caribbean in that the former are drawn by undeveloped natural resources—forests, wildlife preserves, national parks, pristine rivers, lakes, and islands—rather than by beach resorts. Ecotourists are often satisfied with more rustic lodging facilities and interact more closely with the domestic economy than traditional tourists. Although ecotourism still represents a very small portion of the huge international tourism market, its share is growing. For certain Caribbean countries (e.g., Costa Rica), ecotourism is already a significant force, and other countries in the region are attempting to follow suit (Honey 1994; Quinn and Strickland 1994; Holston 1995).

Endowed with significant potential for ecotourism (Simon 1995, 35), Cuba has made efforts to tap into this international market. A national working group for ecotourism was created in 1992 "to make tourism work for the environment—and vice-versa" (Cepero 1992). Among the sites initially identified for ecotourism development were the Sierra del Rosario, a UNESCO-declared biosphere reserve, and the Viñales Valley in the western region of the island; the Zapata Swamp and the Topes de Collantes Hills in the central region; and the Turquino National Park and Saetia Cay in the eastern region, with other areas such as the city of Baracoa and the island's northern cays to be developed subsequently (Cepero 1992). Reportedly, scores of relatively small ecotourism projects were in various stages of development on the island in the mid-1990s (Honey 1996, 10), a departure from the tourism strategy of the 1980s that favored development of large tourist enclaves. In these ecotourism sites, according to then–Tourism Minister Osmany Cienfuegos, "what we've tried is to incorporate the natural environment and the local community. The idea is that the tourists and the community together participate in this" (Honey 1996, 10).

Although ecotourism projects are intended to harmonize with the environment, some have encroached on the country's natural resources and habitat:

▶ To allow access by tourists to beaches in the numerous pristine small cays that surround the island, particularly on its northern coast, Cuban tourism authorities have constructed causeways (or stone embankments) bridging barrier islands to the mainland and to one another called *pedraplenes* (see Map 10.1). These *pedraplenes* block the movement of water in the intra-coastal waters, exacerbating contamination and destroying coastal and marine habitats (Espino 1992, 335). Many of these semi-enclosed water bodies are already subject to weak circulation regimes and high organic matter contents (Alcolado 1991). Examples of *pedraplenes* deemed to have caused substantial harm to the environment and fishing resources include the one joining the islands of Turiguanó and Cayo Coco and others in Caibarién and the northern region of Ciego de Avila Province (Solano 1995). These *pedraplenes* affect over 1,760 square kilometers of fragile ecosystems of the Sabana-Camagüey sub-archipelago, leading to the destruction of natural fauna and flora and the creation of lagoons of stagnant, polluted water devoid of wildlife (Wotzkow 1998, 138–44)

▶ A site marketed by Gaviota, S.A., as an ecotourism site is Cayo Saetia, in eastern Cuba. Gaviota imported some 10,000 Indian and African animals, including zebra, ostrich, gazelle, antelopes, bulls, and even white rhino, and offered camera and hunting safaris (Honey 1996, 12). According to a Cuban scientist, the Cuban military, operating through the enterprise Gaviota, S.A., placed more than 3,000 white-tailed deer in Cayo Saetia to create a hunting preserve for tourists (Wotzkow 1998, 242).

▶ Cayo Coco, site of several hotels and tourism facilities developed with foreign investment, is the most significant breeding ground of the roseate spoonbill and the greater flamingo, species that have been severely affected by human activity (Silva Lee 1996, 111). Reportedly, several colonies of flamingos that used to nest in the Sabana-Camagüey sub-archipelago have left this area because of the destruction of their habitat resulting from tourism facilities and *pedraplenes* and settled in the Bahamas, much to the delight of the tourism industry of that nation (Wotzkow 1998, 144).

▶ Cayo Largo, the site of a very large tourism facility being developed with foreign investment, and neighboring Cayo Majaés are the only habitats of *Mysateles garridoi,* a hutia (a small rodent) believed to be extinct until two individuals were captured in 1989 (Berovides Alvarez and Comas González 1991).

According to an expert, "over-hunting and environmental and social problems produced by the tourist industry have caused some discontent. Conservationists argue that for the ecotourist industry to thrive, regulations protecting the environment must be enforced" (Santana 1991, 15).

Notwithstanding the effects of ecotourism, it is clear that traditional "sun and sand" tourism is the engine of the Cuban tourism industry during the special period: "The reality is that in Cuba today, large-scale beach and urban tourism is capturing the international contracts and bringing in the bulk of the hard currency. Many Cubans openly worry that the island is on a slippery slope back towards the Caribbean-style, foreign-owned tourism. There is a debate over whether ecotourism is a catalyst for creating a new greener tourism industry or whether it is, as one Cuban intellectual put it, simply "'window dressing over the island's mushrooming conventional tourism. [Tourism Minister Osmany] Cienfuegos admits that ecotourism currently accounts for 'very little' of Cuba's overall tourism market" (Honey 1996, 10). Marc Frank, a U.S. economist residing in Cuba, has discussed the tourism–environmental protection dilemma as follows: "In general, I think Cuba is very serious about ecology, very serious about pro-

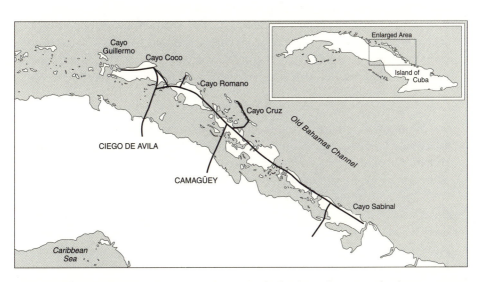

Map 10.1 Stone embankments built to link the island of Cuba with tourism development centers located in offshore cays

tecting the natural environment.... At the same time, Cuba does need to develop mass tourism, with all its negative impact, and the best they can do is try to make it as healthy as possible, but they can't stop its development because they need it in order to survive" (Cuba's Ecotourism 1995). Intense efforts to develop the international tourism industry can only exacerbate pressures on natural resources and have an adverse impact on biodiversity.

Industry

Through 1995, Cuban and foreign enterprises had entered into fifty-six joint venture arrangements in the industrial sector. These arrangements provided idle Cuban industries with funds to import essential raw materials and capital equipment, technology (including management expertise), and export markets for their output. Among the specific industries where joint ventures have been concluded are textiles, footwear, chemicals, glass, cement, pharmaceuticals, and tobacco.

Because of the financial pinch, Cuba has bought cheap oil in the world market to fuel its power plants; this low-quality oil generates dense clouds of pollution that embrace La Habana and other areas near the power plants (McGeary and Booth 1993, 44). As discussed earlier, domestically produced crude oil, used extensively in power plants, is of very low quality and has a very high sulfur content, thus generating a great deal of air pollution. Special period financial strains have put on hold Cuba's efforts to build a nuclear electricity generation plant, a source of electricity that would be more environmentally friendly—in terms of carbon dioxide emissions—than thermoelectric plants, although it would raise other environmental concerns associated with nuclear radiation (see Chapter 8).

Forestry

Shortages of imported lumber and of commercial fuels for home cooking (electricity, kerosene, gas) during the special period have placed additional pressure on Cuban forest resources. Shortages of cooking fuels have been partially addressed by increasing supplies of domestically produced firewood and charcoal. The indiscriminate cutting of trees and bushes for firewood has added to the already-serious problem of soil erosion. To cope with the fuel shortages, every major population center has been assigned brigades of workers whose task is to provide firewood to state-owned facilities, including factories, hospitals, and bakeries (Ecología/Cuba 1996).

Particularly damaging has been the cutting down of mangroves to produce charcoal (Solano 1995). The problem is reported to be particularly acute in the Zapata Swamp, an area populated by mangrove forests and sawgrass prairies, which is a haven for migratory birds from North and South America as well as for some of Cuba's rarest species. Residents of the area have illegally cleared forest remnants for charcoal-burning. According to a Cuban scientist, "Zapata's mangroves are rapidly being fed to the flames, and even its sawgrass is being used as fuel" (Dewar 1993, 6A). Indiscriminate cutting of trees during the special period is contributing to the desertification of the southern plains of Pinar del Río Province, which is advancing eastward at the rate of 11 kilometers per year (Peraza Linares 1997). A 1997 press report noted that using fuelwood for domestic consumption amounts to a misguided use of forest resources. The increase in the use of fuelwood has been attributed by Minister of the Economy and Planning José Luis Rodríguez to the fact that as of April 1997, the government had been able to distribute to the public only 47 percent of the home fuel required by the population. According to Rodríguez, the home cooking fuel necessary to meet the needs of the population for a single day costs $200,000 in hard currency, a sum that the country simply cannot afford (Dirigentes analizan 1997).

Agriculture

Although the environmental effects of Cuba's strategy to overcome the economic crisis appear to be largely benign on the agricultural sector, several developments merit continued attention. Cuba has launched an all-out effort to substitute chemical pesticides with biological agents (Shishkoff 1993). Although these initiatives appear at this time to be meritorious from an environmental point of view, they may carry long-term dangers not readily appreciated because they are being implemented on a wide scale without prior adequate experimental study. Of note are some biotechnology products used as substitutes for imported pesticides and fertilizers. A foreign expert who interviewed Cuban scientists (Perfecto 1994, 103) states: "The accelerated implementation of new technology also has a negative side, however. Some Cuban researchers are concerned about the lack of appropriate testing before a technology is implemented, a consequence of the food crisis during the 'Special Period.' This is an interesting reversal of the common phenomenon of scientists becoming frustrated with bureaucratic impediments that delay the implementation of a new technology." Related concerns have been raised by observers in other countries worried about potential unanticipated adverse consequences resulting from the introduction of biotechnology products in agriculture.

Transportation

Cuba' transportation stock is getting older and, because of poor maintenance and lack of spare parts, can only become more polluting and environmentally unfriendly. As long as the economic crisis continues, Cuba will be unable to modernize its fleet of cars, trucks, and buses (other than for those few vehicles serving the tourist industry). The used buses donated to Cuba by Canada, Spain, and other countries as a rule have been retired after years of service, tend to be obsolete, and do not have the latest environmental control technology.

Public Health

The special period has had considerable negative impacts on the health of Cubans and on the vaunted public health system. Some of these problems have been caused in part by measures introduced to cope with the deteriorated economic conditions. For example, bicycle-related deaths rose by 78 percent from 1989 to 1993, whereas motor vehicle-related deaths fell by 28 percent over the same period (Garfield and Santana 1997, 17). Shortages of imported fuel and spare parts have taken a toll on the public health, for example, limiting the ability to pump and distribute underground water and water from above-ground reservoirs. The dearth of foreign exchange has also affected the ability to import materials and products needed to ensure clean water. The country's ability to produce chlorine declined, reducing the populations covered by chlorinated water systems from 98 percent in 1988 to 26 percent in 1994, contributing to a significant increase in diarrheal diseases (Garfield and Santana 1997, 16). Donations from foreign governments and international organizations have made it possible for Cuba to reinstate water chlorination programs (Zurek 1997).

Urban health problems have escalated because of poor sanitation. Because of shortages of fuel and imported spare parts, many sanitation trucks in urban areas have been taken out of service. Garbage collection schedules are irregular and often carried out with animal-powered vehicles. Landfills have been located closer to populated areas because of the lack of fuel to transport refuse to faraway locations, and heavy equipment used to spread topsoil over the refuse are idle due to a lack of spare parts (Basureros 1997). The shift of urban workers to rural areas to increase agricultural production has created sanitation pressures in areas that are not equipped to handle the larger groups of workers. Sanitation has also been adversely affected by the common practice of raising chickens and pigs in urban homes, a practice encouraged by the government as a means to cope with food shortages.

During 1991–1993, Cuba suffered from the outbreak of a mysterious disease that temporarily blinded those affected by it (Ordúñez-García et al. 1996). After considerable speculation regarding the nature of the illness, an international team of physicians supported by the Pan American Health Organization determined that 50,000 people suffered from optic neuropathy (characterized by decreased visual acuity and color vision, optic-nerve pallor, and decreased sensitivity to vibration and temperature in the legs) and a peripheral form (characterized by diminished ankle reflexes and decreased sensitivity to vibration, pinprick, and light touch). No patients died, and the condition of most patients improved after treatment with vitamins, including vitamin B_{12}. The international team concluded that the epidemic of optic and peripheral neuropathy was linked to reduced nutrient intake caused by the deteriorating economic situation; tobacco use, particularly cigar smoking, was associated with an increased risk of optic neuropathy. From April to May 1993 the Cuban Ministry of Public Health initiated oral supplementation with multiple vitamins in the general population, with the incidence of the disease decreasing substantially after June 1993 (Cuba Neuropathy 1995).

Indeed, the special period austerity has had a significant effect on the nutrition of Cubans. In the 1980s about half of all protein and calories intended for human consumption in Cuba were imported. With imported foodstuffs declining by about 50 percent from 1989 to 1993, per capita protein and calorie availability from all sources declined by 25 percent and 18 percent, respectively, with low-cost rationed goods providing only about 1,200 calories per day. Special protections for children, women, and the elderly have meant that the burden of calorie and protein deficits has fallen predominantly on adult men, whose caloric intake fell from 3,100 in 1989 to 1,863 in 1994. Despite the protections for women, the proportion of newborns weighing under 2,500 grams rose 23 percent between 1989 and 1993, reversing ten years of gradual progress in this area (Garfield and Santana 1997, 16). Data from another report indicate that the percentage of underweight women entering pregnancy increased from 8.7 to 10 percent between 1990 and 1993. There have also been reports of significant increases in mortality among residents of nursing homes in 1993 (Johnson 1993). The gravity of the dietary changes is summarized in the same report: between 1989 and 1993 the daily per capita nutrient intake has declined by 40 percent in proteins, 64 percent in fats, 67 percent in vitamin A, 62 percent in vitamin C, 22 percent in iron, and 19 percent in calcium (U.S. Blockade Causes 1995, 4).

Cuba's advanced system of health care delivery has not been spared by the economic crisis and the end of subsidies from the former Soviet Union and the socialist countries. According to an analyst who visited Cuba in the autumn of 1992, the country's health care system was already "disintegrating" (Barrett 1993, 1). This analyst observed a shortage of imported medical products such as over-the-counter and prescription medications, anesthetics, sutures, surgical gloves,

X-ray plates, diagnostic kits; also in short supply were soap, detergents, other personal hygiene items, chlorine to treat the public water supply, and pesticides to control insects.

Sharp reductions in imports of fats from the former Soviet Union used as input by the domestic industry producing soap and related products brought about a sharp shortage of these products. The yearly per capita ration of soap in 1993 and 1994 was four small bars. The deterioration in the availability of soap and other personal hygiene products led to an epidemic of pediculosis and scabies, which peaked in 1994. Soap substitutes made with caustic soda and other chemicals normally not found in the home resulted in a large increase in burns and poisonings (Garfield and Santana 1997, 17).

More recent evidence suggests that the public health situation continues to be critical, with basic supplies, from antibiotics to sterilizing detergents, hard to find (Stix 1995, 32). As noted in a report to the UN by Roberto Robaina, Cuba's foreign minister, "it has been impossible to procure the necessary resources to maintain the levels of performance of medical services, availability of medicines, and nutrition achieved by Cuba in past decades" (U.S. Blockade Causes 1995, 4).

To conclude, there is fragmentary evidence suggesting that the economic difficulties of the 1990s may be taking a toll on the health status of the population. Mortality rates from several causes of death known to be highly sensitive to environmental factors increased markedly between 1989 and 1993. Death rates from infectious and parasitic diseases, for example, rose by 68 percent, and those from acute diarrhea rose by 152 percent. Similarly, the incidence of low birth weights rose 23 percent. It must be remembered, however, that these increases resulted from a very low base. As discussed earlier, an optical neuropathy epidemic attributed to nutritional deficiencies affected 50,000 persons in the early 1990s. These public health setbacks are likely to have resulted from a severe deterioration in nutritional intake and from the inability—due to lack of financial resources to finance imports—to treat contaminated sources of water.

Science Infrastructure

Cuba's severe economic crisis has forced its government to cut back drastically on its expenditures across all areas, including scientific endeavors. The budget for environmental research has been curtailed and so has scientific training (Cole and Domínguez 1995, 4). The combination of budget cuts and more attractive work opportunities in sectors of the economy where hard currency can be earned (e.g., tourism) has prompted scientists and other professionals to leave their field of specialization and take menial positions as waiters or taxi drivers. The pressure on research institutions to become self-financing has created intense competition on these institutions to market their products independently, thus severing cooperative relations among them (Feinsilver 1995, 119–20).

The economic crisis also threatens Cuba's scientific collections of flora and fauna, which are valuable not only for Cuba but for scientists and environmentalists around the globe. Funds for preservations and maintenance are sorely lacking. Cuban environmental officials Roberto Acosta Moreno and Orlando Rey Santos (1997, 24) describe the critical situation faced by these collections: "An example of this [the adverse impact of the economic crisis] is the precarious state of Cuban scientific collections, brought about by the scarcity of funds and resources for their adequate preservation. The adverse consequences of this situation transcend mere national interests. It ought to be a cause for concern given that the gathering, management, and preservation of this collection is indispensable for any endeavor related to the study, management, and preservation of biodiversity."

The Special Period and the Environment: Myths and New Developments

The special period has had a profound effect on Cuba's environmental situation. It has, on the one hand, arrested a secular process of environmental degradation in the agricultural sector that was vastly aggravated by the adoption of a development model predicated on the heavy use of chemical inputs, mechanization, and irrigation that contributed to the deterioration of Cuba's soils and the contamination of its waters. The economic collapse has also reduced overall levels of industrial pollution. Socialist Cuba's capital stock, including its oil refineries and cement and chemical plants, largely imported from the former Soviet Union and Eastern European countries, is a significant source of air, land, and water pollution.

On the other hand, some of the economic emergency measures introduced during the special period could represent long-term threats to the environment. Particularly alarming are those associated with the speedup of development of the mining and tourism industries. The socialist government, in its zeal to promote the development of the latter industry, appears to be pursuing policies that have led to ecological degradation of most insular Caribbean countries. The economic crisis has also brought about a deterioration of the public health system and an increase in mortality rates from several causes of death known to be highly sensitive to environmental factors.

It is incontrovertible that the decline in imports, particularly of fuel, raw materials, and spare parts, as well as the general economic collapse have brought about sharp declines in pollution, but at a substantial economic cost. The economic-environmental interactions are complex. For example, the reduction in the volume of imported fuels has stimulated the use of firewood and charcoal as cooking fuels, with adverse implications for forest resources and soil erosion. The long-term environmental effects of these developments remain to be seen. Another open question is the permanence of the environment-friendly policies of the 1990s.

The Greening of the Revolution?

Similarly, there is very little information on the impact of Cuba's new agricultural model on ecosystems. One exception is a reported increase in bird species and in the size of their populations in rice fields where chemical products have been discontinued (Díaz González and Muñoz 1995, 51). However, there may be negative environmental consequences from, for example, the short span between the availability of a biopreparation and its extensive use, or from the combined effect of several biopreparations (Díaz González and Muñoz 1995, 51).

Changes that Cuba has made to its productive structure and methods of production to survive the austerity associated with the special period have been hailed by some as the "greening of the revolution," stating that Cuba's commitment to economic development not at the expense of environmental concerns is "profound and sincere" (Eco-Cuba 1993). For example, Richard Levins (1993a, 25) argues that the leftward shift in Cuban politics in the 1980s—during the so-called Rectification Process—contributed to the development of ecological rationality in agriculture, bucking the trend in other socialist countries toward market-oriented policies in agriculture and "creating new forms of organization to resist them, emphasizing collective and social motivations through which people empower themselves and gain their own expertise, aspire to changes in human relations, and place politics, not economics, in command." Elsewhere he writes (Levins 1993b, 59): "The commitment of Cuba to develop toward an ecological society was precipitated by the economic emergency of the special period. But its roots go further back. The socialist economic model starts with the premise that the use of nature must be the result of conscious choices rather than the outcome of individual profit-seeking behaviors. In the absence of private interests favoring the increased use of external inputs (fertilizers, pesticides) in agriculture, debates about the technological strategy are clashes of opinion, not interest. This makes it possible for the ecological option to win." The implication is that the "greening" of the Cuban revolution is not a transitory phenomenon associated with the special period and represents instead a well-calculated, philosophical change.

Has the special period broken the dependence of the Cuban agricultural sector on imported fertilizers and pesticides and on mechanization? How likely is it that recidivism to input-intensive production will set in as soon as the economic situation improves and it is possible to finance additional imports? It is too early to evaluate the long-term environmental effects of special period policies and to answer definitively the question of whether the policy changes are permanent or transitory. Despite the slight economic recovery reported by Cuban officials in 1995–1997, Cuba is still immersed in a severe economic crisis, and the end of the special period is not yet in sight.

The observed behavior of some key export-oriented agricultural sectors puts into question the Cuban commitment to a new environmentalism, however. During 1989–1993, production of export-oriented agroindustrial and agricultural products such as sugar (–44.3 percent) and tobacco (–52.2 percent), dropped significantly. To stimulate production—and exports—of these products, Cuba has entered into a variety of arrangements with foreign firms; for example,

▶ Cuba secured short-term loans estimated at $130–300 million from foreign banks and other institutions (mostly European) to finance the 1995–1996 sugarcane crop. Financing terms were very stiff, with interest rates reported to be in the range of 14 to 18 percent per annum (Mesa-Lago 1996, 5). The loans were to be used to "pre-finance" imports of essential inputs such as oil and fertilizers. For the 1996–1997 crop, the Cuban Ministry of Sugar sought $60 million in additional financing (Ross 1996, 165).

▶ German automobile producer Mercedes-Benz entered into an arrangement in mid-1995 to install internal combustion engines in 200 sugarcane harvesting machines, with payment to be made from the revenues obtained from the sugar crop (Pérez-López 1995a, 18).

▶ Cuba entered into joint ventures with two state tobacco monopolies, Tabacalera of Spain and Seita of France, whereby the foreign concerns would provide lines of credit to finance purchases of fertilizers, pesticides, fuels, lubricants, water pumps, and spare parts needed by the Cuban tobacco industry (Ross 1996, 164–65; Pérez-López 1995a, 13–14).

Clearly, the incentives to return to input-intensive production methods to obtain short-term production increases are very strong, thus raising serious questions about the long-term commitment to "green" agricultural practices.

An important factor to take into account is that the vast majority of Cuban agricultural engineers and technicians were trained under the heavy-input agricultural model. To them, modern agriculture implied the application of science and technology and higher output yields. Thus, "green" agricultural techniques are at best temporary solutions to a crisis situation but do not represent an alternative to the traditional approach (Díaz 1997, 39). This concern has been expressed by a Cuban expert (Coyula Cowley 1997, 60): "Equally troublesome is the trend that began to be noticed at the close of 1995 among certain policy makers. After several years during which the Cuban economy worsened continuously, the halt in the decline made them think about returning to technologies and techniques that had been identified during the special period as taxing, wasteful, and prone to dependency."

According to Ramón Castro Ruz, an advisor to the Ministry of Agriculture and Fidel Castro's brother, the poor results of the 1996–1997 sugarcane harvest can in part be attributed to "technical problems and the indiscriminate use of herbicides, defoliants, and fertilizers" (Zafra Azucarera 1997).

When economic and environmental priorities are placed on the balance, the former wins. When decisions had to be made regarding the construction of causeways linking Cuba to Cayo Coco to facilitate the development of tourism centers, Cuban scientists proposed that the road link be constructed using bridges spanning gaps between naturally existing landmasses, thereby permitting water circulation vital to the survival of marine species. The description of what transpired is instructive:

> The construction of these bridges was deemed too expensive by the Cuban government. Because an agreement could not be reached within COMARNA's meeting, the decision was deferred to the Council of Ministers. An important official then intervened by pointing out that the planned road/bridge structure did not follow a direct route. He then proceeded to take out his pen and draw a straight line from Cayo Coco to the nearest point on the mainland.
>
> The design was adopted, and a straight, bermed road with intermittent underwater tunnels was constructed. Scientists argued that there were too few underwater tunnels to maintain natural water flows. Through negotiation, they were able to double the number of passages, a small victory considering their original opposition to the plan.
>
> This example illustrates that while COMARNA was able to settle minor issues itself, it did not possess the authority to make a final deci-

sion regarding controversial matters. When a project was deemed highly attractive, the environmental protection system could be manipulated to serve a more important agenda. In this case, development and the need to attract foreign investment prevailed. (Collis 1995, 2)

Although some efforts are being made to develop ecologically friendly or ecotourism, focusing on small sites and strong interactions with communities, it is clear that the leadership is thinking in terms of mass tourism that maximizes short-term revenue. It is worth quoting at length a statement made by President Fidel Castro in 1990 on this subject:

> Our country has enormous tourism resources. Enormous! It not only has Varadero, which one day will generate 500 million dollars [of revenue per annum]. The country has many Varaderos, uncountable beaches similar to Varadero, some totally virgin. Between Santa Lucía and Cayo Santa María, near Caibarién, along the northern cays of Sabinal, Romano, Cayo Coco, there are more than 100 kilometers of beaches some of them as good and, in my view, even better than Varadero. There are three of them in particular which I have visited. Thus, I am not speaking about second-hand knowledge. Three are comparable to Varadero and others have other advantages: Totally virgin areas where there is nothing built, where everything can be programmed, can be planned, where master plans based on the newest concepts in tourism can be developed in contrast to how it used to be before. Before, each individual grabbed a piece of land and built near the water, harming natural conditions. I have been told about many beaches where individuals began to build without proper organization and the natural conditions were harmed. We have the privilege of being able to program, as perfectly as possible, the development of these areas, determining how much beach to leave untouched, when to build and what. This is very important, very important! There are hundreds of kilometers of virgin beaches and shore which we are now connecting, where possible, to the mainland. (Castro 1993, 103–04)

Environmental Organizations

Cuba has the dubious achievement among socialist countries of not permitting independent environmental organizations to operate openly. This is so despite the fact that the Cuban Constitution of 1992 recognizes the freedom of

Cuban citizens to associate freely and allows the formation of mass and social organizations;[3] the Constitution also guarantees the right of mass organizations to exist[4] and to own property.[5]

There are reports about grassroots environmental groups and organizations in Cuba that have operated or are operating outside of authorized channels. For example, in the 1980s Cuban citizens held informal meetings to discuss ecological matters and environmental problems at a farm located near the town of Guanabacoa, La Habana Province, owned by the Asociación Naturista Vida (Life Naturist Association), an organization created in the 1930s by Spanish Republican exiles. The Sunday sessions, called "Green Sundays" (Domingos Verdes),[6] brought together scores of young people interested in environmental matters. The farm where the group met was expropriated by the government in 1988, presumably because the land was needed to accommodate railroad expansion, and the buildings and library razed. Subsequently, the Association was officially dissolved by the Cuban government (Alfonso 1991).

Orlando Polo and Mercedes Páez, two of the leaders of the Life Naturist Association, founded an organization in 1988 called the Movimiento Ecopacifista Sendero Verde (Green Path Ecopacifist Movement). The objective of the organization was to restructure the Cuban political system to enhance ecological principles; among specific policies supported by the Green Path Ecopacifist Movement were returning land to farmers and using solar rather than nuclear energy for electricity generation (Santana 1992). The group protested peacefully against, among other things, Cuban military involvement in Angola, the government's "destruction" of the island's environment, and the location of a garbage dump that threatened water supplies (Balmaseda 1991). Founder Polo has stated that the group's preferred approach was to work within the Cuban political and administrative systems, presenting its complaints to the Cuban government and appealing to international organizations only as a last resort (Alfonso 1991). Polo and Páez were arrested several times by the Cuban government for their activities and left the island in early 1991; it is not clear whether Sendero Verde still exists on the island, although its founders claim that it does (Santana 1992).

All Cuban mass media are owned and operated by the Cuban government, the Cuban Communist Party, or affiliated organizations (Nichols 1982, 75). The Cuban revolutionary government took over most newspapers and broadcast stations in 1959 because they became insolvent (having lost advertising revenue after the government nationalized businesses), they published news or commentaries "which might tend to impair the independence of the nation or provoke the non-observance of the laws in force," a civil crime since 1959, or the workers of a particular institution denounced the management as counterrevolutionary (MacGaffey and Barnett 1965, 332). In February 1961 only six daily newspapers were being published in La Habana, compared to sixteen in 1959 (MacGaffey and Barnett 1965, 333). After 1963, *Granma*, the organ of the Central Committee of the Cuban Communist Party, modeled after the Soviet Union's *Pravda*, became the only newspaper in the country (Nichols 1982, 77). Other government-operated newspapers were created in the 1970s and 1980s to address special groups (youth, farmers, military personnel) or specific geographic areas; most of these were closed in the 1990s when the economic crisis forced a severe reduction in imports of newsprint. The government also owns and operates all broadcast stations (radio and television) on the island. It is fair to say that since 1960 the Cuban government has tightly controlled the information that is available to Cuban citizens.

Recently, independent journalists within the island have begun to challenge the government's monopoly over information. Although they are still unable to disseminate their dispatches within the island since the official media outlets (newspapers, radio and television stations) are all owned by the government, these independent journalists file their reports with the foreign press mostly through telephone communication with collaborators in other countries (Ackerman 1997). Reports by Cuban independent journalists are disseminated through the World Wide Web by CubaNet, an internet organization established outside the island for this purpose, and also appear in different media (see, e.g., Desde Cuba 1997). Thus, Cuban independent journalist Olance Nogueras Rofes has reported on industrial pollution in the bay of Cienfuegos (see Chapter 7) and on concerns about the safety of the nuclear power plant under construction in that area (see Chapter 8).[7]

In April 1997 CubaNet announced that it would henceforth disseminate news stories and reports on environmental matters produced by a recently established independent environmental news service based in Camagüey, the Agencia Ambiental Entorno Cubano (AAMEC, Cuban Environs Environmental Agency). Short reports on environmental problems prepared by AAMEC contributors have been posted on the Internet—in Spanish and English—under the title CubaEco. AAMEC reported in July 1997 that the authorities in San Juan y Martínez, Pinar del Río Province, had arrested environmentalist Raúl Pimentel, a member of the environmental group Alerta Verde (Green Alert). This arrest, AAMEC claimed, ran contrary to Principle 10 of the Río Declaration on Environment and Development,[8] which encourages the participation of concerned citizens in dealing with environmental issues (Declaration 1997). No further information is available about Alerta Verde.

An interesting sidebar to this issue is the lack of support from the international environmental movement to independent Cuban environmentalists. In 1988 a high-level delegation from Germany's Green Party visited Cuba with an agenda that reportedly included Cuba's nuclear energy policy and the use of alternative sources of energy, biotechnology, genetic engineering, environmental protection, and human rights. The Cuban press reported that the delegation held about thirty meetings with Cuban Communist Party and government leaders, including with President Castro, but apparently not with any grassroots environmental organizations. The head of the Green Party delegation stated that her party did not have any differences with the Cuban Communist Party except for the use of nuclear energy, with the Green Party espousing the view that nuclear energy is risky and renewable sources of energy—such as solar, eolian, and biomass—are preferable (Colina 1988).[9]

As far as we can tell, like their European counterparts, U.S. environmentalists who have visited Cuba have not made contact with grass-roots environmental groups either. Some of them have essentially given Cuba a pass from the criticism they usually level upon governments, choosing instead to shower praise for the so-called organic agriculture policies that Cuba is pursuing during the special period and overlooking policies and actions that pollute the environment and waste natural resources. In a statement in a piece published in 1993 that invites disbelief as it calls into question the very basic independence from the state of environmental movements, U.S. ecologist Richard Levins (1993a, 22) says: "There is now a growing ecological movement in Cuba. But it is not an ecological movement in the sense of those in Europe or North America. It is not a distinct political movement such as the Greens, nor is it an opposition movement confronting a resistant government and corporations; nor is it yet an 'official' movement of the sort set up by a government to say yes. Cuban ecology activists are political, committed revolutionaries who see their struggles for ecologically sound policies as part of the duty of communists in building a new society with its own relation to nature."

Elsewhere, Levins (1993b, 53) has stated that Cuba's environmental movement "is neither 'official' in the sense of being sponsored by the national leadership, nor, with some few exceptions such as the Sendero Verde group, 'dissident.' It consists of scientists in or out of the Party, professional and amateur associations, journalists, individuals, and networks who educate the public, the Party, and the government on ecological issues, press for an ecological focus in planning, and help develop the research plans." Another view on the Cuban environmental movement is offered by journalist Karen Wald (1991). She describes Cuba during the special period as an "ecologist's dream": greening of cities, national reforestation, green (herbal) medicine, green (organic, home-grown) food, shift from automobiles, buses, and other pollution-generating transportation to bicycles, oxen, and horse-drawn carts. There is only one black mark in this otherwise idyllic situation: nuclear power. Wald (1991, 26) writes: "If the Cubans were not holding onto their belief that nuclear energy might resolve some of their energy problems, the scenario could well be one that any of the world's Green parties could adopt as a platform." Wald had sharp words for Cuba's independent environmental groups. Recounting a public forum held in La Habana to discuss the creation of parks and other green areas in the city, according to her the first public environmental forum ever in the island, Wald writes:

> Representatives of the electric and telephone companies (the major tree-fellers in the absence of any lumber industry) were invited guests, but they failed to attend. Also notably absent were members of Cuba's only opposition "green" group, *Sendero Verde* (Green Path), a tiny organization announced a few years ago by a wandering dissident named Orlando Polo. In 1988, Polo attended press conferences with some of Fidel Castro's most outspoken opponents, describing himself to the foreign press as an "eco-pacifist." If *Sendero Verde* actually exists, it missed a unique opportunity to make itself heard and felt among the like-minded, highly placed individuals and organizations that attended this event. (1991, 26)

Summary and Implications

The environmental consequences of the special period have been mixed. The inability to import agricultural and industrial inputs could be considered a blessing if one were to ignore the devastating consequences of this development for the standard of living of the population. Many of the emergency economic policies instituted since the onset of the crisis in 1990 have been championed by environmentalists abroad, since in their view they are consistent with the preservation of the environment. These initiatives, however, have almost solely responded to economic necessity rather than to a newly discovered fondness for nature on the part of the political leadership. The evidence suggests that if the leadership had the option, it would again prefer to pursue the input-intensive development model that was embraced for three decades but that had to be abandoned due to the collapse of Soviet and Eastern European socialism and the end of foreign subsidies. Thus, scarce resources—obtained through foreign loans at exorbitant interests rates—have been used to obtain fertilizers and pesticides for the sugar and tobacco industries. Furthermore, there is strong evidence that environmental concerns are being sacrificed whenever there is a reason to do so.

Balancing economic and environmental priorities remains a sensitive issue that will only be resolved once the country's post-socialist economic and political course is set. Under the systemic failings of socialism and the current economic crisis, the environmental future appears to be bleak. A transition to a more open society and a capitalist economic system will not necessarily ensure environmentally benign development policies, unless a proper regulatory framework is put in place and enforced. Cuba's environmental tomorrow will depend on the development model the country pursues in years to come, and on the extent to which this model takes into account economic–environmental tradeoffs regarding production and consumption decisions.

Notes

1. The fate of the Centro de Estudios sobre América (CEA, Center for American Studies) provides an illustration of the problem faced by researchers within the island who exercise some independence and stray from the government's line. CEA was dismembered in 1996 after a purge that began in 1995 and culminated in public criticisms by Raúl Castro, who accused some of the researchers of being "agents of imperialism" and "fifth columnists" (agents operating within the ranks of the enemy to undermine its cause). See Giuliano 1998.

2. See, for example, Oro 1992; Lezcano 1994a,b, and 1995; and Wotzkow 1998. Their names and those of other scientists appear in the preface and are also cited throughout the text.

3. Castro's views on these issues seem to waver depending on the context in which they were expressed and the audiences to which they were directed. While making impromptu angry remarks in response to political tensions with Hungary as socialism was crumbling in Eastern Europe, he directly attributed pollution to products received from the socialist bloc by claiming that Hungarian buses polluted the city of La Habana by releasing large quantities of smog (Castro Discusses 1990, 15). In other cases he has embraced a more orthodox view by stating that "the ultimate responsibility for the cumulative environmental deterioration in the Third World as a whole belongs to the developed capitalist world" (Castro 1993, 22). The latter remarks, taken from an edited text distributed to the international gathering assembled in Rio de Janeiro during the 1992 Earth Summit, were obviously intended to gather favor with representatives from Third World countries. Yet, in the same text, which was prepared as a speech addressed to the assembled representatives but never delivered (heads of states speaking to the assembly were only allotted five minutes), Castro stated that "of the 10 countries which generate most of the gas emissions that cause the greenhouse effect, five are highly industrialized. If the former Soviet Union is included, this group would be responsible for over 40 percent of the total emissions" (20). He then went on to add that "the principal producers of pesticides, fertilizers, and other noxious chemical products, even after they have been banned, continue to be the developed countries" (22). The latter remarks are particularly interesting since Castro knows full well that the former Soviet Union and Eastern European socialist countries produced prodigious amounts of these chemicals, including some banned in the West, many of which were exported to Cuba.

4. It is particularly instructive in this regard to examine the issues of *Voluntad Hidráulica*, the Cuban technical hydraulic journal, published in the 1960s and 1970s. Numerous articles were published in the journal that raised cautionary flags about salinization, problems associated with the construction of water projects, and the haphazard development of large-scale regional hydraulic development projects in Cuba and acknowledged that these plans were often drawn in the absence of even the most essential geological information.

5. It is also noteworthy that in the water sector at least there was full awareness of the environmental dangers associated with poorly managed projects. The same issues of *Voluntad Hidraúlica* mentioned in note 4 raised the alert about the mismanagement of water resources, in Cuba as well as in the Soviet Union.

6. Núñez Jiménez, who died in 1998, succeeded Fidel Castro in 1959 as president of the National Institute for Agrarian Reform. He also served as vice minister of culture and president of the Cuban Academy of Sciences, deputy to the National Assembly of People's Power and member of the Central Committee of the Cuban Communist Party.

7. This became the name of the technical hydraulic journal cited earlier and published in La Habana during the 1960s, 1970s, and 1980s.

Notes to Chapter 2

1. The data in Table 2.1 after 1990 are from the low-variant UN projections.

2. In the other large island-nations of the Caribbean, according to the UN, population density increased between 1950 and 1995 as follows: Dominican Republic, from 48 inhabitants per square kilometer in 1950 to 161 in 1995; Haiti, from 118 to 259; Jamaica, from 128 to 223; and Puerto Rico, from 249 to 413.

3. The Cuban censuses of 1970 and 1981 define as urban the "resident population in population centers with 2,000 and more inhabitants, as well as those living in centers with fewer than 2,000 but more than 500 people whenever this last category had the following characteristics: public power network, paved roads, water supply lines, a sewage system, sewers, medical assistance services and an educational center. Also centers with a population between 200 and 500 were included whenever they complied with the six above mentioned characteristics (some towns, created by the Revolution, with these urban characteristics but with very few inhabitants were included)." The rural population was defined as the "population that lives in places with fewer than 500 inhabitants and population centers between 500 and 2,000 inhabitants which had less than four of the above-mentioned urban characteristics."

4. The first Cuban central bank, the Banco Nacional de Cuba, began operations in 1950, followed by the Banco de Fomento Agrícola y Industrial de Cuba, created in 1951 to provide credit for the promotion of agricultural diversification and new industries; the Fondo de Seguro de Depósito, created in 1952 to stimulate personal savings and insure time deposits against the failure of commercial banks; Financiera Nacional de Cuba, created in 1953 to finance self-liquidating public and semipublic works; the Banco Cubano del Comercio Exterior (BANCEX), created in 1954 to coordinate and promote Cuban exports either by aiding private exporters or by its own efforts; and the Banco de Desarrollo Económico y Social (BANDES), created in 1954 to provide credit for the promotion of economic development in general.

5. The establishment of the Banco Nacional marked the beginning of systematic collection and preparation of national accounts in Cuba. Prior to that time, macroeconomic data are esti-

mates whose reliability is questionable. For this reason, this overview of the economy of pre-revolutionary Cuba focuses on the decade of the 1950s.

6. For example, Carranza Valdés (1992, 142) has estimated that the GSP, the broadest measure of economic activity for the Cuban economy under the economic accounting system used by centrally planned economies, declined by 24 percent in 1991 and by 15 percent in 1992, for a two-year reduction of 40 percent. Combining this estimate with those by other economists yields a reduction in GSP over the period 1989–1993 of about 45 percent (Mesa-Lago 1996, 4).

Notes to Chapter 4

1. From a broad perspective soil degradation entails "a deterioration in quality and capacity of the life-supporting processes of the land," whereas from a farming perspective it refers to "a reduction in the land's actual or potential uses," often resulting from the process of crop production itself (as quoted in Pagiola 1994, 22).

2. Another typology reported in the literature indicates that 8.0 percent of Cuba's soils have "very high fertility" and a further 26.2 percent "high fertility." Of the remaining soils, 41.1 percent have "low" fertility and 21.6 percent "very low" fertility. This typology reportedly resulted from "an ambitious program to reclassify, evaluate, and map the country's soil in great detail, and to interpret the maps for sustainable management." Although it is not stated, apparently this effort was begun after 1990 (Gersper et al. 1993, 17).

3. *Marabú* (*Dischrostachys nutants*) is a widespread leguminous plant, generally regarded as a pest but with the ability to regenerate exhausted agricultural soils by fixing nitrogen in its roots. *Marabú* is very thorny and difficult to eradicate.

4. Mesa-Lago included in one of his publications a quotation by Fidel Castro that powerfully reveals the basis for the traditional Marxist bias against small farm holdings (Mesa-Lago 1988, 63; as quoted by Pryor 1992, 48). This quotation, consistent with socialist agricultural organizational viewpoints, helps explain the vigor of collectivization efforts in Cuba for over three decades. At a time when several tens of thousands of small farms remained, Castro felt that "working with them is much more difficult [than with cooperatives]; it is terrible, virtually insolvable because one must discuss and make plans with tens of thousands of them." Had these discussions been held, not only with small farmers but also with state farm and agricultural cooperative members, they would have been able to alert the authorities about the potential adverse environmental impacts of centrally dictated agricultural policies.

5. Yields for most crops began to decline in 1990; these trends continued well into 1994 (see Food and Agriculture Organization 1995).

6. The World Bank has noted that "[soil] degradation figures [for Central America and Caribbean countries] quoted in the literature are often extrapolated from very limited data and may exaggerate the problem because they often consider 'moved soil' as 'lost soil,' even though much of it may be deposited on other agricultural land." This same source notes that aggregate soil degradation figures must be taken with a great deal of skepticism since generally they have a rather weak empirical basis (Lutz et al. 1994, 4, 16). We suspect that similar problems affect the estimates of soil degradation in Cuba, and thus they should be interpreted cautiously.

7. In comparison (similar caveats apply), in 1984, 42 percent of the agricultural land in Costa Rica was "thought to have been affected by erosion"; in Panama 1.2 million hectares in 1981 were

affected by soil degradation; in Haiti "soil erosion has become the dominant environmental problem"; and in Mexico "80 % of the land surveyed . . . is eroded at least slightly" (Lutz et al. 1994, 40, 81, 98, 106).

8. Pryor (1992, 220), in a discussion of political versus economic leadership in farm management in socialist countries, includes a most revealing quote in which Castro said, "Who should manage a farm? A revolutionary. What are his pre-requisites? That he be a revolutionary."

9. To make matters even worse, farm managers often were not even aware of the intent of the organizational or productive directives they received and were forced to follow. While conducting field observations in Cuba an associate of Pryor found that blanket organizational changes related to the deployment and functions of labor brigades "took little account of local circumstances . . . farm managers faithfully followed instructions in making such changes, but they were unable to provide . . . any convincing justification for them or to explain how they would improve farm performance" (Pryor 1992, 172–73).

10. This is an interesting point since it raises several questions regarding recent claims by Cuban officials that Soviet advisors forced upon Cuba the adoption of the socialist agricultural model (see, for example, Rosset and Benjamin 1993, 22). Although the Soviets indeed appear to have disregarded the crop rotation issue, most Eastern European countries, and hence their advisors in Cuba, probably did not. The Cubans probably chose to follow Soviet practices, since these were more in accord with their own views regarding the role of technology, machinery, and chemical inputs in agricultural development (see Chapter 1).

11. This section draws heavily from Atienza Ambou et al. 1995.

12. Unforested areas adjacent to dams and microdams are also significant sources of soil erosion (Medina Pérez n.d.).

13. According to another source, the per hectare use of chemical fertilizers was even higher, reaching 202 kilograms per cultivated hectare in the late 1980s, the country importing at the time 600,000 metric tons of fertilizers (Atienza Ambou et al. 1992, 11). Another source states that in the late 1980s, fertilizer imports reached 800,000 tons (French 1994, 7).

14. Pesticide consumption data are available only for a few other Latin American countries. Pesticide consumption in 1989 appears to have reached levels comparable to those of Cuba (in relation to country size) only in Colombia (20,019 metric tons), Costa Rica (6,264 metric tons), and Trinidad and Tobago (2,303 metric tons). In Paraguay pesticide use reached only 988 metric tons, and in Surinam, only 636 metric tons (WRI 1994, 295).

Notes to Chapter 5

1. A country's hydraulic potential (or renewable water resources) refers "to the average annual flow of rivers and groundwater generated from endogenous precipitation. . . . Annual withdrawals as a percentage of water resources refer to total water withdrawal, not counting evaporative losses from storage basins, as a percentage of internal water resources and river flows from other countries" (the latter, obviously, does not apply to Cuba, an insular country)(WRI 1994, 358).

2. A lower estimate of Cuba's total hydraulic resources of 32.2 cubic kilometers, of which 23.2 cubic kilometers represent surface waters and 9 cubic kilometers represent underground waters, has also been reported in the literature (see Celeiro Chapis 1995, 88).

3. Atienza Ambou et al. (1995, 5) report that the results of several studies and numerous observations indicate that 75 percent of the annual volume of precipitation is lost to evapotranspiration.

4. Some of these aquifers, particularly the Miocene karstic limestones, form vast reservoirs of generally good quality water, having in some parts of Cuba a thickness of as much as 1,300 meters (Cereijo 1992, 75).

5. U.S. drinking water standards, as determined by the U.S. Public Health Service, however, set total dissolved solids limits at 500 milligrams per liter, half of Cuba's current acceptable limits (Cereijo 1992, 75).

6. The estimates on water availability and use we have come across tend to vary considerably from source to source. One estimate cited by Espino (1992, 332) claims that the amount of water being withdrawn in 1990 had increased to 12.7 cubic kilometers. This is higher than the 10 cubic kilometers estimated by Figueras. There are also some disagreements in the literature regarding the distribution of water use by sector. Espino cites figures suggesting that of the water being withdrawn in 1990, 74 percent was for agricultural use, 12 percent for domestic human consumption, and 14 percent for industry. Atienza Ambou and her colleagues (1995) provide a somewhat different distribution of water use in the late 1980s. According to their estimates, 60 percent of water was destined for irrigation, 20–23 percent for urban and industrial use, and the rest was accounted for by waters not used or lost before reaching end users.

7. There seems to be some disagreement regarding how dams and microdams are defined. Figueras (1994, 22–23) claims that between 1959 and 1989, 135 dams and some 1,000 microdams were built. Similar estimates are reported by Atienza Ambou and her colleagues (1995, 6).

8. It is interesting to note that Atienza Ambou and her colleagues (1995) state that the National Institute of Hydraulic Resources was established in 1986, although, as we have noted, this government entity began to function in 1962.

9. To illustrate the relationship between irrigation and yields, Cuban technicians have estimated that nonirrigated citrus plantations yield about 6 metric tons of fruit per hectare, whereas irrigated lands have potential yields of over 40 tons (Riego 1982, 61).

10. The irrigation data, as with many other Cuban statistics, must be used cautiously. By 1985, according to one source, 220,182 *caballerías* (2.95 million hectares) were irrigated (exclusive of land planted with rice), more than a quarter of Cuba's agricultural land (Analizan la situación 1985, 3). In relation to 1983, the amount of land irrigated in 1985 was 40 percent higher. If true, this would constitute a remarkable achievement in just two years. According to other irrigation figures reviewed here, however, the land surface under irrigation prior to the 1990s was far less than the 2.95 million hectares claimed in this source. The *1989 Anuario Estadístico* (Comité Estatal 1989, 215) indicates that the total amount of land irrigated in the state sector reached 896,000 hectares, an estimate consistent with the 900,000 hectares reported by other sources we consulted.

11. Although Atienza Ambou and her associates (1995, 12) claim that the urban water sector was not assigned high priority, it was allocated a substantial amount of resources. It received 30 million pesos in 1985 and 246 million pesos in the 1986–1990 quinquennium. According to the last budget prepared before the special period, it was slated to receive 385 million pesos in the 1991–1995 quinquennium.

12. As related (on 22 March 1997) to one of the authors by Arturo Pino, an agricultural engineer with many years of experience in Cuba and a former staff member of the Inter-American Development Bank.

13. Personal communication (March 1997) from José Ramón González, an international irriga-

tion consultant and 1997 president of the Miami-based Colegio de Ingenieros Agrónomos y Azucareros.

14. Estimates of amount of agricultural land vary from source to source, in part because the amount of actual agricultural land changes as a function of land use. During some periods, for example, forests may be converted to agricultural land, or marginally productive land may be improved and brought into cultivation. For more on these definitional issues, see Chapter 4.

15. In Indonesia, a country with geographical features akin to Cuba's, comparable problems with saltwater intrusions have been reported. According to the World Bank (1994, 58), "in the northern parts of Jakarta, the salinized area is expanding rapidly (at a rate of 0.5–1.0 km/year), and now extends 15 km from the coast." The extensive salinization has been attributed to overexstraction of underground water.

16. The adverse environmental consequences of the Dique Sur have been hinted at, but not mentioned directly, by the government-controlled Cuban press. Castro himself, the reputed force behind the final design and construction of the dike (several eyewitness accounts related to us by hydrologists and geographers who were involved in the design of the dike and left Cuba in the 1990s, claim that Castro personally overruled technical counsel offered regarding the original design of the dike), mentioned in a 29 December 1992 speech that "they are putting in canals around Batabanó. They are putting in, or should put in, some canals around now in Alquizar, since the South Dike has made the water table rise. It is necessary to drain the area. . . . They should start to put in those canals to protect some tens of thousands of *caballerias* [a *caballeria* equals 13.4 hectares] of bananas and other crops" (Castro Delivers 1992, 7). We were alerted to these problems by Manuel J. Acevedo, a hydrologist who left Cuba in the early 1990s (letter to Sergio Díaz-Briquets dated 3 October 1993). This account was also confirmed by José Ramón González, an agricultural engineer who served as 1997 president of the Colegio de Ingenieros Agrónomos y Azucareros, a Miami-based organization, and also recounted by several other Cuban hydrologists. Lezcano (1994a) has also noted that the Dique Sur has led to the explosive appearance of insect-borne plagues, mosquitoes in particular.

Notes to Chapter 6

1. Agroforestry is defined as "'on-farm tree establishment and management,' including seeding, planting, and managing natural regrowth and trees planted as borders or interplanted in agricultural crops, woodlots, home gardens, and so on. Agroforestry does not involve reforestation in its traditional sense" (Current et al. 1995b, 1).

2. Bucek (1986, 15, 16), for instance, provides a forested area estimate for the mid-nineteenth century citing results of the 1852 colonial census. His estimates classify 30.8 percent of the land area as forests, 17.6 percent as arid soils, and 8.6 percent as fruit groves. The reliability of Bucek's estimates, like any other historical estimate, is open to question given the then-constrained access to many of Cuba's less developed regions and limited cartographical development.

3. There is no precise definition of *montes* in Cuban parlance. What is obvious is that it is a broader concept than forests and that it is generally used to refer to most land surfaces not occupied or being intentionally altered by people. The term encompasses forests and other areas naturally covered by vegetation (whether undisturbed or recolonized), such as swampland and *marabú* patches. Some of the sources we have consulted implicitly seem to include

swamps and *marabú* fields in their definition of *montes*, whereas others, particularly in more recent times, classify them separately.

4. For periods prior to 1900, we reconstructed the series primarily by consecutively adjusting percentage estimates of forested land area in terms of the original estimate of forested land area (in hectares) provided by Bucek in 1986. Twentieth-century estimates were derived by evaluating and selecting from different land use data point estimates found in several sources. In the text we indicate those estimates that in our judgement appear to be more credible. In one particular instance in which conflicting estimates are shown for closely spaced dates (1923, 1929, 1945, 1959, and 1962) we judge our adjusted 1945 estimate and the 1959 estimate as the most reliable.

5. It is uncertain how these sources treat swampland. Various estimates suggest that swampland accounts for about 5 percent of the national territory (see Dinerstein et al. 1995, 100). Atienza Ambou et al. (1992, 7) estimate that 500,000 hectares of the Cuban archipelago are swamplands, which is consistent with the 5 percent figure.

6. The estimate of between 790,000 and 1,188,000 hectares forested in nonfarm land assumes that the remaining 1,188,000 to 1,585,000 hectares were occupied by human settlements, roads and other human activities, swamps and marshes (about 500,000 hectares), as well as nonforested coastal areas and inland water bodies. It must be noted that the 1945 census classified an additional 1,652,000 hectares as in "other uses." This figure suggests that the amount of land forested may in fact have been higher than estimated by us here.

7. These official estimates are inconsistent with data provided by the World Resources Institute in its 1996–1997 report (WRI 1996, 219), which indicates that in 1990 only 17.1 percent of Cuba was forested. According to the World Resources Institute estimates, natural forests accounted for 15 percent of the total and plantation forests for 2.1 percent. This source, in addition, notes the presence of an additional 1.3 million hectares classified as "other wooded land" (in tropical countries this definition "encompasses forest fallows [closed and open forests] and shrubs"). Although ongoing deforestation of natural forests continues to occur, the report also notes a rapid increase in tree plantations (increasing between 1981 and 1990 at an annual rate of 12.3 percent).

8. The ratios of harvesting to reforestation are 2:15 for Cuba and 2:75 for Chile.

9. According to the Foreign Policy Association (1935, 466), Decree-law No. 753 prohibited "the cutting down of trees, no matter to whom the wood belongs, without authorization. Every owner or renter of land dedicated to pasturage or cultivation must set aside at least 15 percent of the total area for a permanent wooded area. Persons whose land is traversed by rivers, ravines, brooks, etc., where shade trees have been destroyed must annually plant quick-growing trees to a depth of 50 meters on the banks. The Secretary of Agriculture will furnish gratis trees for replanting, and will pay from 30 to 50 cents a tree planted."

10. This estimate is based on the assumption that it takes at least twenty years for a tree plantation to mature and be ready for harvesting and on COMARNA figures cited by Reed (1993, 38) indicating that, in the early 1990s, 6,500 hectares of trees were being harvested annually.

11. Our hypothetical densities are based on recommended tree plantation densities provided by Poore and Fries for different species (1985, 12). They range from 472 (for *Tectona grandis*) to 1,678 (for *Shorea robusta*) trees per hectare. The recommended density for eucalyptus hybrids is 1,658 trees per hectare. For our purposes, we assume 1,500 trees per hectare as the most likely tree density, although we also consider a higher density of 2,000 trees per hectare.

12. Personal telephone communication by Sergio Díaz-Briguets with Dr. Nancy Diamond, a

forestry expert, on 7 March 1996. Preeg (1996, 27), in his discussion of reforestation rates in Haiti during the 1980s, considers survival rates of 50 percent as "exceptionally high."

13. Associated field preparations included removing large boulders and stones and in some instances even contouring the soils (e.g., by filling small ravines and streams with soil).

Notes to Chapter 7

1. This section draws heavily on Pérez-López (1991).

2. The plant at Las Camariocas is often referred in the literature as "CAME-I." CAME is the Spanish acronym for the CMEA or COMECON (Council for Mutual Economic Assistance), the now-defunct common market of socialist states. Financing for the Las Camariocas plant was provided by a group of CMEA member countries (the former Soviet Union, Romania, Bulgaria, Czechoslovakia, and East Germany), with one-half of production (15,000 tons per annum) earmarked for delivery to these countries.

3. Using the reported level of particulates emitted by the kilns of the Nuevitas plant and some assumptions, an estimate can be made of the amount of cement dust launched into the atmosphere by the plant prior to the installation of the filters. Assuming gas emissions of 400,000 cubic meters per hour—the reported capacity of the filters—it can estimated that each kiln of the Nuevitas plant emitted an estimated 4.8 metric tons of cement dust per hour (400,000 cubic meters x 12 grams of cement dust per cubic meter) or 38.4 tons per 8-hour work shift. When operating together, the three kilns (or "lines") at the Nuevitas plant would therefore emit over 100 tons of cement dust per 8-hour shift. Production capacity of the plant has been reported as 600,000 metric tons per annum or 200,000 metric tons per annum per kiln. Assuming that the plant operated for 250 days at full rated capacity, each kiln would produce 800 tons of cement per day, or 4.8 percent of output.

4. Oro (1992, 57) reports that Cuba's worst oil spill was in Cienfuegos Bay, where, because of human error, 2,300 metric tons of oil spilled into the bay, damaging all sea life. Oro's reference is probably to the 1986 accident mentioned here, although the magnitude of the oil spill he reports differs from that reported officially.

5. At the time, the Cabaiguán refinery was already under the control of the Cuban government. It had been processing Soviet crude since April 1960.

6. In this calculation, electricity produced by "thermoelectric plants" also includes production by gas turbine plants and diesel plants. Cuban statistics distinguish among these plants on the basis of the fuel they use, but they are all thermoelectric plants for the purpose of our discussion.

7. Gébler (1988) and Garcia (1995) provide information and pictures about the ancient fleet of U.S.-made cars operating in Cuba.

Notes to Chapter 8

1. Independent journalist Nogueras Rofes—a member of the Bureau of Independent Press of Cuba—was jailed by Cuban authorities shortly after the publication of this article and accused of "dangerousness," "usurping the role of a journalist," and "spreading false news that attempt against international peace." Police seized his notes, tape recorder, and files and offered to release him from jail if he agreed to leave the country (Cuban Independent 1996). He was later released and remained in Cuba until 1997 (see later discussion and Chapter 10).

2. It appears that the original location selected for the plant was on the shore of the Arimao River, in the area of Caonao, east of Cienfuegos. The location of the plant was shifted to Juraguá because geological irregularities were found in Caonao (Pérez-López 1979, 13).

3. According to Oro (1992, 70–71), the two nuclear plant complexes being planned were (1) "Jururú," in the municipality of Antilla, Holguín Province; and (2) "Occidente," in the municipality of Viñales, Pinar del Río Province. Each of these complexes was conceived to operate with four 440-megawatt nuclear reactors.

4. Section 111 of the Cuban Liberty and Solidarity Act (commonly known as the Helms-Burton Act), which entered into force in March 1996, directs the president of the United States to withhold foreign assistance for any country in an amount equal to the value of assistance and credits provided by that country, or an entity in that country, in support of completion of the nuclear facilities at Juraguá. Section 106 of the same act directs the president to report to the appropriate congressional committees on progress toward the withdrawal of personnel (including advisors, technicians, and military personnel) of any independent state of the former Soviet Union from the Cienfuegos nuclear facility.

5. There have been no announcements of new construction of VVER-440s during 1995–1998, and therefore the list is current.

6. These include (1) design issues: in-core monitoring, decay heat removal, reliability of active components, service water systems, main steam line isolation, primary circuit pressure relief, emergency core cooling system (ECCS) capability, ECCS redundancy and physical separation, confinement function, ventilation/cooling system; (2) component integrity: reactor vessel embrittlement, vessel stress analysis, applicability of leak-before-break, primary circuit inservice inspection, primary circuit stress analysis, vessel support integrity; (3) instrumentation and control: quality and performance, redundancy, separation and independence, control room support, equipment qualification, equipment and power supply classification, signal priority, control room habitability and remote shutdown panels; (4) electric power supplies: redundancy, separation and independence, quality and performance, diesel generation loading, battery capacity and d.c. system design, emergency protection signals; and (5) accident analysis: analysis of loss-of-coolant accidents, evaluation of modifications, fire protection (Ripon 1992, 69–73).

7. The specific problems identified were in the areas of management involvement, safety culture, design modification control, work control, equipment material condition, quality assurance, operating procedures, limits and conditions, radiation protection, emergency operating procedures, training, emergency planning, and operational experience feedback.

8. The G7 consists of the major industrialized countries: Canada, France, Germany, Italy, Japan, the United Kingdom, and the United States.

9. The assessment was made in 1988 by the international insurance group Müncher Rückversicherungs-Gesellschaft of Munich, Germany.

Note to Chapter 9

1. In the same publication a different set of statistics is also reported for 1992: drinking water for 82.9 percent of the urban population and 77.8 percent of the rural population, sanitary excreta disposal for 96 percent of the urban population, and sewerage services for 39 percent of the urban population (PAHO 1994, vol. 2, 164). These differences may arise in part from differences in definitions, but they are indicative of the inconsistencies in Cuban statistics even when published by international organizations.

Notes to Chapter 10

1. The fuel import figure cited by Lage for the period prior to the special period may include some quantities of fuel that were re-exported and therefore probably overestimated fuel consumption.

2. The information on joint ventures in this section is drawn primarily from Pérez-López (1995a).

3. Article 54 states: "The rights to assembly, demonstration, and association are exercised by manual and intellectual workers, farmers, women, students, and other sectors of the working people, who have at their disposal the necessary means to serve this purpose. Mass and social organizations have at their disposal the facilities for carrying out said activities, in which their members enjoy the most extensive freedom of speech and opinion, based on the unrestricted right to initiative and criticism" (Constitución 1992, 26).

4. Article 7 states: "The Cuban socialist state recognizes and encourages mass and social organizations, which have emerged in the historic process of our people's struggles, bringing together different sectors of the population, representing their specific interests, and incorporating them into the tasks of construction, consolidation, and defense of socialist society" (Constitución 1992, 5).

5. Article 22 states: "The State recognizes the ownership by political, mass, and social organizations of the assets used to fulfill their objectives" (Constitución 1992, 14).

6. The name "Green Sundays" was a take off on "Red Sundays" (Domingos Rojos), work sessions organized by the government on Sundays to mobilize "volunteers" to work in agricultural or construction tasks.

7. Nogueras Rofes was again arrested in April and in May 1997 because of his journalistic activities and released after a few days in jail (Desaparecido 1997; López Baeza 1997; Alfonso 1997b). In August 1997 he left the country for the United States under pressure from the Cuban government (Corzo 1997; Remos 1997) and is currently a reporter for Miami's *El Nuevo Herald*.

8. Principle 10 of the Rio Declaration on Environment and Development states: "Environmental issues are best handled with the participation of all concerned citizens, at the relevant level. At the national level, each individual should have appropriate access to information concerning the environment that is held by public authorities, including information on hazardous materials and activities in their communities, and the opportunity to participate in decision-making processes. States shall facilitate and encourage public awareness and participation by making information widely available. Effective access to judicial and administrative proceedings, including redress and remedy, shall be provided."

9. A subsequent report of the visit in the Cuban press did not raise any environmental issues and instead concentrated on the points of agreement between the Green Party and the Cuban Communist Party: condemnation of U.S. aggressiveness, especially in Central America; support for liberation movements; and nonpayment of the Third World debt. Even on the issue of political prisoners—where the Green Party advocates the release of all political prisoners everywhere—the visitors "confirmed that prisoners held in Cuba for supposedly political reasons were in fact guilty of acts of terrorism and therefore would be in jail in any other country in the world" (Triana 1988).

References

Abascal López, José. 1985. A tales barcos, tales muelles. *Cuba Internacional* 17:188–90.

Academia de Ciencias de Cuba. 1989. *Nuevo Atlas Nacional de Cuba.* La Habana: Instituto de Geografía.

Academia de Ciencias de Cuba/Academia de Ciencias de la URSS. 1970. *Atlas Nacional de Cuba.* La Habana: Academia de Ciencias de Cuba/Academia de Ciencias de la URSS.

Ackerman, Elise. 1997. Periodistas libres luchan por quitarse la mordaza. In *Desde Cuba con valor,* 21–41. Madrid: Editorial Pliegos.

Ackerman, Richard. 1991. Environment in Eastern Europe: Despair or Hope? *Transition* 2 (4) (April): 9–11.

Acosta, Dalia. 1996. Cuba-Population: Aging Population Requires Fresh Policies. Interpress Third World News Agency. 29 August.

Acosta León, Maruja, and Jorge E. Hardoy. 1972. La urbanización en Cuba. *Demografía y Economía* 6 (2): 41–67.

Acosta Moreno, Roberto. 1989. Clean Technologies: The Cuban Experience. *UNEP Industry and Technology* 12 (1) (January–March): 42–44.

———. 1995a. An Assessment of Biodiversitry in Cuba. In *The Environment in U.S.–Cuba Relations: Opportunities for Cooperation,* 23–30. Washington, D.C.: Inter-American Dialogue.

———. 1995b. The United States and Cuba: Possibilities of Joint Actions for a Better Environment. In *The Environment in U.S.–Cuban Relations: Opportunities for Cooperation,* 61–68. Washington, D.C.: Inter-American Dialogue.

Acosta Moreno, Roberto, and Orlando Rey Santos. 1997. Frameworks for Cooperation: From the Realm of the Possible to Action. In *The Environment in U.S.–Cuban Relations: Opportunities for Cooperation,* 23–29. Washington, D.C.: Inter-American Dialogue.

Administración de los recursos hidráulicos: Antecedentes, situación actual y perspectiva. 1982. *Voluntad Hidráulica* 19:30–38.

Akiner, Shirin. 1993. Environmental Degradation in Central Asia. In *Economic Development in Cooperation Partner Countries from a Sectoral Perspective,* 255–63. Brussels: NATO Economics Directorate.

Alcolado, Pedro M. 1991. Ecological Assessment of Semi-enclosed Marine Water Bodies of the Archipelago Sabana-Camaguey (Cuba) Prior to Tourism Development Projects. *Marine Pollution Bulletin* 23:375–78.

Alfonso, Pablo. 1991. Ecopacifistas abogan por armonía en la isla. *El Nuevo Herald* (25 March): 1A, 4A.

———. 1992. Destituido de cargo hijo de Castro. *El Nuevo Herald* (18 June): 6A.

———. 1995. Cuba reorganiza gobierno; se aleja de modelo soviético. *El Nuevo Herald* (23 April): 9A.

———. 1997a. Incendio devasta bosques en la provincia de Santiago. *El Nuevo Herald* (21 February): 1B.

———. 1997b. Periodista independiente de nuevo bajo amenaza de seguridad castrista. *El Nuevo Herald* (26 April): 6A.

Altieri, Miguel. 1993. The Implications of Cuba's Agricultural Conversion for the General Latin American Ecological Movement. *Agriculture and Human Values* 10 (3): 91–92.

Altshuler, Igor I., and Iurii N. Golubchikov. 1990. Ecological Semi-Glasnost. *Environmental Policy Review* 4 (2) (July): 1–12.

Alvarez, José, and Lázaro Peña Castellanos. 1995. *Preliminary Study of the Sugar Industries in Cuba and Florida Within the Context of the World Sugar Market.* International Working Paper Series IW95-6. Gainesville, Fla.: Institute of Food and Agricultural Sciences, University of Florida.

Alvarez, José, and William A. Messina, Jr. 1996. Cuba's New Agricultural Cooperatives and Markets: Antecedents, Organization, Early Performance, and Prospects. In *Cuba in Transition,* vol. 6, 175–95. Washington, D.C.: Association for the Study of the Cuban Economy.

Alvarez, José, and Ricardo Puerta. 1994. State Intervention in Cuban Agriculture: Impact on Organization and Performance. *World Development* 22:1663–75.

Alvarez, María Teresa. 1992. Subdesarrollo y deterioro ecológico. *Boletín de Información sobre Economía Mundial* 207 (June): 8–12.

Alvarez Tabío, Fernando. 1981. *Comentarios a la Constitución Socialista.* La Habana: Editorial de Ciencias Sociales.

Alvarez-Vázquez, Luisa. 1993. Trends, Recent Levels, and Determinant Factors of Abortion in Cuba. HRP/WHO (August). La Habana: Center for Research in Human Reproduction.

Amaro, Nelson. 1996. Decentralization, Local Government, and Citizen Participation in Cuba. In *Cuba in Transition,* vol. 6, 262–82. Washington, D.C.: Association for the Study of the Cuban Economy.

Analizan criticamente el trabajo sobre protección de plantas y animales, en reunión del Consejo Científico Nacional. 1978. *Granma* (20 March): 1.

Analizan la situación actual del riego en la rama agropecuaria. 1985. *Granma* (27 March): 1, 3.

ANPP Ninth Regular Session Ends. 1997. Habana Tele Rebelde and Cuba Vision Networks (11 July). Foreign Broadcast Information Service FBIS-LAT-97-195 (13 July).

Aplicación del Programa 21: Examen de los Adelantos Realizados Desde la Conferencia de las Naciones Unidas Sobre el Medio Ambiente y Desarrollo, 1992. 1997. <http://www.un.org/dcsd/earthsummit/cuba-cp.htm>

Aprobadas por la Asamblea 3 nuevas leyes. 1997. *El Nuevo Herald Digital* (13 July).

Aprobados los informes del Ministerio de la Industria Pesquera y de la Asamblea Provincial de Granma. 1985. *Granma* (10 July): 1–4.

Area Handbook for Cuba. 1971. Washington, D.C.: Foreign Area Studies, American University.

Assessment of Juraguá Nuclear Power Plant Situation. 1995. Prensa Latina Service (12 July). Foreign Broadcast Information Service FBIS-LAT-95-137 (18 July): 9–10.

Atienza Ambou, Aída. 1996. Criterios para una estrategia ambiental. *Cuba: Investigación Económica* 2 (1): 61–68.

Atienza Ambou, Aída, Anicia García Alvarez, and Oscar U. Echevarría Vallejo. 1992. Repercusiones medio ambientales de las tendencias de desarrollo socioeconómico en Cuba. La Habana: Instituto Nacional de Investigaciones Económicas. Mimeo.

Atienza Ambou, Aída, Carmen A. Pérez Castañeda, and María J. Tutusaus. 1995. Principales aspectos del desarrollo hidráulico en Cuba. La Habana: Instituto Nacional de Investigaciones Económicas. Mimeo.

Atoms for Peace. 1990. Milestone Documents in the National Archives Series. Washington, D.C.: National Archives and Records Administration.

Avella, Amparo E. 1995. The Process of Environmental Impact Assessment in Cuba. In *Cuba in Transition,* vol. 5, 396–98. Washington, D.C.: Association for the Study of the Cuban Economy.

Babbitt, B. 1980. The Russians' Nuclear Power Plants. *The New York Times* (2 January): A23.

Bai, Eugenio. 1995. Robaina en Moscú. *El Nuevo Herald* (28 June): 10A.

Balmaseda, Liz. 1991. Cuba's Hippies: Now that They've Touted Eco-Pacifism in the U.S., Activists Won't Pay Fees to Get Back Home. *The Miami Herald* (13 September): E1.

Banco Nacional de Cuba (BNC). 1994. *Economic Report 1994.* La Habana: Banco Nacional de Cuba (August 1995).

———. 1995. *Economic Report 1995.* La Habana: Banco Nacional de Cuba (May 1996).

Barba, Ralph B., and Amparo E. Avella. 1995. Cuba's Environmental Law. In *Cuba in Transition,* vol. 5, 276–80. Washington, D.C.: Association for the Study of the Cuban Economy.

Barquinero, Eduardo, et al. 1981. La contaminación causada por la industria de la pulpa y el papel en Cuba. *Revista ICIDCA* 15 (1) (January–April): 3–11.

Barrett, Kathleen. 1993. The Collapse of the Soviet Union and the Eastern Bloc: Effects on Cuban Health Care. Cuban Briefing Paper, no. 2. Washington, D.C.: Center for Latin American Studies, Georgetown University.

Barrio Menéndez, Emilio del. 1979. Comenzó a materializarse proyecto para investigar y controlar la contaminación marina en Cuba. *Granma* (28 November): 1.

Basureros: Bomba de tiempo ante el verano. 1997. *El Nuevo Herald* (1 July) 6A.

Batista, Carlos. 1997. Persiste grave problema de distribución de agua. *El Nuevo Herald* (23 August): 6A.

Batista Valdés, Pastor. 1997. Los residuales no irán al Cauto. *Trabajadores* (13 January): 10.

Bendoyro, Radalberto. 1984. Broad Soil Reconditioning Plan. *Granma* (11 April): 8. English version.

Benesch, Susan. 1992. Cuban Warns of Risky Reactors. *Washington Times* (6 May): A1, A9.

Benjamin-Alvarado, Jonathan. 1996a. Cuban Nuclear Developments. *Monitor,* Center for International Trade and Security, University of Georgia, 2 (1–2) (winter–spring): 1, 5–7.

———. 1996b. The Quest for Power: Analyzing the Costs and Benefits of Cuba's Nuclear Energy Policy. In *Cuba in Transition,* vol. 6, 440–49. Washington, D.C.: Association for the Study of the Cuban Economy.

Benjamin-Alvarado, Jonathan, and Alexander Belkin. 1994. Cuba's Nuclear Power Program and Post–Cold War Pressures. *Nonproliferation Review* (winter): 18–26.

Bennett, Hugh H., and Robert V. Allison. 1928. *The Soils of Cuba.* Washington, D.C.: Tropical Plant Research Foundation.

Bennett, Leonard, and Malik Derrough. 1996. Electricity, Health, and the Environment: Selecting Suitable Options. *IAEA Bulletin* 38 (1): <http://www.iaea.org/worldatom/inforesources/bulletin>.

Benson, Robert W. 1992. An Island on the Way to Ecotopia. *Los Angeles Times* (29 March): M5.

Berg, Thomas F. 1991. Cuban Nuclear Plant Assailed for Safety Flaws. *Public Utilities Fortnightly* (15 July): 32–33.

Berovides Alvarez, Vicente, and Antonio Comas González. 1991. The Critical Condition of Hutias in Cuba. *Oryx* 25 (4) (October): 206–208.

Bianchi, Andres. 1964. Agriculture. In *Cuba: The Economic and Social Revolution*, edited by Dudley Seers, 65–157. Chapel Hill: University of North Carolina Press.

Bianchi Ross, Ciro. 1985. Con dinamita, señor, con dinamita, *Cuba Internacional* 16 (182) (January): 20–27.

Bohemia Article on Effect of Tourism in Key Areas. 1997. Havana Prensa Latina (30 April). Foreign Broadcast Information Service FBIS-LAT-97-125 (5 May).

Bohi, Douglas R. 1994. Foreword. In *Pollution Abatement Strategies in Central and Eastern Europe*, edited by Michael A. Toman, ii–viii. Washington, D.C.: Resources for the Future.

Bonner, Raymond. 1993. Armenia Amidst Wars: Environmental Puzzles. *New York Times* (17 August): A2.

Borges Hernández, Teresita, and Cristóbal Díaz Morejón. 1997. Cuba: Política ambiental a tono con los nuevos tiempos. *Temas* 9 (January–March): 13–19.

Borhidi, A. 1991. *Phytogeography and Vegetation Ecology of Cuba*. Budapest. Akadémia Kiadó.

Brañes, Raúl. 1991. *Institutional and Legal Aspects of the Environment in Latin America, Including the Participation of Nongovernmental Organizations in Environmental Management*. Washington, D.C.: Inter-American Development Bank.

Brundenius, Claes. 1979. Measuring Income Distribution in Pre- and Post-revolutionary Cuba. *Cuban Studies* 9 (2): 29,44.

Bucek, Antonio. 1986. Aseguramiento territorial de la estabilidad ecológica y sus condiciones en Cuba. In *Unidad Hombre-Naturaleza*, Instituto de Geografía, 9–24. La Habana: Editora de la Academia de Ciencias de Cuba.

Bush, Keith. 1972. Environmental Problems in the USSR. *Problems of Communism* 21 (4) (July–August): 21–31.

———. 1974. The Soviet Response to Environmental Disruption. In *Environmental Deterioration in the Soviet Union and Eastern Europe*, edited by Ivan Volgyes, 8–36. New York: Praeger Publishers.

Cabrera Trimiño, Gilberto Javier. 1994. Interrelación población-ambiente en Los Palacios: Cuba. Paradigma emergente de demografía ambiental. Paper presented at the Eighteenth International Congress of the Latin American Studies Association.

———. 1997. *Economía ecológica, demografía ambiental y desarrollo*. La Habana: Editorial de Ciencias Sociales.

Capital en ruinas es meca que atrae a miles. 1997. *El Nuevo Herald Digital* (14 May).

Carlos Lage Comments on Economy. 1992. Havana Radio Rebelde Network (7 November). Foreign Broadcast Information Service FBIS-LAT-92-219 (12 November): 2–14.

Carlos Lage Interview on Economic Situation. 1993. Havana Radio Rebelde Network (3 November). Foreign Broadcast Information Service FBIS-LAT-93-216A (10 November): 2–15.

Carney, Judith A., ed. 1993. Low-Input Sustainable Agriculture in Cuba. *Agriculture and Human Values* 10 (3). Special issue.

Carranza Valdés, Julio. 1992. Cuba: Los retos de la economía. *Cuadernos de Nuestra América* 9 (19) (July–December): 131–58.

Carter, F. W., and D. Turnock. 1993. Problems of the Pollution Scenario. In *Environmental Problems in Eastern Europe*, edited by F. W. Carter and D. Turnock, 188–219 . London: Routledge.

Castañeda Gómez, José. 1991. Los mesías miran al sur. *Nucleus* 11:25–29.

Castro, Fidel. 1953. *History Will Absolve Me*. London: Jonathan Cape, Ltd., 1969.

———. 1974. Address to Construction Workers on Builders Day. *Granma* (7 December): 1–4.

———. 1980. Discurso pronunciado en la clausura del II período de sesiones de la Asamblea Nacional del Poder Popular, el 27 de diciembre de 1980. *Granma* (29 December): 2.

———. 1984. Address by President Fidel Castro on July 26. *Granma* (28 July): 1–4.

———. 1986. Respuesta de Fidel. *Granma* (25 August): 1.

———. 1992. *Ecología y Desarrollo: Selección Temática 1963–1992.* La Habana: Editora Política.

———. 1993. *Tomorrow Is Too Late.* Melbourne: Ocean Press.

Castro Díaz-Balart, Fidel. 1985. La energía nuclear en Cuba: Sus perspectivas y las realidades del mundo de hoy. *Cuba Socialista* 5 (2) (May–June): 38–93.

———. 1986. *La energía nuclear en la economía nacional de la República de Cuba.* Moscú: Secretariado del Consejo de Ayuda Mutua Económica.

———. 1990a. *Energía nuclear y desarrollo: Realidades y desafíos en los umbrales del Siglo XXI.* La Habana: Editorial de Ciencias Sociales.

———. 1990b. Nuclear Energy in Cuba: An Indispensable Link Toward Development. *IAEA Bulletin* 32 (1): 49–52.

Castro, Miriam, and Mercedes Ramos. 1984. Milagro del polvo rojo. *Prisma Latinoamericano* 10 (147) (November): 22–23.

Castro, Raúl. 1996. Maintaining Revolutionary Purity. Excerpts from a Report Presented to the Central Committee of the Cuban Communist Party of Cuba, March 23, 1996. In *Cuba: Political Pilgrims and Cultural Wars*, 31–37. Washington, D.C.: Free Cuba Center of Freedom House.

Castro Delivers Speech at Guines Cooperative. 1992. Havana Television and Radio Networks (27 December). Foreign Broadcast Information Service FBIS-LAT-92-250 (29 December): 4–8.

Castro Discusses Revolution, USSR, Nicaragua. 1990. Havana Domestic Radio and Television Services (7 March). Foreign Broadcast Information Service FBIS-LAT-90-046 (8 March): 1–17.

Castro Says Son Was Fired for Incompetence. 1992. *Chicago Tribune* (31 July): B7.

Castro's Second September Address to Juraguá Power Plant Workers on Its Closure. 1992. Cuba Visión Network (10 September), British Broadcasting Corporation, *Summary of World Broadcasts* (12 September).

Celeiro Chapis, Maira. 1995. Problemas actuales de los recursos hídricos y climáticos de Cuba. In *Proceedings of the Nineteenth Annual Caribbean Studies Association Conference*, Mérida, Mexico, 87–92. Mérida, Mexico: Facultad de Ciencias Antropológicas, Universidad Autónoma de Yucatán.

Central Intelligence Agency (CIA). 1994. *Cuba: Handbook of Trade Statistics, 1994.* ALA 94-10011. Washington, D.C. Central Intelligence Agency.

———. 1996. *Cuba: Handbook of Trade Statistics, 1996.* ALA 96-10002. Washington, D.C.: Central Intelligence Agency.

Centro de Estudios Demográficos. 1976. *La Población de Cuba.* La Habana. Editorial de Ciencias Sociales.

Centro Latinoamericano de Demografía (CELADE). 1989. *Boletín Demográfico* 22:43.

———. 1995. America Latina: Proyecciones de población urbana-rural. *Boletín Demográfico* 28:56.

Cepero, Alina. 1992. Ecotourism: New Option for Cuba. *Cuba Update* (September): 43.

Cepero, Endel. 1997. Contaminación química en la bahía de Nuevitas. CubaEco/AAMEC (May). <http://members.xoom.com/aamec>.

Cereijo, Manuel, ed. 1992. *The Cuban Economy: Blueprint for Reconstruction.* Miami, Fla.: Endowment for Cuban American Studies, Cuban American National Foundation.

Chakraborty, S. 1995. Safety of Nuclear Power Reactors in the Former Eastern European Countries. *Nuclear Safety* 36 (1) (January–June): 46–53.

Cinco preguntas sobre la rama nuclear en Cuba. 1976. Interview with Tirso W. Sáenz, president of the National Committee for the Peaceful Utilization of Atomic Energy. *Granma* (11 March): 6.

Clark, Ismael. 1989. Problemas ambientales en Cuba y en el mundo. *Granma* (5 June): 4.

Colaboración para el uso pacífico de la energía atómica entre Cuba y la URSS. 1967. *Granma* (15 January): 5.

Cole, Sally, and Jorge I. Domínguez. 1995. U.S.-Cuban Environmental Cooperation: Shared Interests, Problems, and Opportunities. In *The Environment in U.S.-Cuban Relations: Opportunities for Cooperation*, 1–8. Washington, D.C.: Inter-American Dialogue.

Colina, Cino. 1988. Comments on First Contact with Cuban Reality. *Granma* (12 June): 8.

Colina, Cino, and Rosa Elvira Peláez. 1976. El reino consolareño del arroz. *Granma* (5 July): 4.

Collis, David S. 1995. Environmental Implications of Cuba's Economic Crisis. Cuba Briefing Paper, no. 8. Washington, D.C.: Center for Latin American Studies, Georgetown University.

———. 1996a. *The Caribbean Environment: Issues of Mutual Concern.* Conference Summary Document. Washington, D.C.: Caribbean Project, Georgetown University.

———. 1996b. Tourism/Ecotourism in Cuba. In *Cuba in Transition,* vol. 6, 451–55. Washington, D.C.: Association for the Study of the Cuban Economy.

Comisión Económica para la América Latina y el Caribe (CEPAL). 1997. *La economía cubana: Reformas estructurales y desempeño en los noventa.* Mexico: Fondo de Cultura Económica.

Comisión Nacional de Protección del Medio Ambiente y el Uso Racional de los Recursos Naturales (COMARNA). 1992. *Informe nacional a la conferencia de Naciones Unidas sobre medio ambiente y desarrollo Brasil/1992. Resumen.* La Habana: COMARNA.

Comité Estatal de Estadísticas. 1988. *Compendio estadístico de energía 1988.* 1989. La Habana: Comité Estatal de Estadísticas.

———. 1989. *Anuario Estadístico de Cuba 1989.* La Habana: Comité Estatal de Estadísticas.

Committee on Selected Biological Problems in the Humid Tropics 1982. *Ecological Aspects of Development in the Humid Tropics.* Washington, D.C.: National Academy Press.

Constitución de la República de Cuba. 1976. La Habana: Editorial de Ciencias Sociales, 1995.

Constitución de la República de Cuba. 1992. *Granma* (22 September): 3–10.

Consultores Asociados, S.A. (CONAS). 1994. *Cuba: Inversiones y negocios 1994–1995.* La Habana: CONAS.

Contaminación afecta a tres ríos orientales. 1997. *El Nuevo Herald Digital* (9 May).

Contaminación de la presa Guirabo de la ciudad de Holguín y otras. 1990. Letter received at Radio Martí in Washington, D.C., from Holguín, Cuba (dated 20 April).

Contamination Affects Fishing Industry, Economy. 1988. Havana Radio Rebelde Network (1 August). Foreign Broadcast Information Service FBIS-LAT-88-151 (5 August): 3.

Convenio sobre Cooperación entre el Gobierno de la República de Cuba y el Gobierno de los Estados Unidos de América relativo al empleo civil de la energía atómica. 1958. *Gaceta Oficial* (6 January): 129–32.

Cook, Edward. 1988. Prospects for Polish Agriculture in the 1980s. In *Socialist Agriculture in Transition*, edited by Joseph C. Brada and Karl-Eugen Wadekin, 131–45. Boulder, Colo.: Westview Press.

Copa, V. 1981. Nuclear Power Station. *Direct from Cuba* (31 July): 13–14.

Corzo, Cynthia. 1997. Periodistas: Castro acosa a disidentes. *El Nuevo Herald* (7 August): 1A, 6A.

Coyula Cowley, Mario. 1997. Ambiente urbano y participación en un socialismo sustentable. *Temas* 9 (January–March): 54–61.

Crítica la situación del transporte público en Cuba. 1993. *Diario las Américas* (8 May): 7A.

Cuba: La Habana promueve reflexion ambiental. 1997. *InterPress Service, Rec APC* (12 September).

Cuba Holds 2 Ships, Cargo in Spill. 1980. *Oil Daily* (29 February): 8.

Cuba Intent on Finishing Plant. 1997. *Journal of Commerce* (14 July): 10A.

Cuba Neuropathy Field Investigation Team. 1995. Epidemic Optic Neuropathy in Cuba: Clinical Characterization and Risk Factors. *New England Journal of Medicine* 333 (18): 1176–82.

Cuban Commission Views Valley, Mountain Municipalities. 1996. Foreign Broadcast Information Service FBIS-LAT-96-024 (5 February): 11–12.

Cuban Constitution and the Environment, The. 1993. *Cuba Update* (February–March): 40.

Cuban Economic Research Project. 1965. *Cuba: Agriculture and Planning, 1963–1964.* Miami, Fla.: University of Miami Press.

Cuban Independent News Media Under Attack. 1996. *Freedom House Cuba Brief* 1:1–2.

Cuban Official on Turquino Manati Plan Implementation. 1996. Foreign Broadcast Information Service FBIS-LAT-96-023 (2 February): 4.

Cuban Radio Program Discusses Internal Migration to Havana. 1996. Foreign Broadcast Information Service FBIS-LAT-96-014 (17 January), 12–14.

Cuba's Ecotourism. 1995. Radio broadcast, Living Earth Network (1 March).

Current, Dean, Ernst Lutz, and Sara J. Scherr. 1995a. The Costs and Benefits of Agroforestry to Farmers. *World Bank Research Observer* 10 (2) (August): 151–80.

———, eds. 1995b. Costs, Benefits, and Farmer Adoption of Agroforestry: Project Experience in Central America and the Caribbean. World Bank Environment Paper, no. 14. Washington, D.C.: World Bank.

Dahmén, Erik. 1971. Environmental Control and Economic Systems. In *The Economics of Environment,* edited by Peter Bohm and Allen V. Kneese, 44–52. London: Macmillan.

Danilov-Danilian, Victor. 1993. Problemas ecológicos en la Federación Rusa. *Cuadernos del Este* 10:33–40.

Darling, Juanita. 1997. Cubanos inmigrantes en su propio país. *El Nuevo Herald Digital* (15 June).

Dávalos, Fernando G. 1984a. Ostión Cero I. *Granma* (13 December): 5.

———. 1984b. Los principales focos de contaminación (Ostión Cero II). *Granma* (18 December): 3.

———. 1984c. Nosotros tambien somos dolientes (Ostión Cero III). *Granma.* (19 December): 2.

———. 1984d. La responsabilidad (Ostión Cero IV). *Granma* (20 December): 4.

———. 1984e. En tres años la bahía de Nipe será esteril, si no se detiene la contaminación. *Granma* (11 September): 3.

Dawson, Frank G. 1976. *Nuclear Power: Development and Management of a Technology.* Seattle: University of Washington Press.

de la Cuesta, Leonel-Antonio, ed. 1974. *Constituciones Cubanas.* New York: Ediciones Exilio.

de la Osa, José A. 1977. Creada Comisión Nacional para la Protección del Medio Ambiente y la Conservación de los Recursos Naturales. *Granma* (3 March): 3.

———. 1978. Cuba se une a la conmemoración del Día Mundial del Medio Ambiente trabajando en la protección de nuestro medio y sus recursos naturales. *Granma* (3 June): 3.

DeBardeleben, Joan, ed. 1991. *To Breathe Free: Eastern Europe's Environmental Crisis.* Baltimore, Md.: Johns Hopkins University Press.

———. 1991. Introduction. In *To Breathe Free: Eastern Europe's Environmental Crisis*, edited by Joan DeBardeleben, 1–21. Baltimore, Md.: Johns Hopkins University Press.

Declaration of the Cuban Environmental Agency. 1997. CubaEco (July): <http://members.xoom.com/aamec/>.

Decreto No. 52. 1981. *Gaceta Oficial* (20 November): 137–38.

Decreto No. 56. 1982. Para la regulación del uso pacífico de la energía nuclear. *Gaceta Oficial* (26 May): 95–100.

Decreto No. 142. 1988. Reglamento para el trabajo con sustancias radioactivas y otras fuentes de radiaciones ionizantes. *Gaceta Oficial* (30 April): 29–44.

Decreto No. 174. 1992. Contravenciones de las regulaciones para el control y el registro del ganado mayor y de razas puras. *Gaceta Oficial* (31 October): 137–38.

Decreto No. 175. 1992. Regulaciones sobre calidad de las semillas, y sus contravenciones. *Gaceta Oficial* (31 October): 138–40.

Decreto No. 176. 1992. Protección a la apicultura y a los recursos mieliferos, y sus contravenciones. *Gaceta Oficial* (31 October): 140–42.

Decreto No. 179. 1993. Protección, uso y conservación de los suelos, y sus contravenciones. *Gaceta Oficial* (26 February): 41–44.

Decreto No. 199. 1995. Contravención de las regulaciones para la protección y el uso racional de los recursos hidraúlicos. *Gaceta Oficial* (11 May): 216–17.

Decreto-Ley No. 128. 1990. Estructura, organización y funcionamiento del Sistema Nacional de Protección del Medio Ambiente y del Uso Racional de los Recursos Naturales, y su órgano rector. *Gaceta Oficial* (18 January): 1–7.

Decreto-Ley No. 138. 1993. De las aguas terrestres. *Gaceta Oficial* (2 July): 121–27.

Decreto-Ley No. 147. 1994. De la reorganizacion de los organismos de la administracion central del estado. *Gaceta Oficial* (21 April): 3–5.

Deere, Carmen Diana, Niurka Pérez, and Ernel Gonzalez. 1994. The View from Below: Cuban Agriculture in the "Special Period in Peacetime." *Journal of Peasant Studies* 21:194–234.

Defector Calls Nuclear Plant "Technical Disaster." 1991. Madrid EFE (4 June). Foreign Broadcast Information Service, FBIS Daily Report—Latin America (6 June): 5–6.

del Aguila, Juan M. 1988. *Cuba: Dilemmas of a Revolution*. Rev. ed. Boulder: Westview Press.

del Barrio Menéndez, Emilio. 1990. Plan para restaurar lagunas costeras. *Granma* (15 August): 3.

Demeritt, David. 1991. Boards, Barrels, and Boxshooks: The Economics of Downeast Lumber in Nineteenth-Century Cuba. *Forest & Conservation History* 35 (July): 108–21.

Desaparecido periodista independiente cubano. 1997. Buro de Periodistas Independientes Cubanos (23 May) CubaNet: <http://www.cubanet.org>.

Desde Cuba con valor. 1997. Madrid: Editorial Pliegos.

Dewar, Heather. 1993. Unlocking the Mysteries of Cuba's Rare Wildlife Species. *The Miami Herald* (26 October): 1A, 6A, 12A.

Díaz, Beatriz. 1995. Biotecnologia agricola: Estudio de caso en Cuba. Paper presented at the Nineteenth Annual International Congress of the Latin American Studies Association. Washington, D.C.

———. 1997. El desarrollo agrícola sustentable en Cuba. *Temas* 9 (January–March): 33–41.

Díaz-Briquets, Sergio. 1988. Regional Differences in Development and Living Standards in Revolutionary Cuba. *Cuban Studies* 18:45–63.

Díaz-Briquets, Sergio, and Lisandro Pérez. 1981. Cuba: The Demography of Revolution. *Population Bulletin* 36:1.

———. 1982. Fertility Decline in Cuba: A Socioeconomic Interpretation. *Population and Development Review* 8 (3): 513–37.

Díaz González, Beatriz, and Marta Rosa Muñoz. 1995. Agricultural Biotechnology and the Environment in the Cuban Special Period. *CartaCuba*, 43–52.

Dictan sentencia contra acusados que provocaron el vertimiento de petróleo en la Bahía de Cienfuegos. 1987. *Granma* (19 May): 3.

Dierksmeier, Gustavo. 1996. Pesticide Contamination in the Cuban Agricultural Environment. *Trends in Analytical Chemistry* 15 (5): 384–89.

Dinerstein, Eric, David M. Olson, Douglas J. Graham, Avis L. Webster, Steven A. Primm, Marnie P. Bookbinder, and George Ledec. 1995. *A Conservation Assessment of the Terrestrial Ecoregions of Latin America and the Caribbean*. Washington, D.C.: World Bank and World Wildlife Fund.

Dirigentes analizan quejas de la población. 1997. *El Nuevo Herald Digital* (13 April).

Dlott, Jeff, Ivette Perfecto, Peter Rosset, Larry Burkham, Julio Monterrey, and John Vandermeer. 1993. Management of Insect Pests and Weeds. *Agriculture and Human Values* 10 (3) (summer): 9–15.

Dorticós, Pedro Luis. 1982. Aprovechamiento de los recursos hidraúlicos. *Voluntad Hidraúlica* 19:6–19.

Dos requisitos básicos: Formación de cuadros y rigor científico. 1982. *Voluntad Hidráulica* 19:71–80.

Drovichev, G. 1966. Nuevo impulso a la irrigación y al mejoramiento de los terrenos en la URSS. *Voluntad Hidráulica* 3 (12) (August): 46–52.

D'Silva, Emmanuel, and S. Appanah. 1993. Forestry Management for Sustainable Development. Economics Development Institute, EDI Policy Seminar Report, no. 32. Washington, D.C.: World Bank.

Dumont, Rene. 1970a. *Cuba ¿Es socialista?* Caracas: Editorial Tiempo Nuevo.

———. 1970b. *Cuba: Socialism and Development*. New York: Grove Press.

Dunleavy, M. P., and Adam Penenberg. 1993. Castro's Green Revolution. *Green Magazine* 4 (5): 12–17.

Durham, William H. 1979. *Scarcity and Survival in Central America*. Stanford, Calif.: Stanford University Press.

Eckstein, Susan. 1981 . The Debourgeoisement of Cuban Cities. In *Cuban Communism*, edited by Irving Louis Horowitz, 119–40. 4th ed. New Brunswick: Transaction Books.

Eco-Cuba: Making the Best of a Bad Thing. 1993. *Cuba Update* (September–October): 21.

Ecología/Cuba. 1996. Crisis del medio ambiente. CubaNet: <http://www.cubanet.org> (7 September).

Economic Commission for Europe (ECU). 1991. *Energy Reforms in Central and Eastern Europe: The First Year*. New York: United Nations.

Economic Commission for Latin America and the Caribbean. 1995. Cuba: Evolución Económica durante 1994, LC/MEX/R.524, Mexico (23 May).

Editorial. 1988. *Voluntad Hidráulica* 25 (77): 2.

Effects of U.S. Blockade on Environment Noted. 1995. Foreign Broadcast Information Service FBIS-LAT-95-186 (26 September): 8.

Egorov, S. V., and José R. Luege. 1967. *Hidrogeología de Cuba*. La Habana: Instituto Nacional de Recursos Hidráulicos and Instituto Cubano de Recursos Minerales.

El paisaje se transforma. 1982. *Voluntad Hidráulica* 19:52–58.

Emprende ciudad de la Habana el saneamiento del río Almendares. 1985. *Granma* (5 April): 3.

En Cauto–El Paso quedarán las aguas de la primavera. 1991. *Trabajadores* 32:5 (2 February): 1.

Encalló petrolero soviético en la bahía de La Habana. 1970. *Granma* (23 June): 3.

Entra en vigor decreto de migración en Cuba. *El Nuevo Herald Digital* (12 May).

Environmental Protection and Foreign Aid: Institutional Reform. 1984. Report of the Second Seminar on International Environmental Issues. Medford, Mass.: Tufts University, Department of Urban and Environmental Policy.

Environmental Summary. 1997. CubaEco (July): <http://members.xoom.com/aamec/>.

Escobar Casas, Reinaldo. 1987. Satisfacer las demandas, pero con calidad. *Cuba Internacional* 18:206 (January): 16–21.

Espino, María Dolores. 1992. Environmental Deterioration and Protection in Socialist Cuba. In *Cuba in Transition,* vol. 2, 327–42. Washington, D.C.: Association for the Study of the Cuban Economy.

———. 1994. Tourism in Cuba: A Development Strategy for the 1990s? In *Cuba at a Crossroads,* edited by Jorge Pérez-López, 147–66. Gainesville: University of Florida Press.

Estudio para utilizar suelos con salinidad. 1991. *Trabajadores* (15 July): 12.

EU defiende ayuda de agencia a central nuclear de Juraguá. 1997. *El Nuevo Herald* (6 February): 2B.

Evenson, Debra. 1994. *Revolution in the Balance: Law and Society in Contemporary Cuba.* Boulder, Colo.: Westview Press.

Exhorta el partido a llegar a conclusiones sólidas en las investigaciones que permitan detener el avance de la salinización en Guantánamo. 1985. *Granma* (1 July): 1.

Farah, Douglas. 1992. Cubans Are Feeling Unempowered. *Washington Post* (21 December): A15.

Feinsilver, Julie M. 1995. Cuban Biotechnology: The Strategic Success and Commercial Limits of a First World Approach to Development. In *Biotechnology in Latin America,* edited by N. Patrick Peritone and Ana Karina Galve Peritone, 97–125. Wilmington, Del.: Scholarly Resources Books.

Fernández, Damián J. 1992. Opening the Blackest of Black Boxes: Theory and Practice of Decision-Making in Cuba's Foreign Policy. *Cuban Studies* 22:53–78.

Fernández Soriano, Armando. 1997. Movimientos comunitarios, participación y medio ambiente. *Temas* 9 (January–March): 26–32.

Feshbach, Murray. 1991. Economics of Environment: Costs, Needs, and Realities. In *The Soviet Economy Under Gorbachev,* 223–46. Brussels: NATO Economics Directorate.

———. 1993. Environment: Improved Knowledge, Expanded Concern. In *Economic Development in Cooperation Partner Countries from a Sectoral Perspective,* 233–40. Brussels: NATO Economics Directorate.

———. 1995. *Ecological Disaster: Cleaning Up the Hidden Legacy of the Soviet Regime.* New York: Twentieth Century Fund.

Feshbach, Murray, and Alfred Friendly, Jr. 1992. *Ecocide in the USSR.* New York: Basic Books.

Fialka, J. 1978. Soviets Think They've Solved Atom Safety Problem. *The Washington Star* (1 October).

The 50-Year War on the Everglades. 1997. *New York Times* (20 April): 14.

Figueras, Miguel Alejandro. 1994. *Aspectos estructurales de la economía cubana.* La Habana: Editorial de Ciencias Sociales.

Food and Agriculture Organization (FAO). 1993. *Tropical Forest Action Programme Update.* Rome: UN Food and Agriculture Organization.

———. 1994. Mangrove Forest Management Guidelines. FAO Forestry Paper, no. 117. Rome: UN Food and Agriculture Organization.

———. 1995. *FAO Yearbook. Production.* Vol. 48. Rome: UN Food and Agriculture Organization.

Foreign Policy Association. 1935. *Problems of the New Cuba.* New York: Foreign Policy Association.

Fortescue, Stephen. 1986. *The Communist Party and Soviet Science.* Baltimore, Md.: Johns Hopkins University Press.

Frank, Mark. 1997. U.S. Law Stalls Power Plant, Cuba Says. *Journal of Commerce* (9 January): 7B.

French, Howard W. 1993. Cuba's Ills Encroach on Health. *The New York Times* (16 July): A3.

———. 1994. Harvest of Headaches: Cuba's Sugar Crop Woes. *The New York Times* (20 February), 3:7.

Fultz, Kenneth O. 1995. *Concerns with the Nuclear Power Reactors in Cuba.* Testimony Before the Subcommittee on the Western Hemisphere, Committee on International Relations, House of Representatives. GAO/T-RCED-95–236. Washington, D.C.: U.S. General Accounting Office (1 August).

García Azuero, Francisco. 1994. Crisis en Cuba amenaza el medio ambiente. *El Nuevo Herald* (17 February): 1A, 6A.

Gardner, Gary T. 1994. *Nuclear Nonproliferation: A Primer.* Boulder, Colo.: Lynne Rienner Publishers.

Garfield, Richard, Julia Denin, and Joy Fausey. 1995. The Health Impact of Economic Sanctions. *Bulletin of the New York Academy of Medicine* 72:454–69.

Garfield, Richard, and Sarah Santana. 1997. The Impact of the Economic Crisis and the U.S. Embargo on Health in Cuba. *American Journal of Public Health* 87 (1): 15–20.

Gébler, Carlo. 1988. *Driving Through Cuba.* New York: Simon and Schuster.

Gersper, Paul L., Carmen S. Rodríguez-Barbosa, and Laura F. Orlando. 1993. Soil Conservation in Cuba: A Key to the New Model for Agriculture. *Agriculture and Human Values* 10:16–23.

Giuliano, Maurizio. 1998. *El Caso CEA: Intelectuales e Inquisidores en Cuba.* Miami, Fla.: Ediciones Universal.

Global Partnership for Environment and Development: A Guide to Agenda 21, Post Rio Edition. 1993. New York: United Nations.

Goldman, Marshall I. 1972. Externalities and the Race for Economic Growth in the USSR: Will the Environment Ever Win? *Journal of Political Economy* 80 (2) (March–April): 314–27.

———. 1973. Pollution Comes to the U.S.S.R. In *Soviet Economic Prospects for the Seventies,* Joint Economic Committee, 56–70. Washington, D.C.: U.S. Government Printing Office.

———. 1979. The Convergence of Environmental Disruption. *Science* 170 (2 October): 37–42.

Gómez, Manuel R. 1981. Occupational Health in Cuba. *American Journal of Public Health* 71 (5) (May): 520–24.

Gómez, Orlando. 1979. Esos verdes pulmones de Cuba se multiplicaron muchas veces. *Granma* (23 June): 5.

———. 1985. Se inició la batalla por la descontaminación del río Luyanó; mejora la situación del Almendares. *Granma* (17 December): 3.

González, Eduardo, and Guivi Gagua. 1979. Nuevo estudio sobre la evaporación en Cuba. *Voluntad Hidráulica* 16 (51): 23–34.

González, Humberto. 1991. Mercury Pollution Caused by a Chlor-Alkali Plant. *Water, Air, and Soil Pollution* 56:83–93.

González, Humberto, Martin Lodenius, and Mirta Otero. 1989. Water Hyacinth as Indicator of Heavy Metal Pollution in the Tropics. *Bulletin of Environmental Contamination and Toxicology* 43:910–14.

González, Humberto, Ivis Torres, and Martha Ramírez. 1995. Contaminación por metales pesados en la Bahía de Levisa, Cuba. *Minería y Geología* 12:41–43.

González, Maby. 1992. Medio ambiente y desarrollo capitalista. *Boletín de Información sobre Economía Mundial* 207 (June): 3–7.

González Báez, Arturo, and Sigilfredo Jiménez Hechevarría. 1988a. La protección sanitaria a los acuíferos cársicos cubanos: Un problema actual (primera parte). *Voluntad Hidráulica* 25 (77): 3–18.

———. 1988b. La protección sanitaria a los acuíferos cársicos cubanos: Un problema actual (segunda parte). *Voluntad Hidráulica* 25 (78): 3–25.

Gort, Sergio A., Tomas Escobar, José Izquierdo, Modesto Correoso, and Narciso Singh. 1994. *Isla de la Juventud: Su Naturaleza*. Editorial Cientifico-Técnica. La Habana: Instituto Cubano del Libro.

Granma Province Mill Sued over Toxic Waste Dumping. 1997. La Habana Radio Rebelde Network (11 February). Foreign Broadcast Information Service FBIS Electronic Service, PA1802143797 (18 February).

Grave sequía afecta a todas la isla. 1997. *El Nuevo Herald Digital* (1 May).

Greenblat, Sara Reva. 1993. Internal and External Perspectives of Energy Security Issues in Eastern Europe and the Former Soviet Union. In *Economic Development in Cooperation Partner Countries from a Sectoral Perspective*, 245–54. Brussels: NATO Economics Directorate.

Gregerse, H. M., et al. 1995. Valuing Forests: Context, Issues, and Guidelines. *FAO Forestry Paper* 127. Rome: UN Food and Agriculture Organization.

Gregory, Paul R., and Robert C. Stuart. 1974. *Soviet Economic Structure and Performance*. New York: Harper & Row.

Griñán, Arnoldo. 1986. Operación combustible. *Granma* (25 July): 2.

Grupo Cubano de Investigaciones Económicas. 1963. *Un estudio sobre Cuba*. Coral Gables: University of Miami Press.

———. 1965. *Cuba: Agriculture and Planning*. Coral Gables, Fla.: University of Miami Press.

Guevara, Ernesto. 1961. En Punta del Este, Uruguay: Dos Discursos. In *Obra Revolucionaria*, 413–47. México: Ediciones Era, 1967.

———. 1964. Cuba: Su economía, su comercio exterior, su significado en el mundo actual. In *Obra Revolucionaria*, 616–26. México: Ediciones Era, 1967.

Gugler, Josef. 1980. "A Minimum of Urbanism and a Maximum of Ruralism": The Cuban Experience. *Studies in Comparative International Development* 15 (2): 27–44.

Guma, José Gabriel. 1989. Para que el Almendares fluya como en el pasado. *Granma* (20 December): 3.

Gunn, Gillian. 1995. Cuba's NGOs: Government Puppets or Seeds of Civil Society? Cuban Briefing Paper Series, no. 7. Washington, D.C.: Center for Latin American Studies, Georgetown University.

Gutiérrez, Pedro Juan. 1991. ¿Hay polución en Cuba? *Bohemia* 83:37 (13 September): 26–30.

Gwiazda, A. 1991. The Decline of Comecon and Its Impact on the Crude Oil Foreign Trade of Eastern European Countries. *OPEC Review* 15:69–90.

Habashi, Fathi. 1993. Nickel in Cuba. In *Extractive Metallurgy of Copper, Nickel and Cobalt*. Vol. 1: *Fundamental Aspects*. Proceedings of the Paul E. Queneau International Symposium, 1165–78. Warrandale, Pa.: Minerals, Metals, and Materials Society.

Heffter, Jerome L., and Barbara J. B. Stunder. 1992. Transport and Dispersion from a Potential Accidental Release of Radioactive Pollutants from the Nuclear Reactor at Cienfuegos, Cuba. Mimeo. Washington, D.C.: U.S. National Oceanic and Atmospheric Administration (August).

Henderson, James M., and Richard E. Quandt. 1971. *Microeconomic Theory: A Mathematical Approach*. New York: McGraw-Hill.

Hernández, Gregorio. 1982. Detener la contaminación de nuestras aguas marinas. *Bohemia* 74 (18) (30 April): 28–31.

Hernández Castellón, Raúl. 1988. *La Revolución Demográfica en Cuba.* La Habana: Editorial de Ciencias Sociales.

Hernández Cruz, S. 1987. El ahorro de energía: Nueva fuente de recursos energéticos. *Revista Estadística* 9:5–42.

Hernández Pardo, Héctor. 1975. La contaminación de las aguas, el problema del río Quibú, plausible iniciativa del MINAZ en la lucha por la protección de los cursos fluviales. *Granma* (28 November): 4.

Hertzman, Clyde. 1995. *Environment and Health in Central and Eastern Europe.* A Report for the Environmental Action Programme for Central and Eastern Europe. Washington, D.C.: World Bank.

Hiatt, Fred. 1995. Armenia's Nuclear Risk. *Washington Post* (28 May): A35, A36.

Hollerbach, Paula E., and Sergio Díaz-Briquets. 1983. *Fertility Determinants in Cuba.* Committee on Population and Demography. Washington, D.C.: National Academy Press.

Holston, Mark. 1995. Trails of Survival. *Américas* 47 (2): 36–43.

Honey, Martha. 1994. Paying the Price of Ecotourism. *Américas* 46 (4): 40–47.

———. 1996. Tourism and the Environment. In *The Caribbean Environment: Issues of Mutual Concern,* edited by David S. Collis, 10–12. Washington, D.C.: Caribbean Project, Georgetown University.

Illán, José M. 1964. *Cuba: Facts and Figures of an Economy in Ruins.* Miami, Fla.: Editorial AIP.

Informe acerca de cuestiones relacionadas con el drenaje agrícola en Cuba, leído por René Peñalver. 1979. *Granma* (28 December): 3.

Informe de Cuba a la Comisión de Desarrollo Sostenible. 1994. <http://community.wow.net/eclac/carlinks/cubalink.htm>.

Instituto Cubano de Geodesia y Cartografía. 1978. *Atlas de Cuba.* La Habana: Instituto de Geodesia y Cartografía.

Inter-American Dialogue. 1995. *The Environment in U.S.-Cuban Relations: Opportunities for Cooperation.* Washington, D.C.: Inter-American Dialogue.

———. 1997. *The Environment in U.S.-Cuban Relations: Recommendations for Cooperation.* Washington, D.C.: Inter-American Dialogue.

International Atomic Energy Agency (IAEA). 1996. Power Reactors in Operation and Under Construction at the End of 1995. IAEA WorldAtom Internet database: <http://www.iaea.org/worldatom>.

International Labor Organization (ILO). 1988. *Summaries of International Labour Standards.* Geneva: International Labour Organization.

———. 1991. *The International Labor Organization: Facts for Americans.* Washington, D.C.: International Labour Organization.

———. 1992. *International Labour Conventions and Recommendations, 1919–1991.* Geneva: International Labour Organization.

———. 1995. *Lists of Ratifications by Conventions by Country (as of December 31, 1994).* Geneva: International Labour Organization.

Isla produce 31,000 barriles diarios de crudo. 1996. *El Nuevo Herald* (27 October): 2B.

Jankowski, John E. 1983. Energy Use and Conservation in Developing Country Industries. *Natural Resources Forum* 7 (2): 145–60.

Jatar-Hausmann, Ana Julia. 1996. Through the Cracks of Socialism. In *Cuba in Transition,* vol. 6, 202–18. Washington, D.C.: Association for the Study of the Cuban Economy.

Johnson, Tim. 1993. Crisis cubana cobra alto precio a ancianos. *El Nuevo Herald* (12 July): 1A, 5A.

Kabala, Stanley J. 1992. Environment and Development in the New Eastern Europe. Occasional Paper, no. 3. Middlebury, Vt.: Geonomics Institute.

Kaufman, Holly. 1993. From Red to Green: Cuba Forced to Conserve Due to Economic Crisis. *Agriculture and Human Values* 10 (3): 31–34.

Kelley, Donald R. 1976. Environmental Policy-Making in the USSR: The Role of Industrial and Environmental Interest Groups. *Soviet Studies* 28:4 (October): 570–89.

Kneese, Allen V., Robert U. Ayres, and Ralph C. D'Arge. 1970. *Economics and the Environment.* Washington, D.C.: Resources for the Future.

Kneese, Allen V., and Clifford S. Russell. 1987. Environmental Economics. In *The New Palgrave Dictionary of Economics,* edited by John Eatwell, Murray Milgate, and Peter Newman, vol. 2, 159–64. London: Macmillan.

Knox, Paul. 1995. Sherritt Breathes Life into Cuban Mine. *Globe and Mail* (Toronto) (31 July).

Kornai, János. 1992. *The Socialist System: The Political Economy of Communism.* Princeton, N.J.: Princeton University Press.

Kramer, John M. 1974. Environmental Problems in the USSR: The Divergence of Theory and Practice. *Journal of Politics* 36 (4) (November): 886–99.

La industria de los fertilizantes fertiliza su futuro. 1985. *Bohemia* 77 (29) (19 July): 33.

Laird, Roy D. 1965. The Politics of Soviet Agriculture. In *Soviet Agriculture: The Permanent Crisis,* edited by Roy D. Laird and Edward L. Crowley, 147–58. New York: Praeger Publishers.

Laird, Roy D., and Edward L. Crowley, eds. 1965. *Soviet Agriculture: The Permanent Crisis.* New York: Praeger Publishers.

Laird, Roy D., and Betty A. Laird. 1990. Perestroika in Agriculture: Gorbachev's "Rural Revolution." In *Communist Agriculture,* edited by Karl-Eugen Wadekin, 107–19. London: Routledge.

Latin America and the Caribbean Region. 1996. *Environment Matters* (fall): 20–23.

Lee, Susana. 1994. Cuba Expands. *Granma International* (6 July): 14.

Lehmann, David. 1982. Agrarian Structure, Migration, and the State in Cuba. In *State Policies and Migration,* edited by Peter Peek and Guy Standing. London: Croom Helm.

LeVine, Steve. 1995. Despite U.S. Protests, Armenia Gets A-Plant. *New York Times* (24 October): A6.

Levins, Richard. 1990. The Struggle for Ecological Agriculture in Cuba. *Capitalism, Nature, Socialism* 5:121–41.

———. 1993a. "Developmentalism" and the Struggle for Ecological Agriculture in Cuba. *Cuba Update* (February–March): 22–30.

———. 1993b. The Ecological Transformation of Cuba. *Agriculture and Human Values* 10 (3): 52–60.

Lewin, J. 1977. The Russian Approach to Nuclear Reactor Safety. *Nuclear Safety* (July–August): 438–50.

Ley No. 33. 1981. De protección del medio ambiente y del uso racional de los recursos naturales. *Gaceta Oficial* (12 February): 255–65.

Ley No. 73. 1994. Del sistema tributario. *Gaceta Oficial* (5 August): 35–44.

Ley No. 77. 1995. *Gaceta Oficial* (6 September): 5–12.

Ley No. 81. 1997. Del Medio Ambiente. *Gaceta Oficial* (11 July): 47–68.

Lezcano, José Carlos. 1994a. SOS de la naturaleza en Cuba. *Boletín Informativo.* Miami, Fla.: Colegio Nacional de Ingenieros Agrónomos y Azucareros Cubanos (April–June): 66.

———. 1994b. Capital en ruinas. Paper presented at the Congreso Internacional de Derechos Humanos. Miami, Fla.: Florida International University.

———. 1995. Aspectos esenciales sobre la mitigación de los desastres naturales en Cuba. In *Cuba in Transition,* vol. 5, 399–406. Washington, D.C.: Association for the Study of the Cuban Economy.

Lineamientos económicos y sociales para el quinquenio 1981–1985. 1981. La Habana: Editora Política.

Lineamientos económicos y sociales para el quinquenio 1986–1990. 1986. La Habana: Editora Política.

Livernash, Roberet, and Eric Rodenburg. 1998. Population Change, Resources, and the Environment. *Population Bulletin* 53 (March): 1.

Llovio-Menéndez, José Luis. 1988. *Insider: My Secret Life as a Revolutionary in Cuba*. New York: Bantam Books.

Lofstedt, Ragnar E., and Allen L. White. 1990. Chernobyl: Four Years Later, the Repercussions Continue. *Environment* 32 (3) (April): 2–5.

López Bueza, Ana Luisa. 1997. Detenido en Cuba el periodísta Nogueras. *El Nuevo Herald*. (25 April): 6A.

López Bueza, Ana Luisa, and Julian Caldecott, eds. 1996. *Decentralization and Biodiversity Conservation*. Washington, D.C.: World Bank.

López Bueza, Ana Luisa, Stefano Pagiola, and Carlos Reiche, eds. 1994. Economic and Institutional Analyses of Soil Conservation Projects in Central America and the Caribbean. World Bank Environment Paper, no. 8. Washington, D.C.: World Bank.

Luzón, José Luis. 1987. *Economia, Población y Territorio en Cuba (1899–1983)*. Instituto de Cooperación Iberoamericana. Madrid: Ediciones Cultura Hispánica.

———. 1988. Housing in Socialist Cuba: An Analysis Using Cuban Censuses of Population and Housing. *Cuban Studies* 18:65–83.

MacEwan, Arthur. 1981. *Revolution and Economic Development in Cuba*. New York: St. Martin's Press.

MacGaffey, Wyatt, and Clifford R. Barnett. 1965. *Twentieth Century Cuba: The Background of the Castro Revolution*. New York: Anchor Books.

McGeary, Johanna, and Cathy Booth. 1993. Cuba Alone. *Time* (6 December): 42–54.

McIntyre, Robert J., and James R. Thornton. 1978. On the Environmental Efficiency of Economic Systems. *Soviet Studies* 30 (2) (April): 173–92.

———. 1974. Environmental Divergence: Air Pollution in the USSR. *Journal of Environmental Economics and Management* 1:109–20.

———. 1981. Environmental Policy Formulation and Current Soviet Management. *Soviet Studies* 33 (1) (January): 146–49.

MacLachan, A. 1978. Soviet Nuclear Experts Scoff at Safety American-style. *Energy Daily* (20 October): 1.

Magraw, Daniel Barstow. 1990. Legal Treatment of Developing Countries: Differential, Contextual, and Absolute Norms. *Colorado Journal of International Environmental Law* 1 (1) (summer): 69–99.

———, ed. 1991. *International Law and Pollution*. Philadelphia: University of Pennsylvania Press.

Malinowitz, Stanley. 1997. Public and Private Services and the Municipal Economy in Cuba. *Cuban Studies* 27: 68–69.

Marrero, Dania. 1985. Los primeros tocororos. *Bohemia* 77 (16) (19 April): 32–33.

Marrero, Leví. 1950. *Geografía de Cuba*. La Habana: La Moderna Poesía.

Martí, Agenor. 1980. Contaminación en el Caribe. *Mar y Pesca* 6:26–31.

Martín Alonso, Nelson J. 1976. Estudio de la salinidad en las zonas arroceras de la granja Candelaria, en Pinar del Río. *Voluntad Hidráulica* 13 (40): 64–70.

Mateo Rodríguez, José M. 1996. Situación medioambiental de Cuba y perspectivas de aplicación de los principios del desarrollo sustentable. *Estudios Geográficos* (Madrid) 57 (223): 219–43.

Medina Pérez, Heriberto. n.d. Erosión de los suelos limítrofes a los embalses: Un desastre a mitigar. CubaEco: <http://www.cubanet.org>.

Medvedev, Zhores. 1978. *Nuclear Disaster in the Urals.* New York: Norton.

Memorias Inéditas del Censo de 1931. 1978. La Habana: Editorial de Ciencias Sociales.

Mesa-Lago, Carmelo. 1969. Ideological Radicalization and Economic Policy in Cuba. *Studies in Comparative International Development* 5 (10): 203–16.

———. 1971. Economic Policies and Growth. In *Revolutionary Change in Cuba,* edited by Carmelo Mesa-Lago, 277–338. Pittsburgh, Pa.: University of Pittsburgh Press.

———. 1981. *The Economy of Socialist Cuba.* Albuquerque: University of New Mexico Press.

———. 1988. The Cuban Economy in the 1980's: The Return of Ideology. In *Socialist Cuba: Past Interpretations and Future Challenges,* edited by Sergio G. Roca, 59–101. Boulder,Colo.: Westview Press.

———. 1994. *Are Economic Reforms Propelling Cuba to the Market?* Coral Gables, Fla.: North-South Center, University of Miami.

———. 1996. The State of the Cuban Economy: 1995–96. In *Cuba in Transition,* vol. 6, 4–7. Washington, D.C.: Association for the Study of the Cuban Economy.

Ministerio de Agricultura. 1951. *Memoria del Censo Agrícola Nacional 1946.* La Habana: Ministerio de Agricultura.

Ministerio de Agricultura (MINAGRI). 1984a. Evaluación cualitativa de las tierras para uso agropecuario. Departamento de Evaluación y Recomendaciones. La Habana: Centro de Información y Distribución Agropecuario.

———. 1984b. *Manual de interpretación de los índices físico-químicos y morfológicos de los suelos cubanos.* La Habana: Editorial Científico-Técnico.

———. 1984c. *Evaluación cualitativa de las tierras para uso agropecuario.* Departamento de Evaluación y Recomendaciones. La Habana: Centro de Información y Distribución.

Ministerio de Ciencia, Tecnología y Medio Ambiente. 1995. *Cuba: Medio Ambiente y Desarrollo.* La Habana: Ministerio de Ciencia, Tecnología y Medio Ambiente.

———. 1997a. *Estrategia Ambiental Nacional.* La Habana: Ministerio de Ciencia, Tecnología y Medio Ambiente.

———. 1997b. *Estrategia Nacional de Educación Ambiental.* La Habana: Ministerio de Ciencia, Tecnología y Medio Ambiente.

Ministerio de Justicia. 1988. *Protección del medio ambiente y uso racional de los recursos naturales.* Informe Central, III Congreso del Partido. La Habana: Ministerio de Justicia.

Ministerio de Salud Pública. 1994. *Anuario Estadístico 1993.* La Habana. Dirección Nacional de Estadística.

Ministerio de Transporte. Instituto de Investigaciones del Transporte. 1985. *Investigación y Control de la Contaminación Marina en la Bahía de La Habana.* Informe Final, Proyecto CUB/80/001, PNUD-PNUMA-UNESCO. La Habana.

Morales, Pedro. 1985. Otro gigante industrial. *Cuba Internacional* 18 (187) (June): 58–65.

Morales Pita, Antonio. 1995. Algunas reflexiones sobre la vinculación economía-medio ambiente. *Economía y Desarrollo* 95 (2): 121–37.

Moran, Theodore H. 1976. The International Political Economy of Cuban Nickel Development. In *Cuba in the World,* edited by Carmelo Mesa-Lago, 257–72. Pittsburgh, Pa.: University of Pittsburgh Press.

Moreno Fraginals, Manuel. 1964. *El Ingenio: El complejo económico y social cubano del azúcar.* La Habana: Editorial de Ciencias Sociales.

Moroney, John R. 1990. Energy Consumption, Capital, and Real Output: A Comparison of Market and Planned Economies. *Journal of Comparative Economics* 14:199–220.

Munasinghe, Mohan. 1992. *Water Supply and Environmental Management.* Boulder, Colo.: Westview Press.

Murray, Douglas L. 1994. *Cultivating Crisis: The Human Cost of Pesticides in Latin America.* Austin: University of Texas Press.

Murray, Douglas L., and Polly Hoppin. 1992. Recurring Contradictions in Agrarian Development: Pesticide Problems in Caribbean Basin Nontraditional Agriculture. *World Development* 20 (4): 597–608.

National Public Radio. 1995. Cuba Struggles to Save Economy and Ecology. Transcript of *All Things Considered.* Washington, D.C.: National Public Radio (30 May).

Nelson, Lowry. 1950. *Rural Cuba.* Minneapolis: University of Minnesota Press.

———. 1972. *Cuba: The Measure of a Revolution.* Minneapolis: University of Minnesota Press.

Nichols, John Spicer. 1982. The Mass Media: Their Functions in Social Conflict. In *Cuba: Internal and International Affairs,* edited by Jorge I. Domínguez, 71–112. Beverly Hills, Calif.: Sage Publications.

Nickel Plant Said to Pollute Sea with Metals, Acids. 1994. Prensa Latina News Service (9 December). Foreign Broadcast Information Service FBIS-LAT-94-238 (12 December): 15–16.

Niehaus, F., and L. Lederman. 1992. International Safety Review of WWER-440/230 Nuclear Power Plants. *IAEA Bulletin* 34 (2): 24–31.

Noble, Jeanne, and Malcolm Potts. 1996. The Fertility Transition in Cuba and the Federal Republic of Korea: The Impact of Organized Family Planning. *Journal of Biosocial Sciences* 28:211–25.

Nogueras Rofes, Olance. 1996. Moscú no cree en lágrimas. *El Nuevo Herald* (24 January): 9A.

———. 1997. Contaminación de bahía cienfueguera es "irreversible." *El Nuevo Herald* (8 March): 1B.

Norniella, José M. 1982a. Contará la fábrica de cemento de Nuevitas para 1985, con tres electrofiltros que evitarán escape de polvo a la atmósfera. *Granma* (27 October): 3.

———. 1982b. Firmarán hoy contrato para la instalación de un electrofiltro en la fábrica de cemento de Nuevitas. *Granma* (26 October): 3.

Nota Informativa. 1986. *Granma* (25 June): 3.

Novak-Decker, Nikolai. 1965. Soviet Efforts to Introduce Intensive Farming. In *Soviet Agriculture: The Permanent Crisis,* edited by Roy D. Laird and Edward L. Crowley, 193–99. New York: Praeger Publishers.

Nove, Alec. 1965. Some Thoughts on Soviet Agricultural Administration. In *Soviet Agriculture: The Permanent Crisis,* edited by Roy D. Laird and Edward L. Crowley, 1–13. New York: Praeger Publishers.

———. 1980. *The Soviet Economic System.* 2nd ed. London: George Allen & Unwin.

Nuevos cientos de millones de metros cúbicos a nuestros sistemas hidráulicos. 1969. *Voluntad Hidráulica* 17:1–2.

Núñez Jiménez, Antonio. 1968. *La erosión desgasta a Cuba.* La Habana: Instituto del Libro.

———. 1972. *Geografía de Cuba.* La Habana: Instituto Cubano del Libro.

———. 1980. Defensa de la naturaleza cubana en la Asamblea Nacional. *Granma* (4 July): 4.

Oberg, James E. 1988. *Uncovering Soviet Disasters.* New York: Random House.

Official Urges Measures to End Deforestation 1994. Foreign Broadcast Information Service FBIS-LAT-94-066 (6 April): 3.

Oramas, Joaquín. 1986. Iniciarán cimentación y montaje del primer tanque de 50 mil metros cúbicos de la base de supertanqueros. *Granma* (27 August): 3.

Ordúñez-García, Pedro O., F. Javier Nieto, Alfredo D. Espinosa-Brito, and Benjamin Caballero. 1996. Cuban Epidemic Neuropathy, 1991 to 1994: History Repeats Itself a Century After the Amblyopia of the Blockade. *American Journal of Public Health* 86 (5): 738–43.

Oro, José R. 1992. *The Poisoning of Paradise: The Environmental Crisis in Cuba*. Miami, Fla.: Endowment for Cuban American Studies.

Osterling, Jorge P. 1985. The Society and Its Environment. In *Cuba: A Country Study*, edited by James D. Rudolph, 63–107. Washington, D.C.: American University.

O'Toole, T. 1978. Soviet Approach to Nuclear Power Is Different. *Washington Post* (8 October).

Pagés, Raisa. 1981. Una riqueza natural que debemos preservar. *Granma* (24 February): 4.

Pagiola, Stefano. 1994. Cost-Benefit Analysis of Soil Conservation. In *Economic and Institutional Analyses of Soil Conservation Projects in Central America and the Caribbean*, edited by Ernst Lutz, Stefano Pagiola, and Carlos Reiche, 21–39. World Bank Environment Paper, no. 8. Washington, D.C.: World Bank.

Palazuelos Barrios, Raúl. 1990a. Cuatro hitos de una rehabilitación industrial. *Granma* (12 April): 1–2.

———. 1990b. Avanza rehabilitación de la primera línea de la René Arcay. *Granma* (22 August): 2.

Pan American Health Organization (PAHO). 1994. *Health Conditions in the Americas—1994 Edition*. 2 vols. Washington, D.C.: PAHO.

Parenteau, Patrick. 1977. Promoting Exchange of Environmental Law and Policymaking Experience: Some Observations on the Evolution of American Environmental Law. In *The Environment in U.S.–Cuban Relations: Recommendations for Cooperation*, 31–39. Washington, D.C.: Inter-American Dialogue.

Pearce, David W., and Jeremy J. Wardford. 1993. *World Without End: Economics, Environment, and Sustainable Development*. New York: Oxford University Press.

Pedraza Linares, Héctor. 1997. Alarmante situación ambiental en Pinar del Río. Habana Press (18 June). CubaNet: <http://www.cubanet.org>.

Peñalver, René. 1979. Informe acerca de cuestiones relacionadas con el drenaje agrícola en Cuba. *Granma* (28 December): 3.

Pérez-López, Jorge F. 1979. The Cuban Nuclear Power Program. *Cuban Studies* 9 (1): 1–42.

———. 1981. Energy Production, Imports, and Consumption in Revolutionary Cuba. *Latin American Research Review* 16 (3): 111–37.

———. 1982. Nuclear Power in Cuba: Opportunities and Challenges. *Orbis* 26:2 (summer): 95–516.

———. 1987a. Cuban Oil Reexports: Significance and Prospects. *Energy Journal* 8 (1): 1–16.

———. 1987b. Nuclear Power in Cuba After Chernobyl. *Journal of Interamerican Studies and World Affairs* 29 (2) (summer): 79–117.

———. 1990. Cuba. In *Energy and Security in the Industrializing World*, edited by Raju G. C. Thomas and Bennett Ramberg, 153–81. Lexington: University Press of Kentucky.

———. 1991. *The Economics of Cuban Sugar*. Pittsburgh, Pa.: University of Pittsburgh Press.

———. 1992. Cuba's Transition to Market-Based Energy Prices. *Energy Journal* 13 (4): 79–117.

———. 1994. Economic and Financial Institutions to Support the Market. In *Cuba in Transition*, vol. 4, 292–302. Washington, D.C.: Association for the Study of the Cuban Economy.

———. 1995a. Odd Couples: Joint Ventures Between Foreign Capitalists and Cuban Socialists. Agenda Papers, no. 16. Coral Gables, Fla.: North-South Center, University of Miami.

———. 1995b. Castro Tries Survival Strategy. *Transition* 6 (3) (March): 11–14.

———. 1995c. *Cuba's Second Economy*. New Brunswick, N.J.: Transaction Publishers.

Pérez-López, Jorge F., and Sergio Díaz-Briquets. 1990. Labor Migration and Offshore Assembly in the Socialist World: The Cuban Experience. *Population and Development Review* 15 (2): 273–99.

Pérez Zorrilla, Wilfredo, and Galino Ya Karasik. 1989. El escurrimiento sólido y la erosión hídrica actual de Cuba. *Ciencias de la Tierra y el Espacio* 15–16:67–76.

Perfecto, Ivette. 1994. The Transformation of Cuban Agriculture After the Cold War. *American Journal of Alternative Agriculture* 9 (3): 98–108.

Petinaud Martínez, J., and O. González Quintana. 1986. La verdad sobre Chernobyl y la nucleo-energética. *Bohemia* (16 May).

Petróleo crudo nacional en termoeléctricas. 1986. *Cuba Internacional* 205 (December): 8.

Petróleo nacional provoca problemas. 1987. *El Nuevo Herald* (18 December): 3A.

Pichs, Ramón. 1992. Cuba ante los desafíos ambientales globales. Mimeo. La Habana.

Plan para frenar migración a La Habana. 1997. *El Nuevo Herald Digital* (26 April).

Pollitt, Brian H. 1997. The Cuban Sugar Economy: Collapse, Reform, and Prospects for Recovery. *Journal of Latin American Studies.* 29 (1) (February): 171–210.

Poore, M. E. D., and C. Fries. 1985. The Ecological Effects of Eucalyptus. FAO Forestry Papers, no. 59. Rome: UN Food and Agriculture Organization.

Population Reference Bureau, Inc. 1996. *World Population Data Sheet 1996.* Washington, D.C.: Population Reference Bureau.

Portela, Armando. 1997. Cuba Assesses Damage, Finds It Excessive. *Cuba News* 5 (7) (July): 11.

Postergada indefinidamente la terminación de la planta de Juraguá. 1997. *Diario las Américas* (19 January): 1A, 9A.

Powell, David A. 1971. The Social Costs of Modernization: Ecological Problems in the USSR. *World Politics* 23 (4) (July): 327–34.

Pozo, Alberto. 1980. ¡Anduvo, anda y andará! *Bohemia* 72 (1) (4 January): 16–23.

Prado, Jose Antonio. 1996. Kindling. *Economist* (16–22 March): 8.

Preeg, Ernest H. 1996. *The Haitian Dilemma.* Washington, D.C.: Center for Strategic and International Studies.

Presidió Guillermo García reunión del Consejo Científico para la renovación de la fauna y la flora. 1978. *Granma* (26 September): 1.

Prevén un Chernobyl en Cuba. 1991. *Diario las Américas* (31 May): 1A, 11A.

Prieto, Fidel. 1990. Cleanup Effort Targets at Mining Center. Inter Press Service (19 June).

Pryde, P., and L. Pryde. 1974. Soviet Nuclear Power. *Environment* (April): 26–34.

Pryor, Frederic L. 1992. *The Red and the Green: The Rise and Fall of Collectivized Agriculture in Marxist Regimes.* Princeton, N.J.: Princeton University Press.

Pujol, Joaquín P. 1991. Membership Requirements in the IMF: Possible Implications for Cuba. In *Cuba in Transition,* 1: 91–102. Miami: Florida International University.

Quinn, Norman J., and Richard S. Strickland. 1994. Ecotourism in a Tourist-Based Caribbean Economy. *North-South* 4 (2): 42–47.

Raup, Philip M. 1990. Structural Contrasts and Convergences in Socialist and Capitalist Agriculture. In *Communist Agriculture,* edited by Karl-Eugen Wadekin, 90–104. London: Routledge.

Reed, Gail. 1992. On the Razor's Edge: Deforestation in Cuba. *Cuba Update* 1–2:36–38.

———. 1993. Protecting the Environment During the Special Period. *Cuba Update* (September–October): 29, 31–33.

Rehabilitan lagunas costeras para beneficiar fauna marina. 1991. *Trabajadores* (26 August): 12.

Remos, Ariel. 1992. Dice el Ingeniero Oro que Cuba ocultó intensidad del terremoto. *Diario las Américas* (30 May): 1B.

———. 1994. Explotación petrolera en Cuba podría dañar a playas de la Florida y a Varadero. *Diario las Américas* (24 July): 1A, 13A.

———. 1996. Grave la polución soterrada. *Diario Las Américas* (25 May): 1A, 11A.

———. 1997. Violencia oficial podría disparar en Cuba otro cinco de agosto. *Diario de las Americas* (7 August): 1A, 13A.

Report by the Delegation of the U.S. Association of Former Members of Congress Visit to Cuba, December 9–14, 1996. 1996. Washington, D.C.: United States Association of Former Members of Congress.

Report of Cuba. 1976. United Nations Conference on Water. *Voluntad Hidráulica* 13 (39): 47–59.

Resolución conjunta del Ministerio de Salud Pública, el Instituto Nacional de Recursos Hidraúlicos, y el Ministerio de la Industria Pesquera. 1990. *Gaceta Oficial* (11 July): 460–63.

Resolución No. 21/79 del Instituto Nacional de Desarrollo y Aprovechamiento Forestales. 1979. *Granma* (10 September): 370–72.

Resolución No. 34 del Comité Estatal de Trabajo y Seguridad Social. 1977. *Gaceta Oficial* (3 August): 281–82.

Resolución No. 63/83 del Comité Estatal de Finanzas. 1983. *Gaceta Oficial* (29 November): 1766–67.

Resolución unánime de la Asamblea Nacional aprobando la valiente y firme actuación nacional e internacional del gobierno revolucionario a raiz de los sucesos que se iniciaron en la Embajada de Perú. 1980. *Granma* (4 July): 1, 3–4.

Rev, Istvan. 1993. La naturaleza antiecológica de la centralización. *Cuadernos del Este* 10:9–17.

Rey Santos, Orlando. 1997. Expanding Cooperation Between the United States and Cuba: Legislative Policies and the National Legal Framework. In *The Environment in U.S.–Cuban Relations: Recommendations for Cooperation*, 41–45. Washington, D.C.: Inter-American Dialogue.

Rice, John. 1994. Making Do in Cuba. *Cuba Update* (September–October): 15.

Riego: Sinónimo de altos rendimientos en la agricultura. 1982. *Voluntad Hidráulica* 19:60–64.

Riera, Lilliam. 1998. Will Its Waters Ever Run Clear Again? *Granma Electronic Edition* 3 (March).

Ripon, Simon. 1992. Report Says Model 230 VVERs Need Immediate Fixes. *Nuclear News* (April): 69–73.

Rivero, Raúl. 1997. Los niños no conocen las frutas nacionales. *El Nuevo Herald Digital* (24 June).

Roca, Sergio G. 1986. State Enterprises in Cuba Under the New System of Planning and Management (SDPE). *Cuban Studies* 16:153–79.

———. 1994. Reflections on Economic Policy: Cuba's Food Program. In *Cuba at a Crossroads*, edited by Jorge Pérez-López, 94–117. Gainesville: University Press of Florida.

Rodríguez, José Luis. 1990. *Estrategia del desarrollo económico en Cuba*. La Habana: Editorial de Ciencias Sociales.

Rodríguez, José Luis, and George Carriazo Moreno. 1987. *La erradicación de la pobreza en Cuba*. La Habana: Editorial de Ciencias Sociales.

Rodríguez Castellón, Roberto. 1987. La industria del combustible. *Economía y Desarrollo* 99 (July–August): 188–97.

Rodríguez Mesa, Gonzalo M. 1980. *El proceso de industrialización de la economía cubana*. La Habana: Editorial de Ciencias Sociales.

Rohter, Larry. 1996. A Kennedy-Castro Talk Touched by History. *New York Times* (19 February): A1, A6.

Rose, William. 1992. Sugar: Key to Development and Conservation. *Cuba Update* (September): 39–40.

Rosendahl, Bruce. 1991. Cuban Oil Drilling Could Spoil the Keys. *Miami Herald* (5 May): 1C.

Rosett, Claudia, and José de Córdoba. 1995. Russians Say They Are Coming to Build Cuban Nuclear Plant. *Wall Street Journal* (6 June): A1, A11.

Ross, James E. 1996. Agribusiness Investment in Cuba's Post Embargo Period. In *Cuba in Transition*, vol. 6, 163–68. Washington, D.C.: Association for the Study of the Cuban Economy.

Rosset, Peter, and Medea Benjamin. 1993. *Two Steps Backward, One Step Forward*. San Francisco: Global Exchange.

———. 1994a. Cuba's Nationwide Conversion to Organic Agriculture. *Capitalism, Nature, Socialism* 5 (3) (September): 79–97.

———. 1994b. *The Greening of the Revolution: Cuba's Experiment with Organic Agriculture*. Melbourne, Australia: Ocean Press.

Roundup of Economic Activity Reported 23–28 December 1992. Foreign Broadcast Information Service FBIS-LAT-92-252 (31 December): 2–3.

Roundup of Economic Developments. 1994. Foreign Broadcast Information Service FBIS-LAT-94-109 (7 June): 8–9.

Roy, Joaquín. 1997. The Helms-Burton Law: Development, Consequences, and Legacy for Inter-American and European–US Relations. *Journal of Interamerican Studies and World Affairs* 39:77–108.

Rusia reanudará trabajos en planta nuclear de Jaraguá. 1997. *El Nuevo Herald* (6 June): 6A.

Sáenz, Rodolfo. 1990. *Posibles medidas para controlar o atenuar el deterioro de la calidad microbiologica de los recursos hídricos en la América Latina y el Caribe (ALC)*. Washington, D.C.: Organización Panamericana de la Salud.

Sáez, Héctor R. 1994. The Environmental Consequences of Agricultural Development in Cuba. Paper presented to the Eighteenth International Congress of the Latin American Studies Association.

———. 1995. Technology, Property Rights, and Land Degradation: The Case of Santo Domingo, Cuba. Paper presented at the Nineteenth International Congress of the Latin American Studies Association. Washington, D.C.

———. 1997a. Agricultural Policies, Resource Degradation, and Conservation in Cuba. *Cuban Studies* 27:40–67.

———. 1997b. Property Rights, Technology, and Land Degradation: A Case Study of Santo Domingo, Cuba. In *Cuba in Transition*, vol. 7, 472–85. Washington, D.C.: Association for the Study of the Cuban Economy.

Salazar, Alberto. 1991. Una pelea cubana contra la "salinización." *Bohemia* (3 May): 28–32.

———. 1992. Bacterias que "comen" petróleo. *Bohemia* 84 (48) (27 November): B34–B36.

Sánchez Hernández, E. P. 1992. Torula Yeast Wastewater Treatment by Downflow Anaerobic Filters. *Bioresource Technology* 40 (2): 163–66.

Sanear el Almendares costaría $20 millones. 1996. *El Nuevo Herald* (16 May): 1B.

Santana, Eduardo. 1991. Nature Conservation and Sustainable Development in Cuba. *Conservation Biology* 5 (1): 13–15.

Santana, Maydel. 1992. Grupo ecologista anuncia "plan de erosión de emergencia." *El Nuevo Herald* (23 May, 1996): 3B.

Santiago, Ana E. 1990. Ecología de Cuba avanza al desastre, dice experto. *El Nuevo Herald* (14 November): 1B.

Santiesteban, Argelio. 1987a. Arterias del desarrollo. *Bohemia* 79 (44) (30 October): 48–51.

———. 1987b. Sin fiebre ni titulares. *Bohemia* 79 (22) (29 May): 47–51.

Satre Ahlander, Ann-Mari. 1994. *Environmental Problems in the Shortage Economy*. Aldershot, England: Edward Elgard.

Schlachter, Alexis. 1990. Analizán contaminación marina en zonas de Cuba. *Granma* (22 June): 2.

Se incendia tanquero griego en Matanzas. 1993. *El Nuevo Herald* (28 January): 3A.

Segre, Roberto, Mario Coyula, and Joseph L. Scarpaci. 1997. *Havana: Two Faces of the Antillean Metropolis*. Chichester, England: John Wiley & Sons.

Serradet Acosta, Miguel A. 1995. Programa Nucleoenergético Cubano. Paper prepared for Regional Seminar on Public Information, La Habana (May 17–19).

Shayakubov, B., and Cesar Morales. 1980. Consideraciones para la explotación de acuíferos costeros, en presencia de intrusión marina. *Voluntad Hidráulica* 17 (54): 10–15.

Shishkoff, Nina. 1993. Plant Diseases and Their Control by Biological Means in Cuba. *Agriculture and Human Values* 10 (3): 24–30.

Silva Lee, Alfonso. 1996. *Natural Cuba*. Saint Paul, Minn.: Pangaea.

Simon, Francoise L. 1995. Tourism Develpment in Transition Economies: The Cuba Case. *Columbia Journal of World Business* 30 (1): 26–41.

Simon, Joel. 1997. An Organic Coup in Cuba? *Amicus Journal* (winter). <http://www.nrdc.org/eamicus>.

Simons, Marlise. 1993. Major Powers Back a Fund for Soviet-Design Reactors. *New York Times* (29 January): A2.

Síntesis de la exposición de motivos acerca del Proyecto de Ley sobre la Protección del Medio Ambiente y del Uso Racional de los Recursos Naturales. 1980. *Granma* (27 December): 2.

Sismo causa heridos, daña hospitales y escuelas. 1992. *El Nuevo Herald* (26 May): 1A, 5A.

Sismo pudo producir grandes daños. 1992. *El Nuevo Herald* (27 May): 4A.

Smith, R. Jeffrey. 1993. U.S. Denounces N. Korea for Quitting Nuclear Pact. *Washington Post* (13 March): A1, A18.

———. 1994. U.S. to Dangle Prospect of Reactor at N. Korea. *Washington Post* (7 July): A1, A14.

Smith, Wayne S. 1995. Help Cuba with Nuclear Power. *Bulletin of the Atomic Scientist* 51 (5) (September–October): 4.

Sobell, Vlad. 1990. The Systemic Roots of the East European Ecological Crisis. *Environmental Policy Review* 4 (1): 47–52.

Solano, Rafael. 1995. Catástrofe del ecosistema cubano. *El Nuevo Herald* (22 April): 15A.

Solares, Andrés J. 1994. Situación del medio ambiente en Cuba. In *Desarrollo Agrícola en Cuba*, vol. 2, 61–78. Miami, Fla.: Colegio de Ingenieros Agrónomos y Azucareros.

Soto Hernández, Gerardo. 1994. Estudio de los suelos y el futuro agrícola de Cuba. In *Desarrollo Agrícola de Cuba*, vol. 2, 43–59. Miami,Fla.: Colegio de Ingenieros Agrónomos y Azucareros.

Sotolongo, Raúl, and Ernestino Abreu. 1992. El arroz en Cuba: Presente y futuro. In *Desarrollo Agrícola de Cuba*, vol. 1, 153–70. Miami, Fla.: Colegio de Ingenieros Agrónomos y Azucareros.

Soviet Official Concedes Accidents Have Taken Place at Nuclear Sites. 1979. *New York Times* (23 April): 15a.

Spreen, Thomas H., Armando Nova González, and Ronald P. Murato. 1996. The Citrus Industries in Cuba and Florida. International Working Paper Series, IW96-2 (January). Gainesville, Fla.: International Agricultural Trade and Development Center, University of Florida.

Status Report on Atomic Energy Developments. 1958. Foreign Service Despatch, U.S. Embassy, La Habana (21 October). U.S. Embassy Habana, Decimal Files 1955–59, File 837.1901 (available in the U.S. National Archives).

Steer, Andrew. 1996a. Ten Principles of the New Environmentalism. *Finance & Development* 33 (4) (December): 4–7.

Steer, Andrew. 1996b. The Year in Perspective. *Environment Matters* (fall): 4–7.

Stix, Gary. 1995. Ban That Embargo. *Scientific American* 272 (3): 32–34.

Suárez Salazar, Luis. 1995. Cuba's Foreign Policy in the "Special Period." In *Cuba in the International System: Normalization and Integration*, edited by Archibald R. M. Ritter and John M. Kirk, 84–104. London: Macmillan.

Support for a Democratic Transition in Cuba. 1997. Report of the President of the United States, Washington, D.C., 28 January. <http://www.state.gov/www/regions/wha/helmbu/html>.

Surgen los bosques. 1983. *Prisma Latinoamericano* (September): 26.

Tagliabue, John T. 1991. Auf Wiedersehen, Trusty Little Road Companion. *New York Times* (5 January): 2.

Tapanes, Juan José. 1981. Mediciones de la salinidad; temperatura y contenido de oxígeno en el agua del río Zaza. *Voluntad Hidráulica* 18 (57): 33–39.

Terrero, Ariel. 1994. Tendencias de un ajuste. *Bohemia* 85 (28) (28 October): B30–B39.

Terry Berro, Carmen C. 1997. Impacto ambiental: Primeras experiencias en Cuba. *Temas* 9 (January–March): 42–47).

Tésis y resoluciones del Primer Congreso del Partido Comunista de Cuba. 1976. La Habana: Departamento de Orientación Revolucionaria del Comité Central del Partido Comunista de Cuba.

Thiesenhusen, William C. 1995. Landed Property in Capitalist and Socialist Countries: The Russian Transition. In *Agricultural Landownership in Transitional Economies*, edited by Gene Wunderlich, 27–53. Lanham, Md.: University Press of America.

Thomas, Hugh 1971. *Cuba: The Pursuit of Freedom.* New York: Harper & Row.

Thornton, Judith A. 1978. Soviet Methodology for the Valuation of Natural Resources. *Journal of Comparative Economics* 2 (4): 321–33.

Tiende el proyecto de ley sobre el medio ambiente al aprovechamiento óptimo del potencial productivo nacional. 1980. *Granma* (24 December): 1–2.

Timoshenko, V. P. 1953. Agricultural Resources. In *Soviet Economic Growth*, edited by Abram Bergson, 246–71. Evanston, Ill.: Row, Peterson and Company.

Toman, Michael A., ed. 1994. *Pollution Abatement Strategies in Central and Eastern Europe.* Washington, D.C.: Resources for the Future.

Toman, Michael A., and R. David Simpson. 1994. Environmental Policies, Economic Restructuring, and Institutional Development in the Former Soviet Union. In *Pollution Abatement Strategies in Central and Eastern Europe*, edited by Michael A. Toman, 73–80. Washington, D.C.: Resources for the Future.

Top Soviet Power Minister Discloses Nuclear Accidents. 1979. *The Washington Post* (23 April).

Trabalka, J. R., S. I. Auerbach, and L. D. Eyman. 1980. The 1957–1958 Soviet Nuclear Accident in the Urals. *Nuclear Safety* (January–February): 94–99.

Trabalka, J. R., L. D. Eyman, F. L. Parker, E. G. Struxness, and S. I. Auerbach. 1979. Another Perspective of the 1958 Soviet Nuclear Accident. *Nuclear Safety* (March–April): 206–10.

Triana, Fausto. 1988. Ecologists Stress Points in Common with Cuban Revolutionaries. *Granma Weekly Review* (24 July): 3.

Turner, R. Kerry, David Pearce, and Ian Bateman. 1993. *Environmental Economics: An Elementary Introduction.* Baltimore, Md.: Johns Hopkins University Press.

Umali, Dina L. 1993. *Irrigation-Induced Salinity: A Growing Problem for Development and the Environment.* World Bank Technical Paper, no. 215. Washington, D.C.: World Bank.

United Nations. 1980. The Nickel Industry and the Developing Countries. ST/ESA/100. New York: United Nations.

———. 1985. Investigación y control de la contaminación marina en la bahía de la Habana. Mimeo. La Habana.

———. 1992a. *Statistical Yearbook 1992*. New York: United Nations.

———. 1992b. *Nations of the Earth Report*. Vol 3. Geneva: UN Conference on Environment and Development.

———. 1993. *Statistical Yearbook 1993*. New York: United Nations.

———. 1995a. *World Population Prospects: The 1994 Revision*. Policy Analysis, Population Division, ST/ESA/SER.A/145. New York: Department for Economic and Social Information.

———. 1995b. *World Statistics Pocketbook 1995*. World Statistics in Brief Series V, no. 16. New York: United Nations.

U.S. Agency for International Development. 1995. *LAC Regional Program Strategy for FY 1996–FY 2000 and Action Plan for FY 1996–FY 1997*. Washington, D.C.: U.S. Agency for International Development.

U.S. Blockade Causes Significant Economic Losses in 1994. 1995. Foreign Broadcast Information Service FBIS-LAT-95-141 (28 July): 3–4.

U.S. General Accounting Office. 1991. *Nuclear Power Safety: Chernobyl Accident Prompted Worldwide Actions but Further Efforts Needed*. GAO/NSIAD-92-28. Washington, D.C.: U.S. General Accounting Office.

———. 1992. *Concerns About the Nuclear Power Reactors in Cuba*. GAO/RCED-92-262. Washington, D.C.: U.S. General Accounting Office.

———. 1994a. *Consensus on Acceptable Radiation Risk to the Public is Lacking*. GAO/RCED-94-190. Washington, D.C.: U.S. General Accounting Office.

———. 1994b. *International Assistance Efforts to Make Soviet-Designed Reactors Safer*. GAO/RCED-94-234. Washington, D.C.: U.S. General Accounting Office.

———. 1996. *Implications of the U.S./North Korean Agreement on Nuclear Issues*. GAO/RCED/NSIAD-97-8. Washington, D.C.: U.S. General Accounting Office.

———. 1997. *International Atomic Energy Agency's Nuclear Technical Assistance to Cuba*. GAO/RCED-97-72. Washington, D.C.: U.S. General Accounting Office.

U.S. House of Representatives. 1991. Committee on Foreign Affairs. Subcommittee on Western Hemisphere Affairs. *Hearing on Recent Developments in United States–Cuban Relations: Immigration and Nuclear Power*. 102nd Cong., 1st sess. 5 June.

———. 1995. Committee on International Relations. Subcommittee on the Western Hemisphere. *Hearing on the Cienfuegos Nuclear Power Plant*. 104th Cong., 2nd sess. 1 August.

U.S. Senate. 1986. Committee on Governmental Affairs. Subcommittee on Energy, Nuclear Proliferation, and Governmental Processes. *Hearings on Cuban Nuclear Reactors*. 99th Cong., 2nd sess. 30 June.

———. 1991a. Committee on Environment and Public Works. Subcommittee on Nuclear Regulation. *Hearing on International Commercial Nuclear Reactor Safety*. 102nd Cong., 1st sess. 25 July.

———. 1991b. Committee on Environment and Public Works. Subcommittee on Nuclear Regulation. *Hearing on International Commercial Nuclear Reactor Safety*. 102nd Cong., 1st sess. 21 November.

U.S. War Department. 1900. *Report on the Census of Cuba 1899*. Washington, D.C.: U.S. Government Printing Office.

Unpredictable Consequences of Oil Spill from "Princess Anne Marie," Which Ran Aground in Fishing Areas South of Pinar del Rio. 1980. *Direct from Cuba* 231 (15 April): 13–16.

Valev, E. B. 1973. *Economic Geography of Cuba*. Washington, D.C.: Joint Publications Research Service. Translation of *Ekonomischeskaya Geografiya Kuby*. Moscow: Izd-vo MGU, 1972.

Vandermeer, John, Judith Carney, Paul Gesper, Ivette Perfecto, and Peter Rosset. 1993. Cuba and the Dilemma of Modern Agriculture. *Agriculture and Human Values* 10 (3): 3–8.

Varela Pérez, Juan. 1976. La zafra 76, los activos sindicales proximos a comenzar, el ahorro de agua en la industria, la capacitación, la eficiencia en fábrica, el petróleo y otros temas. *Granma* (12 February): 6.

Vedensky, George. 1965. The Soviet Chemical Fertilizer Industry. In *Soviet Agriculture: The Permanent Crisis*, edited by Roy D. Laird and Edward L. Crowley, 18–192. New York: Praeger Publishers.

Vinculación de los Programas de Desarrollo Economico y Social con el Medio Ambiente: Lineamientos para la Acción. 1993.<gopher://ceniai.inf.cu.>

Visión real de lo maravilloso. 1998. *Granma Digital Edition* (2 April): <http://www.cubaweb.granma>.

Volgyes, Ivan, ed. 1974a. *Environmental Deterioration in the Soviet Union and Eastern Europe*. New York: Praeger Publishers.

———. 1974b. Politics and Pollution in Western and Communist Societies. In *Environmental Deterioration in the Soviet Union and Eastern Europe*, edited by Ivan Volgyes, 1–7. New York: Praeger Publishers.

Volin, Lazar. 1962. Agricultural Policy of the Soviet Union. In *The Soviet Economy: A Book of Readings*, edited by Morris Bornstein and Daniel Fusfeld, 243–76. Homewood, Ill.: Richard D. Irwin.

Vosti, Stephen A., and Thomas Reardon, eds. 1997. *Sustainability, Growth, and Poverty Alleviation: A Policy and Agroecological Perspective*. Baltimore: Johns Hopkins University Press.

Vedensky, George. 1965. The Soviet Chemical Fertilizer Industry. In *Soviet Agriculture: The Permanent Crisis*, edited by Roy D. Laird and Edward L. Crowley, 18–192. New York: Praeger Publishers.

Wadekin, Karl-Eugen, ed. 1990. *Communist Agriculture*. London: Routledge.

Wald, Karen. 1991. Cuba Goes Green. *Earth Island Journal* (winter): 26–27.

Walls, Margaret A. 1994. Motor Vehicles and Pollution in Central and Eastern Europe. In *Pollution Abatement Strategies in Central and Eastern Europe*, edited by Michael E. Toman, 1–14. Washington, D.C.: Resources for the Future.

Ward, Fred. 1978. *Inside Cuba Today*. New York: Crown Publishers.

Weinberg, Bill. 1994. Cuba's Reluctant Environmentalism. *Amicus Journal* 16:18–21.

Weir, David, and Mark Schapiro. 1981. *Circle of Poison: Pesticides and People in a Hungry World*. Oakland, Calif.: Institute for Food and Development Policy.

Weiss, Edith Brown, Daniel Barstow Magraw, and Paul C. Szasz, eds. 1992. *International Environmental Law: Basic Instruments and References*. Dobbs Ferry, N.Y.: Transnational Publishers.

Westoby, Jack. 1989. *Introduction to World Forestry*. London. Basil Blackwell.

Whitefield, Mimi. 1990. Cuba admite problemas con petróleo. *El Nuevo Herald* (20 June): 1A, 5A.

———. 1992. Castro's Son Loses Top Nuclear Job. *Miami Herald* (18 June): 24A.

———. 1993. Blackouts Increase Miseries for Cubans. *Miami Herald* (11 August): 6A.

Willett, Joseph H. 1962. The Recent Record in Agricultural Production. In *Dimension of Soviet Economic Power*, report by the U.S. Joint Economic Committee, 91–113. Washington, D.C.: U.S. Government Printing Office.

Wilson, R. 1995. Nuclear Power Safety in Central and Eastern Europe. *Nuclear Safety* 36 (1): 33–46.

Work to Start Soon on a Site for North Korea Reactors.1997. *Journal of Commerce.* (7 July): 11A.

World Bank (International Bank for Reconstruction and Development). 1951. *Report on Cuba*. Washington, D.C.: World Bank.

World Bank. 1990. *Indonesia: Sustainable Development of Forests, Land, and Water*. Washington, D.C.: World Bank.

———. 1992. *Development and the Environment—World Development Report 1992*. New York: Oxford University Press.

———. 1993. *Jamaica: Economic Issues for Environmental Management*. World Bank Country Study. Washington, D.C.: World Bank.

———. 1994. *Indonesia: Environment and Development*. World Bank Country Study. Washington, D.C.: World Bank.

World List of Nuclear Power Plants. 1995. *Nuclear News* (March): 27–39.

World Resources Institute (WRI). 1992. *World Resources 1992–1993*. New York: Oxford University Press.

———. 1994. *World Resources, 1994–95*. New York: Oxford University Press.

———. 1996. *World Resources, 1996–97*. New York: Oxford University Press.

Wotzkow, Carlos. 1998. *Natumaleza Cubana*. Miami, Fla.: Ediciones Universal.

Wunderlich, Gene, ed. 1995. *Agricultural Landownership in Transitional Economies*. Lanham, Md.: University Press of America.

Yablokov, Andrei V. 1990. The Current State of the Soviet Environment. *Environmental Policy Review* 4 (1): 1–14.

Zafra azucarera no superará la del 96. 1997. *El Nuevo Herald Digital* (3 May).

Ziegler, Charles E. 1980. Soviet Environmental Policy and Soviet Central Planning: A Reply to McIntyre and Thornton. *Soviet Studies* 32 (1): 124–34.

———. 1982. Centrally Planned Economies and Environmental Information: A Rejoinder. *Soviet Studies* 34 (2): 296–99.

———. 1987. *Environmental Policy in the USSR*. Amherst: University of Massachusetts Press.

———. 1991. Environmental Protection in Soviet–East European Relations. In *To Breathe Free: Eastern Europe's Environmental Crisis*, edited by Joan DeBardeleben, 83–100. Baltimore, Md.: Johns Hopkins University Press.

Zoerb, Carl. 1965. The Virgin Land Territory: Plans, Performance, Prospects. In *Soviet Agriculture: The Permanent Crisis*, edited by Roy D. Laird and Edward L. Crowley, 29–44. New York: Praeger Publishers.

Zurek, Julia. 1997. Grupo enviará cloro a Cuba para evitar la contaminación. *El Nuevo Herald* (22 January): 4A.

Zyszkowski, Wiktor. 1993. Improving the Safety of WWER Nuclear Power Plants: Focus on Technical Assistance in Central and Eastern Europe. *IAEA Bulletin* 35 (4): 39–44.

Index

317